LIE GROUPS: HISTORY, FRONTIERS AND APPLICATIONS

VOLUME XI

FIRST WORKSHOP ON GRAND UNIFICATION

New England Center
University of New Hampshire
April 10–12, 1980

Editors

Paul H. Frampton
Sheldon L. Glashow
Asim Yildiz

MATH SCI PRESS
1980

Library of Congress Catalog Card Number:
ISBN: 0-915692-31-7
Library of Congress Cataloging in Publication Data

Workshop on Grand Unification, 1st, University of
New Hampshire, 1980.
First workshop on grand unification.

(Lie groups ; v. 11)
Organized by P. Frampton and others.
Includes bibliographical references.
1. Nuclear reactions--Congresses. 2. Unified field theories--Congresses. I. Frampton, Paul H. II. Glashow, Sheldon L. III. Yildiz, Asim. IV. Title.
QC793.9.W66 1980 539.7 80-25294
ISBN 0-915692-31-7

MATH SCI PRESS
53 JORDAN ROAD
BROOKLINE, MASSACHUSETTS 02146

Printed in the United States of America

TABLE OF CONTENTS

EDITORS' INTRODUCTION

This workshop held at the New England Center provided a timely opportunity for over 100 participants to gather in a unique environment and discuss the present status of the unification of strong and electroweak forces. One reason for the timeliness was perhaps that experiments of the seventies had already lent confirmation to the separate theories of strong and of electroweak forces, so that for the eighties it now seems especially compelling to attempt the grand unification of these two forces. Also, the planned experiments to search for proton decay and the new experiments which are suggestive, though not yet conclusive, of non-zero neutrino rest masses add further stimulus to the theory. Thus, the workshop provided an ideal forum for exchange of ideas amongst active physicists.

The presentations at the workshop covered the present status of both theory and experiment with a strong interplay. Also, there were presentations from the discipline of astrophysics which is becoming very intertwined with that of high-energy physics especially when in the latter one is addressing energies and temperatures that were extant only in the first nanosecond of the universe.

On experiment, we heard a comprehensive coverage of the four United States proton decay experiments. The Brookhaven-Irvine-Michigan experiment in the Morton Salt Mine at Fairport Harbor, Ohio was discussed by LARRY SULAK, while DAVID WINN talked on the Harvard-Purdue-Wisconsin effort in the Silver King Mine, Utah. MARVIN MARSHAK and RICHARD STEINBERG described respectively the Soudan Mine, Minnesota, and the Homestake Mine, South Dakota, experiments. The first three experiments are in construction while the Homestake Mine was able to quote a new lower limit on the proton lifetime of approximately 8×10^{30} years.

Another baryon-number violating effect predicted within some theoretical frameworks is neutron-antineutron mixing and RICHARD WILSON explained his ideas on how to detect this effect.

On the question of neutrino masses and oscillations, the experiment at the Savannah River reactor, South Carolina, was discussed by FRED REINES. [Due to illness, Reines' did not present verbally his written contribution, included in these proceedings; in his place, we heard short contributions from JOHN LO SECCO, PIERRE RAMOND, and MAURICE GOLDHABER]. A survey of related neutrino experiments was given by DAVID CLINE.

The general phenomenological picture was painted by SHELDON GLASHOW who emphasized the increasing importance of passive-type high-energy experiments, as opposed to accelerator experiments; also, the key predictions of the various models were summarized.

The subject of grand unification overlaps naturally with cosmology especially the big-bang model for the universe. GARY STEIGMAN described work on neutrinos in cosmology and MICHAEL TURNER outlined the work of the Chicago group on the baryon asymmetry of the universe. The avoidance of superheavy monopoles in the early universe was explained by PAUL LANGACKER. Limits on heavy unstable particles were imposed by PAUL FRAMPTON.

The nature of symmetry breaking is a central issue in grand unification schemes. For unitary groups this was treated by HENRI RUEGG while the intricacies of exceptional groups were covered in FEZA GURSEY's lecture. Further aspects of symmetry breaking were presented in turn by RICHARD SLANSKY, LEONARD SUSSKIND, HOWARD GEORGI, and TERRY GOLDMAN.

One shortcoming of the simplest grand unification schemes is that they include only one family of quarks and leptons. In a larger symmetry group, the multiple families may receive some more justification. An example based on SU(7) was given by ASIM YILDIZ. An additional related contribution was made by KAMESHWAR WALI. JOHN ELLIS, appealing to extended supergravity, advocated an SU(8) structure.

The theory of massive neutrinos is of great interest because of the recent experimental indications of possible neutrino oscillations. The concept of a Majorana mass was explicated by EDWARD WITTEN while the cosmological ramifications were dramatized by ALVARO DE RÚJULA. PIERRE RAMOND and RABINDRA MOHAPATRA explained how neutrino masses are accommodated in grand unification models.

The most prevalent unification schemes are based on a picture of color confinement. An alternative model with non-confined color was the basis of the talk by JOGESH PATI, whose theory leads to predictions for proton decay which differ in detail from those of the conventional theory.

The expected properties of baryon and lepton nonconserving process were outlined by STEVEN WEINBERG. His analysis is based on general considerations of the structure of effective field theories and the strong and electroweak gauge symmetries. He also explained his general formalism for deriving effective Lagrangians by integrating out superheavy particles, and discussed the implications of the survival of a cosmic baryon number.

Taken all together, we believe the presentations at the workship gave a faithful and comprehensive representation of the state of the art in the physics of unification of strong and electroweak forces. There were, of course, many informal interchanges between the participants, in addition to the items contained in this book.

We would like to thank the participants whose attendance contributed so much to the success of the workshop.

Last but not least, we must thank Mrs. JOYCE CASH and her staff for tireless efforts both during, and for several months preceding, the workshop.

These efforts, together with the cooperation of the University of New Hampshire and of the staff of the New England Center, were essential to the smooth running of the workshop.

P. H. Frampton
S. L. Glashow
A. Yildiz

May 1980

WELCOME

Gordon Haaland
Vice-President for Academic Affairs, University of New Hampshire

My experience with physics was initially limited to a standard college course, but as I pursued my own academic career I became more familiar with physics through the philosophy of science. As I studied psychology, much of the concern that psychology had in terms of both theory and experiment led retrospectively to the literature of the thirties, forties and fifties in areas like the unity of science movement, the whole change in logical positivism, and the impact of contemporary quantum mechanics on the epistemology of science. I was fascinated to find many of the eminent physicists of the day being in the forefront of some of those epistemological issues, particularly the relation between theory and experiment.

Normally, my task is simply to say a few things on behalf of the university and get out of your way, and I don't intend to forget that signal responsibility.

I am fascinated by this conference, and believe that the University of New Hampshire is very honored by your presence here today partly because contemporary physics has continued to be in the forefront of raising the significant epistemological issues of science. It is clear that you are dealing with one of the most significant of our contemporary scientific problems at the forefront of our intellectual endeavors.

This workshop represents one of the real responsibilities and missions of a university, that is to provide a context for discussions on the frontiers of intellectual endeavor. The University of New Hampshire is committed to this type of opportunity. We are interested in, and believe in, the value and importance of a university to provide the opportunities for people such as yourselves to explore the breadth of our world and to explore the ranges of ideas. Consequently, we are committed to research.

We are very pleased that you are here. We hope that you find this to be a productive environment. We hope that the university can provide for you in these next few days the types of resources that will make this an exciting intellectual endeavor for you. We are pleased and honored both by the nature of the topic and the quality and importance of you people as participants.

Welcome to the University of New Hampshire!

We trust that the next several days of this workshop will be fruitful to you. We appreciate your coming. Thank you.

THE NEW FRONTIER

Sheldon Lee Glashow*

Lyman Laboratory of Physics
Harvard University
Cambridge, MA 02138

PARTICLE PHYSICS WITHOUT ACCELERATORS

Pions, muons, positrons, neutrons, and strange particles were found without the use of accelerators. More recently, most developments in elementary particle physics depended upon these expensive artificial aids. Science changes quickly. A time may come when accelerators no longer dominate our field: not yet, but perhaps sooner than some may think.

Important discoveries await the next generation of accelerators. QCD and the electroweak theory need further confirmation. We need to know how b quarks decay. The weak interaction intermediaries must be seen to be believed. The top quark (or the perversions needed by topless theories) lurks just out of range. Higgs may wait to be found. There could well be a fourth family of quarks and leptons. There may even be unanticipated surprises. We need the new machines.

On the other hand, we have for the first time an apparently correct *theory* of elementary particle physics. It may be, in a sense, phenomenologically complete. It suggests the possibility that there are no more surprises at higher energies, at least at energies that are remotely accessible. Indeed, PETRA and ISR have produced no surprises, even at energies many times greater than were previously studied. The same may be true for PEP, ISABELLE, and the TEVATRON. Theorists do expect novel high-energy phenomena, but only at absurdly inaccessible energies. Proton decay, if it is found, will reinforce belief in the great desert extending from 100 GeV to the unification mass of 10^{14} GeV. Perhaps the desert is a blessing in disguise. Ever larger and more costly machines conflict with dwindling finances and energy reserves. All frontiers come to an end.

You may like this scenario or not; it may be true or false. But, it is neither impossible, implausible,nor unlikely. And, do not despair nor prematurely lament the death of particle physics. We have a ways to go to reach the desert, with exotic fauna along the way, and even the desolation of a desert can be interesting. The end of the high-energy frontier in no way implies the end of particle physics. There are many ways to skin a cat. In

*Research supported in part by the National Science Foundation under Grant No. PHY77-22864.

this talk I will indicate several exciting lines of research that are well away from the high-energy frontier. Important results, perhaps even extraordinary surprises, may await us. But, there is danger on the way.

The passive frontier of which I shall speak has suffered years of benign neglect. It needs money and manpower, and it must compete for this with the accelerator establishment. There is no labor union of physicists who work at accelerators, but sometimes it seems that there is. It has been argued that plans for accelerator construction must depend on the "needs" of the working force: several thousands of dedicated high-energy experimenters. This is nonsense. Future accelerators must be built in accordance with scientific, not demographic, priorities. The new machines are not labor-intensive, and must not be forced to be so. Not all high energy physicists can be accommodated at the new machines. The high-energy physicist has no guaranteed right to work at an accelerator, he has not that kind of job security. He must respond to the challenge of the passive frontier.

1. CP PHENOMENOLOGY

Here is a small but important enterprise: the search for the electric dipole moment of the neutron. The theorist is confident that the effect does not vanish, but it has not yet been found. One line of thought requires a dipole moment of order 10^{-24} cm. In another, it is expected to be a million times smaller. Which view is correct, if either, will soon be determined by experiment. It is a result of the greatest theoretical interest. In a similar vein is a precision study of the CP violation in K decay. It is essential to know whether or not there are measurable departures from the superweak model. Both of these examples are in the way of loose ends that have been passed over in the push to higher energies. No great surprises await us here, just important and basic physics. There are many other such examples. In this lecture, we are out for bigger game.

2. NEW KINDS OF STABLE MATTER

It has been suggested that there exists a very strong but unobserved interaction that sets the scale of weak interaction effects. Associated with these new forces are new particles with masses between 100 GeV and 100 TeV. (In these technicolor scenarios, the lower reaches of the desert are made to bloom.) Some of these particles may be reasonably stable, so that the particles or their effects may be seen today.

With lifetimes shorter than 10^{10} years, the heavy particles will have already in large measure decayed. Relic high energy neutrinos or photons would be their only spoor. With longer lifetimes, we might see them decaying in real time. Experiments that have been done put severe constraints on the concentration and lifetime of these hypothetical particles. Experiments that will be done can obtain stronger constraints, or perhaps reveal the new particles. Paul Frampton will address this subject in his talk.

Here, I wish to consider the possible existence of new forms of *stable* matter. Imagine that such matter was produced in the big bang, and that it cohabits with us today. To be specific, I shall speak of singly charged heavy particles of subnuclear size with or without nuclear interactions, called $X^{\pm}$ and with unknown masses somewhere between 100 GeV and 100 TeV. What is the fate of such particles?

The X^- could defend itself against X^+X^- annihilation by binding to a nucleus. Binding to a proton would produce a neutral system. It would be subject to fusion processes to yield ^{4}He X^-, which would behave chemically like a super-heavy Hydrogen isotope. Alternatively, X^- could bind to other common nuclei yielding superheavy atoms of Z one less: superheavy Al or Mn are interesting possibilities. On the other hand, X^+ would be expected to form superheavy hydrogen as an end-product unless the X^+ has strong couplings to nucleons. If it did, ^{4}He X^+ would be anticipated as a superheavy Lithium isotope. If $X^{\pm}$ is discovered to exist, its distribution in nuclei will be important both to establish the properties of X-matter and as a key to the nature of the early universe. Meanwhile, these arguments may be used to suggest plausible sites (like Manganese nodules) wherein to discover superheavy matter. I am aware of no experiments that put useful limits on the terrestrial abundance of X-matter, except for the fact that no one has encountered a nugget of the stuff.

The putative superheavy matter should be easily concentrated by centrifugation. Several possibilities present themselves for detection.

P. Horowitz suggests detection by back-scattered non-relativistic Protons. Back-scattering from heavy nuclei involves a maximum recoil energy loss of 2%. Back-scattering from a superheavy isotope leads to no perceptible energy loss.

Detection may be accomplished by an e/m measurement at a tandem Van de Graaf. Conventional sputter sources should produce superheavy ions effectively. Intervening foils can disrupt polymers and molecules that could mimic superheavies. Detectors must be tuned to the superheavy regime.

Here is an ambitious and risky field of scientific endeavor. What could be more exciting than the discovery of an entirely new kind of stable matter? What enterprise could *seem* less likely to produce a positive result? Do I have any takers?

3. NEUTRINO MASSES AND NEUTRINO MIXING

The only good symmetry is a gauge symmetry. Everything that can happen does happen. Neutrinos should have masses, and they should mix. The only open question is how big the effect is. Particle physics is controlled by the unifying mass $\sim 10^{14}$ GeV and the weak mass $G^{-1/2} \sim 100$ GeV. Neutrino masses, being a $\Delta I = 1$ weak effect, do not arise in lowest order. They depend upon the unifying mass scale. They are suppressed by $\sim 10^{-12}$ compared to ordinary masses. This loose argument gives very small neutrino masses, $\sim 10^{-3}$ eV at best. Perhaps it is wrong by one or more powers of α. Only experiment can tell. It is not implausible that neutrino masses are 1 eV or larger and that there is substantial mixing. This would provide a variety of experimentally measurable parameters: four angles and three masses in a three fermion family picture. The experimenter needs the challenge, the theorist needs the hints. (Remember: no one has plausibly predicted the top quark mass.)

There are limits of various kinds on neutrino mixing effects. There are even some indications of an effect. Many kinds of experiment are relevant:

(1) *Accelerator Experiments*: The original experiment of Lederman, Schwartz, and Steinberger established the existence of two neutrinos, and put weak limits on the mixing. Better experiments have been done and still better ones can be done. A beam enriched in energetic electron neutrinos can search for $\nu_e - \nu_\tau$ mixing. A recent BNL experiment using a beam of essentially pure ν_μ can look more closely for $\nu_\mu - \nu_e$ mixing. FNL bubble chamber experiments put limits on $\nu_\mu - \nu_\tau$ mixing. Beam dump experiments seem to yield curious and unexpected results. Much remains to be done, and many good experiments suggest themselves.

(2) *Reactor Experiments*: Whatever mixing $\bar{\nu}_e$ is subject to will show up as an anomalously small cross section for charged current processes. Uncertainties in flux can be compensated by measurement of neutral-current processes. Several experiments are in progress and will be reported on at this conference.

(3) *Meson Factories*: can provide copious sources of neutrinos. These can be used to study $\bar{\nu}_e - \bar{\nu}_\mu$ mixing, or any variety of ν_e mixing. Published data reveal no indication of any effect, but again, more precise work can be done. A marvelous neutrino beam will soon be available at LAMPF.

(4) *Solar Neutrinos*: Here, there is an indication of a large mixing effect. Any neutrino masses greater than 10^{-6} eV can be relevant. More decisive experiments can be done, but they require large quantities of exotic and expensive materials like Gallium or Indium.

(5) *Beta Decay Physics*: Careful experiments have put a limit of 60 eV on the mass of the electron neutrino, from the study of tritium beta decay. It is not impossible that the tau neutrino mass is of order ~30 eV, and that there is considerable $\nu_e - \nu_\tau$ mixing. In this case, the endpoint region of the tritium Kurie plot should display a curious bimodal behavior, with a "glitch" occurring 30 eV before the endpoint. The detection of such an effect is difficult, but perhaps possible. It would seem worthwhile to try.

Better experiments are being attempted in Guelph and in Moscow.

Searches for neutrinoless double beta decay have put a limit of ~1 KeV on the mass of the electron neutrino. Better experiments may be possible. In particular, we suggest the search for the neutrinoless double K capture process.

(6) *Atmospheric Neutrinos*: Neutrinos produced by interactions of primary cosmic rays in the atmosphere produce a 2/1 admixture of muon and electron neutrinos. This fact has not been verified, and it could be altered by neutrino mixing effects. With upward moving neutrinos taken into account, neutrino masses as small as 3×10^{-3} eV can be detected. Some data will result from planned proton decay experiments. Dedicated neutrino-mixing experiments should also be designed.

(7) *Extraterrestrial Sources*: Other neutrino sources can be imagined. We have mentioned relic neutrinos from the decay of shortlived ($<10^{10}$ years) superheavy particles. Supernovae in our galaxy are another possibility, but nearby events are not frequent. Continuous observation at several permanent underground stations would be desirable. Let us not miss the next nearby supernova! Time distributions of neutrinos from such a source can shed considerable light on neutrino masses and mixing.

4. ASTROPHYSICAL NEUTRINO PHYSICS

This subject is large, growing, and unfamiliar to me. My remarks are incomplete and perhaps irresponsible. There is a well known solar neutrino problem. It may have an orthodox explanation, or it may be the first indication of neutrino mixing. The Chlorine experiment is sensitive to an energetic minority of solar neutrinos which arise from a minor solar nuclear process. The Gallium experiment, which would be sensitive to better understood low energy neutrinos, requires fifty tons of Gallium. This experiment, or an equivalent one, deserves the highest priority. Solar neutrino mixing is sensitive to smaller neutrino masses than any other conceivably observable process. It is a beautiful example of fundamental physics done in the passive mode. Moreover, the Gallium will survive intact and constitute a National Strategic Reserve.

Another astrophysical argument for neutrino masses is suggested in the universe according to Guth and Tye, who conclude that the universe is neither open nor closed, but is flat. (This is an appealing possibility quite independently of Guth and Tye). The visible mass of the universe is insufficient for this: something like ten times more mass is needed. Suppose this missing mass resides in relic neutrinos.

The relic black-body radiation consists of photons, gravitons, and neutrinos. There are about 100 neutrinos/cm^3 of each species. Should the heaviest neutrino weigh about 30 eV, these neutrinos would account for the needed missing mass.

Astronomers have discovered that each galaxy is surrounded by a large and more massive halo of non-luminous matter. Some believe the halo consists of burnt-out stellar remnants. If this is so, why is the luminous matter confined to a smaller volume within the halo? Another possibility is that the halo consists of a gravitationally bound cloud of massive neutrinos which were captured during galaxy formation.

The bound neutrinos must be massive in order that their velocities are below escape velocity from the galaxy and that their densities are within the degeneracy limit. This requires them to exceed ~20 eV in mass. What a convenient place to put the neutrinos that provide the missing mass of the universe!

The heavy neutrinos that dominate our universe may occasionally decay into lighter neutrinos plus a photon. This produces UV photons. Alvaro De Rújula will discuss the detectability of such a signal.

5. MAGNETIC MONOPOLES

The last subject I touch upon is the possible existence of magnetic monopoles. From the point of view of grand unified theories, they should exist in nature with masses of order 10^{16} GeV. Monopoles of such great mass cannot be expected to lie about on the surface of the Earth or the moon. They would fall through! Searches that have been done for ambient monopoles are thus almost entirely irrelevant.

Monopoles may be incident upon us as cosmic rays. Their kinetic energies, according to J. Preskill, should be $\sim 10^{10}$ GeV, hence their velocities are ~300 km/sec. Such slowly moving cosmic monopoles may not ionize. However, they are supersonic and will produce Cerenkov sound radiation. Alternatively, they may be detected as they pass through a superconducting loop and change the magnetic flux. Here is another kind of non-accelerator experiment that can explore the new frontier.

6. PROTON DECAY

This is the subject of the conference, and it is indeed the King of the new frontier. To some, it is a foregone conclusion that proton decay is about to be seen by experiments now abuilding. Afterwards, the branching ratios must be measured in detail. This is stressed in a recent work with A. De Rújula and H. Georgi. Here is a key, not only to the nature of the unifying group, but to the very origin of fermion masses and Cabibbo angles.

GRAND UNIFIED THEORIES WITHOUT SUPERHEAVY MAGNETIC MONOPOLES

Paul Langacker

Institute for Advanced Study
Princeton, New Jersey 08540

and

University of Pennsylvania
Philadelphia, PA 19104

Grand unified theories have many very attractive features, such as the approximately correct prediction of the Weinberg angle, the prediction of proton decay, and the possible explanation of the baryon asymmetry of the universe. However, one serious problem that has received much attention recently[1-6] is that many grand unified theories predict the existence of superheavy magnetic monopoles. These monopoles may have been produced prolifically soon after the big bang and could very well contribute too much to the energy density of the universe. In this talk I will briefly review the subject and then describe a model that I have developed in collaboration with S.-Y. Pi[6] in which the superheavy monopoles would never have been produced.

't Hooft-Polyakov magnetic monopoles[7] are classical configurations of the Higgs and gauge fields in a gauge theory that are topologically stable. This means that the Higgs fields far from the monopole are oriented in their internal symmetry indices in such a way that the fields cannot be smoothly rotated into a parallel configuration, as would be needed for the monopole to decay (Figure 1). Monopole solutions exist in a gauge theory[7] when a gauge group G is spontaneously broken down to an unbroken subgroup $H = h \times U_1$ containing an explicit U_1 factor. The monopole mass and magnetic charge are then of order $m_m \simeq M_X/\alpha_g$, $q_m \simeq 2\pi/g$, where M_X is a typical mass of a gauge boson coupling, and $\alpha_g = g^2/4\pi$. For the Georgi-Glashow SU_5 model,[8] for example, SU_5 is broken down to $SU_3^c \times SU_2 \times U_1$ by an adjoint representation of Higgs. Then, $M_X \simeq 10^{14}$ GeV is the mass of the leptoquark boson which mediates proton decay and $m_m \simeq 10^{16}$ GeV. A further breaking down to $SU_3^c \times U_1$ does not greatly alter this estimate.

Figure 1.

(a) A magnetic monopole. The arrows indicate the orientation of the Higgs field in internal symmetry space.

(b) The Higgs fields in three acausal domains which may form a monopole when they merge.

In an important paper, Preskill[2] has considered: (a) How many monopoles were produced in the very early universe when the temperature T was of order $T_i \simeq M_X$? (b) How many monopole-antimonopole pairs subsequently annihilated? (c) What are the observational limits on the number of monopoles remaining in the present universe?

Taking these questions in reverse order, Preskill argued that cosmic ray searches for monopoles would not have been sensitive to superheavy $(m_m \geq 10^{16}$ GeV) monopoles because they would be non-relativistic and would therefore not ionize strongly. Also, these monopoles would not be found in the earth's crust because they could be gravitationally pulled to the center of the earth. However, Preskill argues that the ratio $r(T) \equiv n_m(T)/n_\gamma(T)$ of monopole to photon density must be less than 10^{-24} at the present time because otherwise the monopoles would dominate the mass density of the universe. The present density could have been affected[2,3] by gravitational condensation effects which may have brought monopoles and antimonopoles closely together where they could rapidly annihilate. This consideration does not affect the limit $r(t = 1 \text{ MeV}) < 10^{-19}$, however, which Preskill obtains by requiring that the monopoles did not dominate the energy density of the universe at the time of helium synthesis.

Preskill also considered the annihilation of $m - \bar{m}$ pairs. He concluded that if the initial density $r(T_i)$ exceeded 10^{-10}, then annihilation would have reduced r to $\simeq 10^{-10}$ at $T = 1$ MeV. If $r(T_i) < 10^{-10}$, however, the annihilation rate would be negligible and $r(T = 1 \text{ MeV}) = r(T_i)$. Preskill therefore concluded[2,9] that the observation limits require that $r(T_i)$ must have been less than 10^{-19}. Steigman has argued[10] that even more stringent limits may be required.

The estimate of the initial density $r(T_i)$ is the most difficult question. Presumably, the universe was initially in a phase in which the SU_5 symmetry was not spontaneously broken when the temperature T was large compared to T_i. As the universe cooled, it underwent a phase transition to a phase in which SU_5 was broken. Preskill[2] and Einhorn, Stein, and Toussaint[3] have argued that if this transition was second order (i.e., continuous in the VEV's of the Higgs fields), as shown in Figure 2a, then monopoles could have been prolifically produced by thermal fluctuations in the fields for $T \leq T_i$, where T_i is the temperature below which the symmetry is spontaneously broken. The fields in different regions of space that were not causally connected would have been oriented in different directions, leading to domains, as indicated in Figure 1b. When the domains collided, it would sometimes be impossible for the fields to continuously rearrange themselves into a parallel configuration. The resulting twists or knots are magnetic monopoles. Preskill[2] and Einhorn et al.,[3] estimated $r(T_i) \simeq 10^{-6}$ in this case,[11] thirteen orders of magnitude too large.

One very interesting solution to this problem, by Preskill,[2] Einhorn et al.,[3] and Guth and Tye,[4] is that the transition to the $SU_3^c \times SU_2 \times U_1$ invariant phase may have been strongly first order, or discontinuous in the VEV's, as shown in Figure 2b. In the Guth and Tye scenario, the SU_5 symmetry is first broken (by a second order transition) to an $SU_4 \times U_1$ invariant phase at $T_1 \simeq 10^{15}$ GeV and then to $SU_3^c \times SU_2 \times U_1$ by a first order transition at $T_2 \simeq 10^{13}$ GeV. (This can occur if a cubic term is added to the Higgs potential.) Below T_2, the transition from the metastable $SU_4 \times U_1$ invariant false vacuum to the true vacuum is supposed to occur by the nucleation and expansion of bubbles. Monopoles can be produced by the coalescence of these bubbles, but Guth and Tye argued that if the nucleation rate is small enough, there will be significant supercooling. In this case it may be possible to have a sufficiently small initial monopole density. Guth has also argued[12] that such a scenario would solve the horizon and critical density problems in cosmology. Bardeen has argued,[13] however, that if the nucleation rate is small enough to suppress monopoles, then the phase transition would never occur.

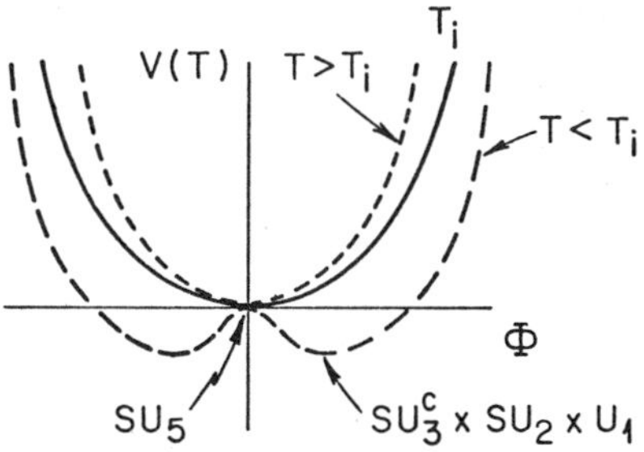

(a) The temperature dependent effective Higgs potential for a second order transition. The minimum varies continuously from an SU_5 invariant phase to an $SU_3^c \times SU_2 \times U_1$ invariant phase as the temperature is lowered.

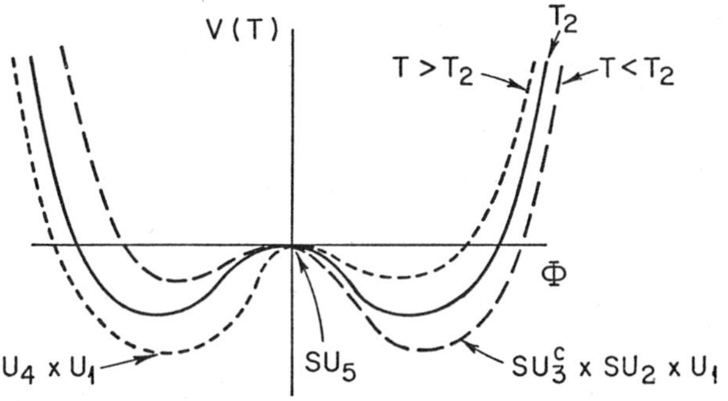

(b) The potential for a first order transition. For $T > T_2$ the $SU_4 \times U_1$ invariant vacuum is the true minimum. For $T < T_2$ this becomes a false vacuum, with the true vacuum $SU_3^c \times SU_2 \times U_1$ invariant.

Figure 2.

Another interesting solution, proposed by Lazarides and Shafi,[5] is that in some grand unified theories (not including SU_5, SO_{10}, or E_6) the monopole antimonopole pairs will be connected by magnetic flux strings of the Z^0 field, allowing for rapid annihilation.

I would now like to describe an alternative scenario for the suppression of monopoles, developed in collaboration with S.-Y. Pi,[6] in which the universe undergoes two or more phase transitions (which can be second order)

$$G \underset{T_1}{\rightarrow} H_1 \underset{T_2}{\rightarrow} \cdots \underset{T_n}{\rightarrow} H_n \underset{T_c}{\rightarrow} SU_3^c \times U_1^{EM} \quad . \tag{1}$$

The critical temperature T_c at which U_1^{EM} appears is $T_c \geq 1$ TeV. Since

$$T_c \lll m_m \simeq 10^{16}\ \text{GeV} \quad ,$$

no monopoles will be produced. U_1^{EM} does not appear an an explicit factor in H_n. Therefore, either U_1^{EM} is spontaneously broken for $T > T_c$, or it is contained in an unbroken subgroup of H_n. The latter possibility would actually lead to the production of light $(m_m \geq T_c)$ monopoles and would probably[14] imply an extremely short lifetime for the proton, which could decay via the interactions in H_n. We therefore choose the first possibility, i.e., that the U_1 of electromagnetism is broken for $T > T_c$. For example, SU_5 could break down to SU_3^c at $T_1 \leq M_X$ and undergo a second phase transition to the higher symmetry group $SU_3^c \times U_1^{EM}$ at $T_c \geq 1$ TeV. I will first display an explicit model which exhibits this behavior and then discuss the cosmological implications.

We consider a model which at $T = 0$ is the standard Georgi-Glashow SU_5 model,[8] with SU_5 broken to $SU_3^c \times SU_2 \times SU_1$, by an adjoint Higgs representation and then to $SU_3^c \times U_1^{EM}$ by one or more five-dimensional Higgs representations. We assume that a hierarchy exists, i.e., that

$$M_{W\pm,Z} \lll M_{X,Y}$$

and that the color triplet components of the Higgs fields have masses $\leq M_X$. For $0 \leq T \ll M_X$ we need only consider the $SU_2 \times U_1$ part of the model. (We assume that SU_3^c is never broken.)

Therefore, consider an $SU_2 \times U_1$ model with n complex Higgs doublets ϕ_i. It will turn out that at least three doublets (or two doublets and a singlet) are required, so we will take $n = 3$. The Higgs potential (at $T = 0$) is

$$V = \sum_{i=1}^{3} [-\mu_i^2\phi_i^\dagger\phi_i + \lambda_i(\phi_i^\dagger\phi_i)^2] \\ + \sum_{i<j} [\sigma_{ij}\phi_i^\dagger\phi_i\phi_j^\dagger\phi_j + \rho_{ij}\phi_i^\dagger\phi_j\phi_j^\dagger\phi_i + \eta_{ij}(\phi_i^\dagger\phi_j)^2 + \eta_{ij}^*(\phi_j^\dagger\phi_i)^2] \tag{2}$$

where we have imposed discrete symmetries under $\phi_i \to -\phi_i$ for simplicity. If the minimum of V occurs when only one doublet (e.g., ϕ_1) has a nonzero VEV $<\phi_i(0)>$, then $SU_2 \times U_1$ is broken down to U_1^{EM} and we can take

$$<\phi_1(0)> = \frac{1}{\sqrt{2}}\begin{pmatrix}0\\v_1\end{pmatrix} \tag{3}$$

If two doublets ϕ_1 and ϕ_2 both have nonzero VEV's, then either

$$<\phi_2(0)> = \frac{1}{\sqrt{2}}\begin{pmatrix}v_2\\0\end{pmatrix} \tag{4}$$

or

$$<\phi_2(0)> = \frac{1}{\sqrt{2}}\begin{pmatrix}0\\v_2\end{pmatrix} \tag{5}$$

which occur for ρ_{12} greater or less than $2|\eta_{12}|$, respectively. $SU_2 \times U_1$ is either completely broken or broken to U_1^{EM} for these two cases, respectively. We want U_1^{EM} to be unbroken at $T = 0$, but broken for $T > T_c$. Therefore, we take

$$\rho_{ij} > 2|\eta_{ij}| \quad , \tag{6}$$

so that the VEV's want to be orthogonal, but we will arrange the other parameters so that

$$<\phi_2(0)> = <\phi_3(0)> = 0 \quad .$$

This occurs for $\mu_1^2 > 0$, $\mu_{2,3}^2 < 0$, and

$$|\mu_i|^2 + \frac{\sigma_{1i}\mu_1^2}{2\lambda_1} > 0 \quad , \qquad i = 2,3 \quad . \tag{7}$$

Then v_1^2 is given by

$$\mu_1^2/\lambda_1 = (\sqrt{2}\,G_F)^{-1} \quad .$$

We also require

$$\lambda_i > 0$$
$$\sigma_{ij} > -\sqrt{\lambda_i \lambda_j} \tag{8}$$

which are sufficient conditions for V to be bounded from below.

For $T > 0$, the VEV's $\langle \phi_i(0) \rangle$ must be replaced by ensemble averages[15,16,17] $\langle \phi_i(T) \rangle$. Weinberg[16] has shown that the $\langle \phi_i(T) \rangle$ can be obtained, at least for sufficiently large T, by minimizing the effective potential

$$V(T) \equiv V + \sum_{i=1}^{3} \frac{T^2}{2} F_i \phi_i^\dagger \phi_i \quad , \tag{9}$$

where

$$F_i = \frac{3g^2 + g'^2}{8} + \lambda_i + \sum_{j \neq i} \left(\frac{\sigma_{ij}}{3} + \frac{\rho_{ij}}{6} \right) \tag{10}$$
$$+ \text{ Yukawa terms}$$

For small fermion masses the Yukawa terms in (10) will generally be negligible. If the F_i are positive, then for $T^2 \gtrsim 2\mu_1^2/F_1$ the coefficient of $\phi_1^\dagger \phi_1$ in $V(T)$ will be positive and the system will undergo a transition to a phase in which $SU_2 \times U_1$ is unbroken ($\langle \phi_i(T) \rangle = 0$). However, Weinberg[16] and more recently Mohapatra and Senjanovic[18] and Zee[19] have emphasized in analogous models that some of the F_i can be negative; in this case the symmetry need not be restored at high T.

It is even possible to have a transition to a state of lower symmetry.[16] We will choose parameters so that (in addition to (6)-(8)),

$$F_{1,2} < 0 \quad . \tag{11}$$

This turns out to require $F_3 > 0$, so that for sufficiently large T we may have a transition to the phase analogous to (4) with $SU_2 \times U_1$ completely broken. As an existence proof that (6)-(8) and (11) can be satisfied, choose

$$\lambda_1 = \lambda_2 \equiv \lambda \gg g^4, |\rho_{ij}|$$
$$\sigma_{12} > -\lambda$$
$$-\sigma_{13} = -\sigma_{23} \equiv \sigma > 3\lambda + \sigma_{12} + 3X \tag{12}$$
$$\lambda_3 > \sigma^2/\lambda$$

where

$$X = \frac{(3g^2 + g'^2)}{8} \simeq 0.16 \quad .$$

The condition $\lambda \gg g^4$ allows us to neglect radiative corrections to V. For a typical set of numbers, chosse

$$\lambda \simeq -\sigma_{12} \simeq g^2 \simeq 0.4 \quad ,$$

$$\sigma \gtrsim 1.3 \, , \quad \lambda_3 \gtrsim 4.1 \quad .$$

The only purpose of introducing ϕ_3 was to lower the energy of ϕ_1 and ϕ_2 at high temperatures because of their coupling to ϕ_3. We see that there is a range of parameters which satisfy conditions (6)-(8) and (11), but a rather large value for λ_3 is required. This value is not so large as to violate tree level unitarity, which would occur[20] for $\lambda_3 \gtrsim 8\pi/3$, but it may lead to serious difficulties with the renormalization group equations[21] for running quartic couplings.[22]

For large T, the effective mass quantities $M_1^2(T)$ and $M_2^2(T)$ defined by

$$M_i^2(T) \equiv \mu_i^2 - \frac{F_i T^2}{2} \quad ,$$

will be positive. V(T) will have an extremum analogous to (3) and (4), with

$$\begin{pmatrix} v_1(T)^2 \\ v_2(T)^2 \end{pmatrix} = \frac{\begin{pmatrix} \lambda_2 & -\sigma_{12}/2 \\ -\sigma_{12}/2 & \lambda_1 \end{pmatrix} \begin{pmatrix} M_1^2(T) \\ M_2^2(T) \end{pmatrix}}{\lambda_1 \lambda_2 - \sigma_{12}^2/4} \qquad (14)$$

and $\langle \phi_3(T) \rangle = 0$. This will be (at least) a local minimum if

$$\begin{aligned} & v_{1,2}^2 > 0 \\ & 2 > \frac{\sigma_{12}}{\sqrt{\lambda_1 \lambda_2}} > -1 \\ & |M_3^2| + \frac{\sigma_{13} v_1^2}{2} + \frac{\sigma_{23} v_2^2}{2} > 0 \quad . \end{aligned} \qquad (15)$$

The second order transition between these phases occurs at T_c such that $v_2(T_c) = 0$. For the special case (11), these conditions are fulfilled if

$$2 > \sigma_{12}/\lambda > -1 \qquad \text{and} \qquad |\mu_2^2| > \mu_1^2 \ .$$

In this case,

$$|F_1| = |F_2| \leq O(\lambda \simeq g^2) \ ,$$

$$F_3 \simeq \lambda_3$$

and the phase transition occurs for

$$T_c = \frac{A\mu_1}{\sqrt{\lambda}} = (246 \text{ GeV})A \ , \tag{16}$$

where

$$A = \left[\frac{[(\lambda|\mu_2|^2)/\mu_1^2] + \sigma_{12}/2}{\frac{1}{2}|F_2|(1 - \sigma_{12}/2\lambda)} \right]^{1/2} \ . \tag{17}$$

A is typically of order unity, but can be made much larger or smaller by adjusting parameters. We will assume $T_c \gtrsim 1$ TeV.

I have therefore demonstrated the existence of a model for which $SU_3^c \times SU_2 \times U_1$ is broken to SU_3^c for $T > T_c$. I will now discuss the cosmological implications for various stages in the evolution of the universe, starting with the present $(T \gtrsim 0)$ and working backwards to earlier times (higher temperature), as indicated in Figure 3.

For $T < T_c$ the symmetry is $SU_3^c \times U_1$, so ordinary physics at $T \gtrsim 0$ and nucleosynthesis at $T \simeq 1$ MeV are unaffected.

Stable superheavy monopoles could, in principle, exist for $T < T_c$, but not for $T > T_c$. However, the number

$$r \simeq \exp(-m_m/T_c) \simeq \exp(-10^{13})$$

expected to be produced by thermal fluctuations is utterly negligible.

For $T_c < T < T_n$ the unbroken symmetry is SU_3^c, with the U_1 of electromagnetism spontaneously broken! During this period, the photon has a mass and electric charge is violated. One might worry that a charge excess produced during this period would become frozen in for $T < T_c$, leading to observable consequences. However, this does not occur to any significant extent.

For $T \gg T_c$ we have

$$v_1(T) \simeq v_2(T) \sim \frac{\sqrt{-F_i T}}{\sqrt{\lambda}} \lesssim T \ ,$$

$$m_1, m_2 \simeq \sqrt{-F_1}\ T < T \tag{18}$$

$$m_3 \simeq \sqrt{F_3}\ T \gtrsim T \quad ,$$

where the four massive gauge bosons (mixtures of $W^{\pm}$, Z, γ) have masses $\simeq g v_i \leq gT$. $m_{1,2}$ are the masses of the Higgs particle eigenstates which are mixtures of ϕ_1 and ϕ_2, and m_3 are the masses of the bosons in ϕ_3 (which do not mix with $\phi_{1,2}$). Fermion masses are of order

$$m_F(T) \sim \frac{m_F(0)}{v_i(0)} v_i(T) \sim m_F(0)\ G_F^{1/2}\ T \ll T \quad . \tag{19}$$

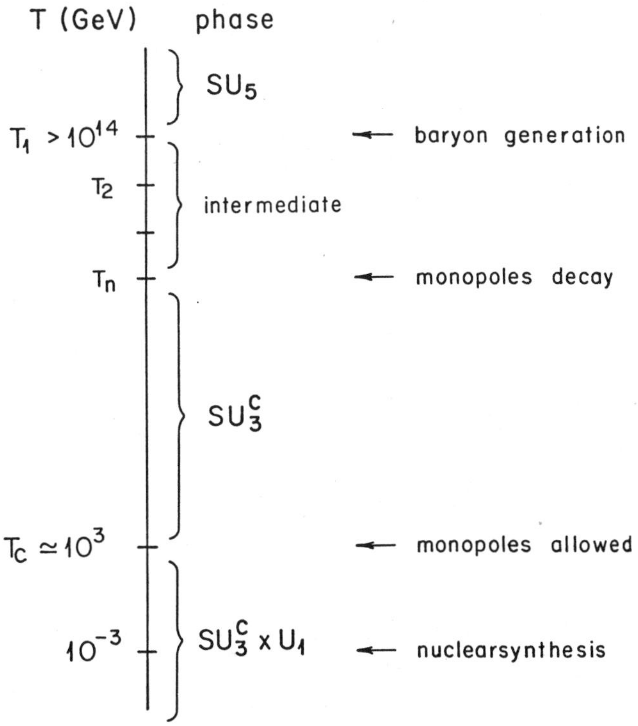

Figure 3. The unbroken symmetry at various times (temperatures) in the evolution of the universe.

The gauge boson masses $M \approx gT$ are negligible compared to the electron plasma frequency

$$\omega_p(T) \sim \left[\frac{4\pi n_e(T) e^2}{m_e(T)}\right]^{1/2} \simeq 400\,T \quad , \tag{20}$$

and can therefore be ignored. The fermion and (hopefully) the Higgs masses are small enough to not be problematic.

The reaction rate for charge violating reactions is[23]

$$\begin{aligned} \Gamma(T) &= \langle \sigma v \rangle_T \, n(T) \\ &\approx 10^{23}\, g\alpha^2 T(\mathrm{GeV})/\mathrm{sec.} \quad , \end{aligned} \tag{21}$$

where we have assumed

$$\langle \sigma v \rangle_T \simeq c \langle \sigma \rangle_T \approx \frac{\alpha^2 v_2(T)^2}{T^4}\, c \approx \frac{\alpha^2 c}{T^2} \tag{22}$$

and a number density

$$n(T) \sim \frac{g n_\gamma(T)}{2} \quad ,$$

with[23]

$$g = \frac{g_B + 7g_F}{8} \gtrsim 100 \quad .$$

$g_{B,F}$ are the number of boson and fermion light degrees of freedom at T. This is large compared to t^{-1}, where

$$t(\mathrm{sec}) = 2.4 \times 10^{-6}\, g^{-1/2} T^{-2}\, (\mathrm{GeV})$$

is the age of the universe $(\Gamma(T)t(T) \sim 10^{14}/T\,(\mathrm{GeV}))$, so the charge violating reactions are in equilibrium for $T \geq T_c$. There will be a small net charge density n_Q in the present universe left over from fluctuations from equilibrium at $T > T_c$. Only charge fluctuations on the scale of the observable universe are distinguishable from the standard scenario, so we will assume a total charge $N_Q \leq \sqrt{N_\gamma}$ in the observable universe (actually, the net charge will probably be much smaller because charged Higgs bosons become massless for $T \sim T_c$; they could be produced prolifically out of the vacuum to neutralize any excess charge produced earlier).[24] With $N_\gamma \simeq 10^{86}$, this implies

$$n_Q < 10^{-43} n_\gamma \sim 10^{-34} n_B$$

in the present universe, where n_B is the baryon density. This is far smaller than the observational limits[25,26] from galaxies and cosmology. Lyttleton and Bondi[25] have estimated that a charge excess of $n_Q/n_B \simeq 10^{-18}$ would be required to eject cosmic ray protons of energy $\leq 10^{19}$ eV from galaxies and that the repulsive electrostatic energy of particles in the universe would equal the attractive gravitational energy only for

$$\frac{n_Q}{n_B} \gtrsim \sqrt{G_N}\,\frac{m_p}{e} \simeq 10^{-18} \quad ,$$

assuming a photon mass m_γ (which is induced by the net charge) of order $m_\gamma \sim R^{-1}$, where R is the radius of the universe. Barnes[26] has shown that this limit is far less stringent if larger photon masses, which lead to a "short range" electrostatic force, are allowed.

For sufficiently early times, the temperature will be so high that superheavy particles can no longer be ignored. These new thresholds (or the T dependence of the coupling constants[18]) will presumably lead to new phase transitions. The most probable situation is that SU_5 will be restored for $T \gtrsim M_X$. There may also be intermediate phases in between the SU_5 and SU_3^c phases, perhaps involving U_1 factors in their unbroken subgroups. Stable monopoles could be produced during these intermediate phases, but they would become unstable and decay (or annihilate) as soon as the temperature dropped below T_n.

Fermion masses are always sufficiently small that the standard scenario for baryon generation is not affected.

We have constructed a model which avoids magnetic monopoles without affecting nucleosynthesis or baryon number generation and without significantly violating the charge neutrality of the universe. The model requires a rather complicated Higgs structure and potentially dangerous large self interactions for the scalars. This aspect and the phase structure near the grand unification scale are under investigation.

One possibly testable consequence of this scenario concerns hadron-hadron collisions in which the center of mass energy exceeds 1 TeV. To the extent that the collision can be described by an equilibrium fireball picture with a temperature greater than T_c, it may be possible to observe the nonconservation of charge in individual reactions.[27]

BIBLIOGRPHY

1. Zel'dovich, Ya. B., and M.Y. Khlopov, *Phys. Lett.* 79B, 239 (1979).

2. Preskill, J.P., *Phys. Rev. Lett.* 43, 1365 (1979).

3. Einhorn, M.B., D.L. Stein, and D. Toussaint, Michigan Preprint UM HE 80-1.

4. Guth, A.H., and S.-H.H. Tye, SLAC-PUB-2448.

5. Lazarides, G., and Q. Shafi, CERN TH 2821.

6. Langacker, P., and S.-Y. Pi, IAS Preprint.

7. 't Hooft, G., *Nucl. Phys.* B79, 276 (1974); A.M. Polyakov, *Pis'ma Eksp. Teor. Fiz.* 20, 430 (1974) [*JETP Lett.* 20, 194 (1974)]. For an introduction, see S. Coleman in *New Phenomena in Subclear Physics, Part A*, A. Zichichi (ed.), Plenum, New York, 1977, p. 297.

8. Georgi, H., and S.L. Glashow, *Phys. Rev. Lett.* 32, 438 (1974). See also A. Buras et al., *Nucl. Phys.* B135, 66 (1978).

9. Annihilations could produce[2] a small enough density if $m_m < 10^{11}$ GeV, but this would lead to an unacceptably short proton lifetime in the standard model. A small unification mass could be tolerated in theories with a stable proton (which can still generate a baryon asymmetry). See P. Langacker, G. Segre, and H.A. Weldon, *Phys. Lett.* 73B, 87 (1978) and *Phys. Rev.* D18, 552 (1978); M. Gell-Mann, P. Ramond, and R. Slansky, *Rev. Mod. Phys.* 50, 721 (1978); G. Segre and H.A. Weldon, Penn. Preprint UPR-0147T.

10. Steigman, G., private communication.

11. A small density could be obtained for an unacceptably large scalar self interaction.

12. Guth, A., to be published.

13. Bardeen, J., private communication.

14. Except for models with a stable proton (Ref. 9).

15. Kirzhnits, D.A., and A.D. Linde, *Phys. Lett.* 42B, 471 (1972) and *Ann. Phys.* 101, 195 (1976).

16. Weinberg, S., *Phys. Rev.* D9, 3357 (1974).

17. Dolan, L., and R. Jackiw, *Phys. Rev.* D9, 3320 (1974).

18. Mohapatra, R.N., and G. Senjanovic, *Phys. Rev. Lett.* 42, 1651 (1979); *Phys. Rev.* D20, 3390 (1979); CCNY Preprint HEP-7916.

19. Zee, A., *Phys. Rev. Lett.* 44, 703 (1980).

20. Lee, B.W., H. Thacker, and C. Quigg, *Phys. Rev.* D16, 1519 (1977), and references therein.

21. Cheng, T.P., E. Eichten, and L.-F. Li, *Phys. Rev.* D9, 2259 (1974).

22. The solutions tend to diverge for large momenta if the initial values are too large. See L. Maiani, G. Parisi, and R. Petronzio, *Nucl. Phys.* B136, 115 (1978) and N. Cabibbo et al., CERN Preprint TH 2683. It might be possible to avoid this problem by fine tuning parameters, introducing more Higgs fields (with smaller $|\sigma_{ij}|$), or by invoking higher order contributions to the equations.

23. See, for example, G. Steigman, in *Ann. Rev. Nucl. Sci.*, Vol. 29.

24. We thank S. Barr for this comment.

25. Lyttleton, R.A., and H. Bondi, *Proc. Roy. Soc. London* A252, 313 (1959).

26. Barnes, A., *Ast. Journal* 227, 1 (1979).

27. We thank G. Kane for this comment.

SYMMETRY BREAKING PATTERNS FOR UNITARY AND ORTHOGONAL GROUPS*

H. Ruegg

Stanford Linear Accelerator Center
Stanford University
Stanford, CA 94305

and

Universite de Geneve
Switzerland

Abstract

We discuss the spontaneous symmetry breaking pattern for SU(n) and O(10). It is based on the exact treatment of the absolute minimum of the Higgs potential as a function of scalar fields belonging to the fundamental *and* adjoint representations of SU(n), the spinor *and* adjoint representations of O(10).

1. INTRODUCTION

The unified model[1] of weak, electromagnetic and strong interactions based on the gauge group SU(5) has many attractive features. However, it still involves too many arbitrary parameters. One way to reduce this number is to imbed SU(5) into higher simple groups. Various schemes have been proposed by many authors,[2] but since no fully convincing solution has been found so far, it seems worthwhile to keep the discussion as general as possible.

In all schemes we have in mind, the Lagrangian is invariant under a given gauge group. The symmetry is spontaneously broken, that is, if a scalar Higgs field transforms as a representation of the group, some of its components develop nonzero vacuum expectation values (VEV). This defines a privileged direction in representation space and determines the pattern of symmetry breaking. The subgroup which leaves these VEV invariant remains unbroken.

The scalar fields may be elementary. In this case, one wants the Higgs potential to be renormalizable, which limits it to a polynomial of degree four. A natural condition for the nonzero VEV is to require that they minimize the Higgs potential. This is the criterion used here. It is also possible to

---

*Joint work with F. Buccella, Instituto di Fisica, Universiata di Roma, INFN Sezione di Roma and C.A. Savoy, Universite de Geneve, Switzerland and Centre de Recherches Nucleaires, Universite Strasbourg, France.

Work supported in part by the Department of Energy, contract DE-AC03-76SF00515, and in part by the Swiss National Science Foundation.

consider the scalar fields as bound states of the fundamental fields.[3,4] In the absence of a satisfactory dynamical theory, it turns out that in this case also it may be necessary to minimize an effective Higgs potential.[4] Hence, a general discussion of the absolute minima of scalar potentials is useful also for dynamical symmetry breaking schemes.

A general group theoretic discussion of the Higgs potential has been given by L.F. Li,[5] who considered scalar fields in various irreducible representations. However, this is insufficient. For example, to break SU(5) down to SU(3) × SU(2) × U(1) and eventually to the exactly conserved SU(3) × U(1) one needs at least two irreducible components. Various particular examples have been discussed in some approximate schemes.[6-8]

Here we consider Higgs fields belonging to the adjoint *plus* the fundamental representations of SU(n) and the adjoint *plus* the spinor representation of O(10). This is still not sufficient, but our result has the merit of being exact, i.e., we do not require any parameter to be small. This may be important if one studies the transition between one symmetry breaking regime to another, the parameters in the Higgs potential varying as functions of energy (or temperature). For example, we shall find that in the SU(5) gauge model, a continuous change in a certain ratio of parameters (see Section 4) changes the conserved subgroup from SU(4) to SU(3) × U(1). This may be relevant to the discussion of monopoles.

To our knowledge, no exact treatment of the spinor Higgs fields has been given so far. This may be due to certain unfamiliar properties of these representations. Generalization to O(n) is hampered by the exponential growth of the number of spinor components.

Results presented in this talk have been discussed in more detail elsewhere.[9,10] Some new features will be shown here.

2. SU(n): HIGGS POTENTIAL

Let the complex field H_i $(i = 1,\dots,n)$ transform as the fundamental, the Hermitean, traceless field ϕ_i^j as the adjoint representation of SU(n). The most general renormalizable potential of degree four, invariant (for simplicity)[15] under the discrete operation $\phi_i^j \to -\phi_i^j$, is, using the notation of Ref. 6:

$$V(\phi,H) = -\frac{1}{2}\mu^2 \operatorname{tr}\phi^2 + \frac{a}{4}(\operatorname{tr}\phi^2)^2 - \frac{1}{2}\nu^2 H^+H + \frac{\lambda}{4}(H^+H)^2 + \alpha H^+H \operatorname{tr}\phi^2 + \frac{b}{2}\operatorname{tr}\phi^4 + \beta H^+\phi^2 H \quad . \tag{2.1}$$

V has to be minimized with respect to all components of H and ϕ. However, we are interested in the symmetry breaking pattern, that is, the unbroken subgroups. This does not depend on the norm of H and ϕ, but only on their direction in representation space. Hence, we will discuss the absolute minimum of

$$F = \frac{b}{2} \operatorname{tr} \phi^4 + \beta H^+ \phi^2 H \tag{2.2}$$

keeping $\operatorname{tr} \phi^2$ and H^+H fixed. Diagonalizing ϕ_i^j

$$\phi_i^j = a_i \delta_i^j \tag{2.3}$$

$$\sum_{i=1}^{n} a_i = 0 \quad , \qquad \sum_{i=1}^{n} a_i^2 = \operatorname{tr} \phi^2 \tag{2.4}$$

F becomes

$$F = \frac{b}{2} \sum_{i=1}^{n} a_i^4 + \beta \sum_{i=1}^{n} |H_i|^2 a_i^2 \tag{2.5}$$

Clearly, if b and β are both positive, each sum in (2.5) has to be minimum; if b and β are both negative, each sum has to be maximum. If the signs are opposite, more discussion is required.

Minimizing with respect to H_i gives:[7]

$$|H_i| = 0 \quad , \qquad i = 1,2,\ldots,n-1 \quad . \tag{2.6}$$

If we choose

$$\begin{aligned} a_n^2 &< a_i^2 \quad (i = 1,2,\ldots,n-1) \qquad \text{when} \quad \beta > 0 \\ a_n^2 &> a_i^2 \quad (i = 1,2,\ldots,n-1) \qquad \text{when} \quad \beta < 0 \end{aligned} \tag{2.7}$$

with (2.6), one gets for F:

$$F = \frac{b}{2} \sum_{i=1}^{n} a_i^4 + \beta H^+ H \, a_n^2 \quad . \tag{2.8}$$

3. SU(n): SYMMETRY BREAKING PATTERN

We first minimize with respect to a_i $(i = 1,2,\ldots,n-1)$. The following Lemma, proved in Ref. 9, is very useful:

<u>Lemma</u>.* The absolute minimum or the absolute maximum of

$$f = \sum_{i=1}^{n-1} a_i^4 \tag{3.1}$$

where a_i are real numbers subject to the constraints

$$\sum_{i=1}^{n-1} a_i = \sigma \quad , \quad \sum_{i=1}^{n-1} a_i^2 = \rho^2 \tag{3.2}$$

for fixed σ and ρ^2 $(\sigma^2 \leq (n-1)\rho^2)$, occurs only if at most two of the $n-1$ variables a_i are distinct. Furthermore, the absolute maximum is obtained if at least $n-2$ variables a_i are equal. This lemma can be applied to the minimization of $\mathrm{tr}\,\phi^4$, where ϕ is a second rank tensor with real eigenvalues.

According to the lemma, one has:

$$n_1 \text{ times } a_1 \quad , \quad n_2 \text{ times } a_2$$

$$n_1 + n_2 = n - 1 \quad . \tag{3.3}$$

The subgroup structure after symmetry breaking is then:

$$SU(n_1) \times SU(n_2) \times U(1) \qquad \text{if } n_1 n_2 \neq 0$$

$$SU(n-1) \qquad \text{if } n_1 n_2 = 0$$

In order to find n_1 and n_2, rewrite (2.4) as

$$n_1 a_1 + n_2 a_2 = - a_n \tag{3.4}$$

$$n_1 a_1^2 + n_2 a_2^2 = \mathrm{tr}\,\phi^2 - a_n^2 \quad . \tag{3.5}$$

Solve for a_1 and a_2 and define

---

*It can be shown[16] that the lemma holds also for the function

$$g = b \sum_{i=1}^{n-1} a_i^4 + d \sum_{i=1}^{n-1} a_i^3 \quad .$$

$$x = \frac{n_1 - n_2}{\sqrt{n_1 n_2}} \,, \qquad \sin^2\theta = \frac{a_n^2}{\mathrm{tr}\,\phi^2} \cdot \frac{n}{n-1} \tag{3.6}$$

$$(0 \leq \theta \leq \pi/2)$$

The function F to be minimized can now be written

$$F = \frac{b}{2}\,\mathrm{tr}\,\phi^2 \left[\sin^4\theta\,(n^2 - 3n + 3) + 6\sin^2\theta\cos^2\theta + n(1 + x^2)\cos^4\theta \right.$$

$$\left. \pm 4x\sqrt{n}\,\sin\theta\cos^3\theta + \frac{\beta}{b}\,2\,\frac{H^+H}{\mathrm{tr}\,\phi^2}\,(n-1)^2\sin^2\theta \right] \frac{1}{n(n-1)} \tag{3.7}$$

The result of the minimization is given in Table I. For $b > 0$, $\beta < 0$, the solution depends on the ratio β/b.

Table I. SU(n) symmetry breaking pattern. The subgroups in the table leave invariant the minimum of the potential $F = (b/2)\ \mathrm{tr}\,\phi^4 + \beta H^+\phi^2 H$, $\phi \in$ adjoint, $H \in$ fundamental representation of SU(n).

A) n even	
$b > 0,\ \beta > 0$	SU(n/2) × SU((n/2) − 1) × U(1)
$b > 0,\ \beta < 0$*	SU(n/2) × SU((n/2) − 1) × U(1) SU((n/2) + 1) × SU((n/2) − 2) × U(1) - - - - - - - - - - - - - - - - - SU(n − 1 − m) × SU(m) × U(1) - - - - - - - - - - - - - - - - - SU(n − 2) × U(1) SU(n − 1)
$b < 0,\ \beta < 0$	SU(n − 1)
$b < 0,\ \beta > 0$	SU(n − 2) × U(1)
B) n odd	
$b > 0,\ \beta > 0$	SU((n − 1)/2) × SU((n − 1)/2) × U(1)
$b > 0,\ \beta < 0$*	SU((n + 1)/2) × SU((n − 3)/2) × U(1) SU((n + 3)/2) × SU((n − 5)/2) × U(1) - - - - - - - - - - - - - - - - - SU(n − 1 − m) × SU(m) × U(1) - - - - - - - - - - - - - - - - - SU(n − 2) × U(1) SU(n − 1)
$b < 0,\ \beta < 0$	SU(n − 1)
$b < 0,\ \beta > 0$	SU(n − 2) × U(1)

*Increasing the ratio β/b.

4. SU(n): DISCUSSION

The characteristics of the subgroup after symmetry breaking are the following:

1) One loses one rank. This is due to H in the fundamental representation. With $\nu = \lambda = \beta = 0$, one would get instead $SU(n-1) \times U(1)$ or $SU(n/2) \times SU(n/2) \times U(1)$ (n even) or $SU((n-1)/2) \times SU((n-1)/2) \times U(1)$ (N odd).

2) The subgroup is the product of at most three factors, with at most one U(1). This follows from the Lemma.

3) For $b > 0$ and $\beta < 0$, as the ratio β/b varies, one find $n/2$, respectively, $(n-1)/2$ different subgroups for n even, respectively, n odd.

4) SU(n-1) is obtained for $b < 0$, $\beta > 0$ and for part of the quadrant $b > 0$, $\beta < 0$. $SU(n/2) \times SU((n/2)-1) \times U(1)$ (n even) is obtained for $b > 0$, $\beta > 0$ and for part of the quadrant $b > 0$, $\beta < 0$.

5) For SU(5), the solution is: SU(4) for $b < 0$, $\beta < 0$ and part of the quadrant $b > 0$, $\beta < 0$. $SU(3) \times U(1)$ for $b < 0$, $\beta > 0$ and part of the quadrant $b > 0$, $\beta < 0$. $SU(2) \times SU(2) \times U(1)$ for $b > 0$, $\beta > 0$.

5. O(10): HIGGS POTENTIAL

We consider the antisymmetric Higgs scalar, $\phi_{ij} = -\phi_{ji}$ $(i,j = 1,\ldots,10)$ belonging to the adjoint representation 45, and the $16 + \overline{16}$ Higgs scalars χ in the spinor representation. We are interested in the most general renormalizable potential of degree four, excluding odd terms by requiring invariance under $\phi \to -\phi$.

Notice that:

1) The product $16 \times 16 \times 16 \times 16$ vanishes.

2) For the symmetric product of two 16's, one gets $(16 \times 16)_s = 10 + 126$ so that one might have only two independent invariants $16 \times 16 \times \overline{16} \times \overline{16}$.

3) From $16 \times \overline{16} = 1 + 45 + 210$ and $(45 \times 45)_s = 1 + 54 + 210 + 770$ one sees that there are only two independent invariants $16 \times \overline{16} \times 45 \times 45$.

From this it follows that the most general Higgs potential is:

$$V(\phi,\chi) = V_0 + V_s + V_a + V_i$$

$$V_0 = a\chi^+\chi + b\ \mathrm{tr}\ \phi^2 + c(\chi^+\chi)^2 + d(\mathrm{tr}\ \phi^2)^2 + e\chi^+\chi\ \mathrm{tr}\ \phi^2$$

$$V_s = \kappa \sum_{i=1}^{10} (\chi^T c\gamma_i\chi)(\chi^T c\gamma_i\chi)^* \tag{5.1}$$

$$V_a = \lambda \operatorname{tr} \phi^4$$

$$V_i = \mu\chi^+ \left(\sum_{i,j=1}^{10} \sigma_{ij}\phi_{ij} \right)^2 \chi$$

where c, γ are defined in Refs. 11, 12, 13.

6. O(10): ORBITS OF THE SPINOR REPRESENTATION AND SYMMETRY BREAKING

A given spinor χ can be considered as a point in a 16 dimensional space. Acting on χ with all the group elements of O(10), one gets a set of points called the orbit of χ. According to Michel and Radicati,[14] all smooth, real, invariant functions are stationary on what they call critical orbits. A fortiori, an absolute minimum of the Higgs potential will occur on a critical orbit.

To characterize these orbits, define the basic states χ_A $(A = 1,\ldots,16)$. They are eigenstates of five mutually commuting generators H_α $(\alpha = 1,\ldots,5)$ belonging to the Cartan subalgebra. The eigenvalues are $\lambda_\alpha^A = \pm\frac{1}{2}$, their product over α being positive for representation 16, negative for $\overline{16}$. Any spinor χ can be written as

$$\chi(c_A) = \sum_{A=1}^{16} c_A\chi_A \quad . \tag{6.1}$$

For a given basic state χ_A, define a $\chi_{\bar{A}}$ by changing the sign of four eigenvalues λ_α^A. For example, if χ_A is given by $(+\frac{1}{2}, +\frac{1}{2}, +\frac{1}{2}, +\frac{1}{2}, +\frac{1}{2})$, then a $\chi_{\bar{A}}$ is given by $(+\frac{1}{2}, -\frac{1}{2}, -\frac{1}{2}, -\frac{1}{2}, -\frac{1}{2})$. There are five possibilities for $\chi_{\bar{A}}$. It can be shown that any $\chi(c_A)$ can be transformed by an O(10) rotation into

$$\chi(\theta) = \chi_A \cos\theta + \chi_{\bar{A}} \sin\theta \tag{6.2}$$

Especially, any basic state χ_A can be transformed into another basic state $\chi_{A'}$. The invariant V_s of Eq. (5.1) becomes

$$V_s = \kappa|\chi^+\chi|^2 \sin^2\theta \cos^2\theta \quad . \tag{6.3}$$

For each value θ one gets a different orbit. The critical orbits correspond to extrema of V_s. For $\kappa > 0$, V_s is minimum for $\theta = 0$ and $\theta = \pi/2$. The corresponding little groups, i.e., the subgroups of O(10) leaving $\chi(0)$ or $\chi(\pi/2)$ invariant are SU(5) groups conjugate to each other. For $\kappa < 0$, V_s is minimum for $\theta = \pi/4$. The little group of $1/\sqrt{2}\,(\chi(0) + \chi(\pi/2))$ is O(7).

The set of orbits for $\theta \neq 0$, $\pi/4$, $\pi/2$ (modulo $\pi/2$) is called[14] generic and not critical. The little group is $O(6) \sim SU(4)$.

Hence, the symmetry breaking due to the spinor representation alone yields the subgroups $SU(5)$ or $O(7)$.

7. O(10): SYMMETRY BREAKING PATTERNS

It remains to minimize V_i and V_a in (5.1). Through an $O(10)$ transformation, one can always rotate the Higgs scalar ϕ^{ij} in such a way that the only nonzero components are $(\alpha = 1,\ldots,5)$

$$\begin{aligned} \phi_{2\alpha-1,2\alpha} &= -\phi_{2\alpha,2\alpha-1} \\ &= a_\alpha \\ \sum_{\alpha=1}^{5} a_\alpha^2 &= \mathrm{tr}\,\phi^2 \quad . \end{aligned} \tag{7.1}$$

Replacing (7.1) in (5.1), we get

$$V_a = \lambda \sum_{\alpha=1}^{5} a_\alpha^4 \tag{7.2}$$

Keeping $\mathrm{tr}\,\phi^2$ fixed, the minimum of V_a is obtained for $a_1 = a_2 = a_3 = a_4 = a_5$ if $\lambda > 0$ and $a_1 = a_2 = a_3 = a_4 = 0 \neq a_5$ if $\lambda < 0$, the unbroken subgroups being $SU(5) \times U(1)$ and $O(8) \times U(1)$, respectively.[5]

With (6.1) and (5.1), one obtains

$$\begin{aligned} V_i &= \frac{\mu}{4} \sum_{A=1}^{16} |c_A|^2 \sigma_A^2 \quad , \\ \sigma_A &= 2 \sum_{\alpha=1}^{5} \lambda_\alpha^A a_\alpha \quad . \end{aligned} \tag{7.3}$$

At fixed values of the a_α's, we define $\sigma^2 = (\mathrm{Min}\ \mu\sigma_A^2)/\mu$ so that the absolute minimum $V_i = (\mu/4)\chi^+\chi\sigma^2$ is obtained for any direction χ_{A_0} such that $\sigma_{A_0} = \sigma$.

Now one has to minimize

$$\begin{aligned} V_i + V_a &= \lambda \sum_{\alpha=1}^{5} a_\alpha^4 + \frac{\mu}{4} \chi^+\chi\sigma^2 \\ \sum_{\alpha=1}^{5} a_\alpha^2 &= \mathrm{tr}\,\phi^2 \quad , \end{aligned} \tag{7.4}$$

with respect to a_α, where

$$\sigma = \sigma_{A_0}$$

$$= \sum_\alpha 2\lambda_\alpha^{A_0} a_\alpha \quad .$$

As noticed before, the basic state χ_{A_0}, associated to $\sigma_{A_0} = \sigma$, can be transformed into any other basic state. Correspondingly, only the signs of the a_α's will be affected, $a_\alpha \to \pm a_\alpha$, in Eq. (7.3). Therefore, we take

$$\chi_{A_0} = \chi_1$$

$$\lambda_\alpha = \lambda_\alpha^1 = +\frac{1}{2} \, , \qquad (\alpha = 1,\ldots,5) \quad ,$$

$$\sigma = \sum_\alpha a_\alpha \quad ,$$

without any loss of generality.

We can now again apply the lemma of Section 3 to the minimum of $\lambda\,\Sigma_\alpha = a^4$ keeping $\Sigma_\alpha a_\alpha^2 = \mathrm{tr}\,\phi^2$ and $\Sigma\, a_\alpha = \sigma$ fixed.

It is then a matter of simple algebra[10] to find the absolute minimum of (7.4). In particular, one finds that $a_\alpha \neq 0$ and $a_\alpha \neq -a_\beta$ $(\alpha,\beta = 1,\ldots,5)$. The result is given in Table II.

Table II. O(10) symmetry breaking pattern. The subgroups in the table leave invariant the minimum of the potential:

$$V = \kappa \sum_{i=1}^{10} \chi^T c\gamma_i \chi (\chi^T c\gamma_i \chi)^* + \lambda\, \mathrm{tr}\, \phi^4 + \mu\chi^+ \left(\sum_{i,j=1}^{10} \sigma_{ij}\phi_{ij} \right)^2 \chi,$$

$\phi \in$ adjoint, $\chi \in$ spinor representation.*

$\lambda > 0$,	$\mu > 0$:	SU(3) × SU(2) × U(1)
$\lambda > 0$,	$\mu < 0$:	SU(5)
$\lambda < 0$,	$\mu > 0$:	SU(4) × U(1)
$\lambda < 0$,	$\mu > 0$:	SU(4) × U(1) or** SU(5)

*Table II gives the result for $\kappa > 0$. For $\kappa < 0$, see text.

**Depending on the ratio μ/λ.

The absolute minimum of the Higgs potential V of Eq. (5.1) is as follows. For $\kappa > 0$, the choice of the spinor χ on the critical orbit with invariance group SU(5) is the proper one to minimize both V_s and $V_i + V_a$. Minimizing with respect to a_i yields the subgroups of SU(5) given in Table II.

If, instead, $\kappa < 0$ and $\lambda\mu \neq 0$, it can be shown that in order to get the absolute minimum of V, χ cannot stay on the orbit with O(7) invariance group, which minimizes V_s. In general, χ will belong to the orbit defined by $\chi_A \cos\theta + \chi_{\bar{A}} \sin\theta$, with θ depending on the parameters of the potential V. If $|\kappa|$ is small enough, $\theta = 0$ (or $\theta = \pi/2$), and the pattern of symmetry breaking will again be given by Table II. Otherwise, the residual symmetry group will be a subgroup of rank 3 of SU(4).

Acknowledgements: H.R. thanks Prof. S.D. Drell for the hospitality at SLAC.

REFERENCES

1. H. Georgi and S.L. Glashow, *Phys. Rev. Lett.* 32, 438 (1974).

2. See for example, H. Georgi and A. Pais, *Phys. Rev.* D19, 2746 (1979); E. Farhi and L. Susskind, SLAC-PUB-2361 (1979); H. Georgi, *Nucl. Phys.* B156, 126 (1979); P. Frampton, *Phys. Lett.* 89B, 352 (1980); M. Claudson, A. Yildiz, and P.H. Cox, HUTP-80/A013 (1980). See also Refs. 8 and 13.

3. S. Coleman and E. Weinberg, *Phys. Rev.* D7, 1888 (1973).

4. S. Raby, S. Dimopoulos, and L. Susskind, Stanford University Preprint ITP-653 (1979).

5. L.F. Li, *Phys. Rev.* D9, 2977 (1972).

6. A.J. Buras, J. Ellis, M.K. Gaillard, and D.V. Nanopoulos, *Nucl. Phys.* B135, 66 (1978).

7. M. Magg and Q. Shafi, CERN Preprint TH 2729 (1979).

8. H. Georgi and D.V. Nanopoulos, *Nucl. Phys.* B128, 506 (1977).

9. F. Buccella, H. Ruegg, and C.A. Savoy, Geneva Preprint UGVA-DPT 1979/11-221.

10. F. Buccella, H. Ruegg, and C.A. Savoy, Geneva Preprint UGVA-DPT 1980/01-232.

11. R. Brauer and H. Weyl, *Am. J. Math.* 57, 425 (1935).

12. A. Pais, *J. Math. Phys.* 3, 1135 (1962).

13. M. Chanowitz, J. Ellis, and M.K. Gaillard, *Nucl. Phys.* B128, 506 (1977).

14. L. Michel and L.A. Radicati, *Ann. Phys.* (N.Y.) 66, 758 (1971).

15. The potential containing in addition terms odd in ϕ has recently been minimized. See H. Ruegg, SLAC-PUB-2518 (May, 1980).

16. See Ref. 15, where the following stronger result has been shown: For values of the variables a_i for which the first derivative of g is zero (stationary points), the matrix of second derivatives has eigenvalues of opposite signs if three variables a_i are distinct (saddle points). See also L. Michel, Ref. TH. 2716-CERN.

THE LIMITS ON UNSTABLE HEAVY PARTICLES*

Paul Frampton
Lyman Laboratory of Physics, Harvard University, Cambridge, MA 02138

In this talk, I will describe some recent work with Glashow[1] on the question: do there exist in our universe unstable heavy particles?

By heavy is meant that the mass M satisfies $M \gtrsim 300$ GeV; unstable means that the lifetime τ is, of course, finite but the principal interest here is in cosmological lifetimes $10^5 y < \tau < 10^{30} y$. Recall that the age of the universe is $a \simeq 2 \times 10^{10} y$. The first question is what limits can be already placed from existing experimental data? The second is whether these limits can be improved by the upcoming proton-decay experiments? The four United States proton-decay experiments are by the Harvard-Purdue-Wisconsin group at the Silver King Mine in Utah[2], by the Brookhaven-Irvine-Michigan group at the Fairport Harbor Mine in Ohio[3], by Minnesotans at the Soudan Mine in Minnesota[4], and by Pennsylvanians at the Homestake Mine in South Dakota.[5] All four are discussed in some detail at this workshop.[2-5]

The reasons for pursuing the question are several in number:

A. Curiosity

B. The energy desert between 100 GeV and 10^{14} GeV occurring in the simplest SU(5) and SO(10) grand unification schemes requires investigation.

C. The viability of certain technicolor phenomenologies rests on the existence of particles in this mass region.

D. Even if a multi-million dollar experiment fails to find proton decay, it is still useful to search for heavy particles and hence constrain theoretical model building. After all, a proton lifetime $\tau_p > 10^{33} y$ would not be disastrous for theory.

It is worth noting that the heavy particle decays discussed here cannot be confused with proton decay itself because the energy involved is much higher than the 1 GeV available there.

Of interest is the 3-dimensional parameter space spanned by the variables M (mass), τ (lifetime) and C (mean concentration or abundance per nucleon). Rather than M, we take a typical decay energy E(~M/3).

For a deep subterranean detector the most relevant scenarios for decay of the heavy particle are

(1) For lifetime τ short compared to the age of the universe a, all such heavy particles have already decayed, leaving only *relic neutrinos* which populate uniformly the present universe.

(2) For lifetimes τ comparable to and larger than a, we must consider *terran neutrinos* arising from present decays of heavy particles residing in the Earth.

(3) From the rock in the geographic vicinity of the detector, *endogenous muons* from heavy particle decay may be detected.

We will discuss these three scenarios in turn.

For scenario (1) we may detect either the relic neutrinos directly, or secondary muons produced by charged interactions near the detector.

Recall the basic parameters:

$$\text{Age of universe, } a = 2 \times 10^{10}\ \text{y}$$
$$\text{Velocity of light in vacuum, } c = 10^{18}\ \text{cm y}^{-1}$$
$$\text{Volume of visible universe} = 10^{85}\ \text{cm}^3$$
$$\text{No. of nucleons therein} = 10^{80}$$
$$\text{Mean nucleon density} = 10^{-5}\ \text{cm}^{-3}.$$

The flux of relic neutrinos is thus given by

$$J_r = 10^{13}\ c\ (1-e^{-a/\tau})\ \ \text{cm}^{-2}\ \text{y}^{-1}. \tag{1}$$

Consider a cubic volume detector of volume $V_d = \ell^3\ \text{cm}^3$ containing $6 \times 10^{23}\ \rho_d V_d$ nucleons where ρ_d = density. Since the neutrino-nucleon cross-section is $\sigma(\nu N) \sim 10^{-38}\ \text{cm}^2\ \text{GeV}^{-1}$, the total cross-section presented by the detector, for water $\rho_d = 1$, is

$$\Sigma_d = 6 \times 10^{-25}\ E'\ \ell^3\ \text{cm}^2. \tag{2}$$

The energy E' is the red-shifted value

$$E' = E\left[\frac{\tau}{a}\ \theta(a-\tau) + \theta(\tau-a)\right]. \tag{3}$$

Now, for an upward-going ν, there is essentially no background so we may conservatively estimate the limit of sensitivity as 5 events/y, that is

$$J_r\ \Sigma_d > 5\ \text{y}^{-1}. \tag{4}$$

Taking ℓ = 2000 cm as representative of the new experiments, this translates into

$$cE\left[\frac{\tau}{a}\ \theta(a-\tau) + \theta(\tau-a)\right]\ (1-e^{-a/\tau}) > 10^{-8}\ \text{GeV}. \tag{5}$$

The old limit may be well approximated by setting ℓ = 200 cm, thus changing the right-hand side of (5) to 10^{-5} GeV. This existing limit is from the South African gold-mine experiment of Reines et al.

The range of a muon in rock is $R \simeq 10^2\ \text{cm GeV}^{-1}$ and hence weakly-produced muons within a rock sphere of radius $R = 10^2\ E'$ cm may reach the detector, taking into account a solid angle factor $\sim \ell^2/R^2$. The mean rock density is 3 g cm^{-3} so secondary muons give a limit

$$J_r\ \Sigma_d \left(\frac{3R}{\ell}\right) > 5\ \text{y}^{-1}. \tag{6}$$

Note that the Earth is transparent for neutrino energies up to 50 TeV so the assumption of isotropy is good.

The new bound, from Eq. (6), translates into

$$CE^2 \left[\left(\frac{\tau}{a}\right) \theta(a-\tau) + \theta(\tau-a)\right]^2 (1-e^{-a/\tau}) > 7 \times 10^{-8}\ \mathrm{GeV}^2 \tag{7}$$

while the old bound is

$$CE^2 \left[\left(\frac{\tau}{a}\right) \theta(a-\tau) + \theta(\tau-a)\right]^2 (1-e^{-a/\tau}) > 7 \times 10^{-6}\ \mathrm{GeV}^2. \tag{8}$$

Next, consider terran neutrinos. Here, we need

$$\text{Volume of Earth} = 1.1 \times 10^{27}\ \mathrm{cm}^3$$
$$\text{Mean density of Earth} = 5.5\ \mathrm{g\ cm}^{-3}$$
$$\text{Surface Area of Earth} = 5 \times 10^{18}\ \mathrm{cm}^2$$
$$\text{Number of nucleons in Earth} = 4 \times 10^{51}.$$

Thus the flux of terran neutrinos at the Earth's surface is

$$J_t = 8 \times 10^{32}\ C\ \tau^{-1}\ e^{-a/\tau}\ \mathrm{cm}^{-2}\ \mathrm{y}^{-1}, \tag{9}$$

and leads to a limit

$$J_t\ \Sigma_d > 5\ \mathrm{y}^{-1}. \tag{10}$$

Exactly as for the relic neutrinos, we may consider secondary muons produced in the endogenous rock; this leads to the limit

$$J_t\ \Sigma_d \left(\frac{3R}{\ell}\right) > 5\ \mathrm{y}^{-1}, \tag{11}$$

where $R = 10^2\ E$ cm.

For heavy particles decaying in the Sun, the neutrino flux is down by a factor of several hundred, while for the Moon and other planets there is an even bigger reduction. Thus, only the particles which have settled in this planet need be considered.

The third and final scenario is where endogenous muons arise from decays in the rock immediately neighboring to the detector. In a sphere of radius $R = 10^2\ E$ cm the number of heavy particles decaying per year is

$$N = \frac{4}{3}\pi R^3\ (6 \times 10^{23})\ (3c)\ \tau^{-1}\ e^{-a/\tau}\ \mathrm{y}^{-1}. \tag{12}$$

The limit of sensitivity is then

$$N\left(\frac{\ell^2}{R^2}\right) > 5\ \mathrm{y}^{-1}. \tag{13}$$

This turns out to provide the best limit if, say, E = 1 TeV for all $\tau > 6 \times 10^8$ y.

Collecting results for the (new) limits together, we have the following inequalities for the domain which may be detected (or excluded by lack of detection!)

[Scenario 1] Relic neutrinos

$$CE \left[\left(\frac{\tau}{a}\right) \theta(a-\tau) + \theta(\tau-a)\right] (1-e^{-a/\tau}) > 10^{-8}\ \mathrm{GeV}.$$

[1A] Secondary muons from relic neutrinos

$$CE^2 \left[\left(\frac{\tau}{a}\right) \theta(a-\tau) + \theta(\tau-a) \right]^2 (1-e^{-a/\tau}) > 7 \times 10^{-8} \text{ GeV}^2.$$

[Scenario 2] Terran neutrinos

$$CE\, \tau^{-1} e^{-a/\tau} > 1.2 \times 10^{-28} \text{ GeV y}^{-1}.$$

[2A] Secondary muons from terran neutrinos

$$CE^2 \tau^{-1} e^{-a/\tau} > 8 \times 10^{-28} \text{ GeV}^2 \text{ y}^{-1}.$$

[Scenario 3] Endogenous Muons

$$CE\, \tau^{-1} e^{-a/\tau} > 1.7 \times 10^{-33} \text{ GeV y}^{-1}.$$

There are, of course, corresponding expressions for the old, existing limit.

To exhibit the results, we choose E = 1 TeV (that is, M ~ 3 TeV) and plot the allowed abundance C versus lifetime τ (see Figure 1). The existing bound is from endogenous muons if $\tau > 6 \times 10^8$ y and from secondary muons produced by relic neutrinos if $\tau < 6 \times 10^8$ y. The new bound is deduced similarly except that if $\tau < 1.4 \times 10^7$ y, the limit now comes from direct detection of relic neutrinos.

There are two main points to stress in concluding. Firstly, the existing limit is quite stringent and should be taken into account in constructing either grand unified theories or technicolor phenomenologies. As an example, suppose that the lifetime scales as M^{-5} and that the proton lifetime is $\tau_p = 10^{30}$ y. Then, an abundance of 10^{-9} per nucleon is excluded for masses M = 300 GeV to 30 TeV. This is not necessarily realistic but only illustrates that the present bound is highly non-trivial.

Secondly, we see that the new proton-decay mine experiments can usefully extend the limits.

References

(1) P. H. Frampton and S. L. Glashow, Phys. Rev. Letters 44, 1481 (1980).

(2) D. Winn, these proceedings.

(3) L. Sulak, these proceedings.

(4) M. Marshak, these proceedings.

(5) R. Steinberg, these proceedings.

*Research supported in part by the U. S. Department of Energy and by the National Science Foundation under Grant No. PHY77-22864.

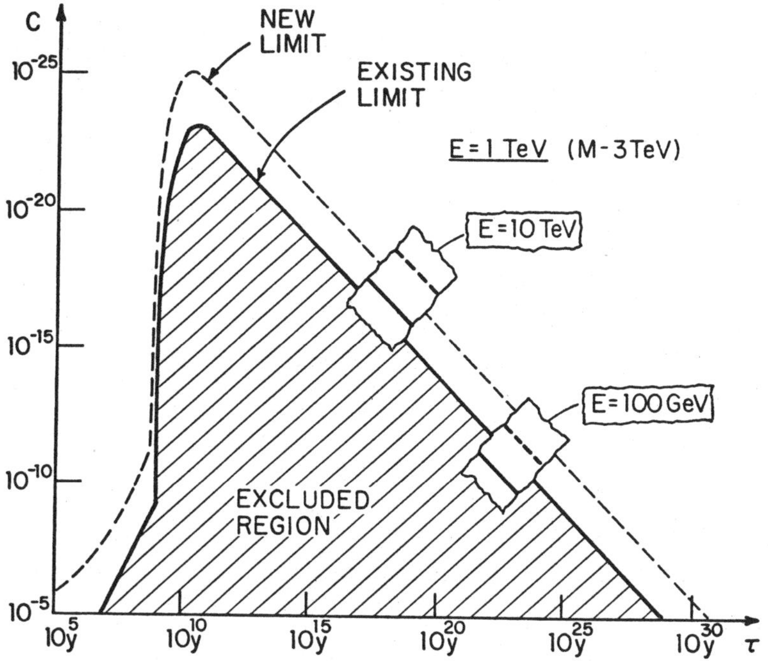

Figure 1: Plot of concentration C against lifetime τ for E = 1 TeV (corresponding to mass ~3 TeV). Shown are the present bound (solid line) and the bound (broken curve) accessible to the new proton decay experiments. In this plot, the existing bound is from endogenous muons ($\tau > 6 \times 10^8$ yr) and from secondary muons produced by relic neutrinos ($\tau < 6 \times 10^8$ yr). The new bound is from endogenous muons ($\tau > 6 \times 10^8$ yr), from secondary muons produced by relic neutrinos (1.4×10^7 yr $< \tau < 6 \times 10^8$ yr), and from direct detection of relic neutrinos ($\tau < 1.4 \times 10^7$ yr). Also shown by "windows" are portions of the curves corresponding to E = 100 GeV (M ~300 GeV) and E = 10 TeV (M ~30 TeV).

SYMMETRY BREAKING PATTERNS IN E_6 *

Feza Gürsey

Physics Department
Yale University
New Haven, CT 06520

Abstract

The two possible ways of using the exceptional group E_6 as a grand unified gauge model are discussed. Natural directions that occur in its breaking are exhibited through a generalization of a method employed by Michel and Radicati. The corresponding Higgs sector is analyzed both for elementary or composite Higgs scalars. The possible enlargement of the electroweak group from $SU(2)_W \times U(1)$ to $SU(2)_W \times SU'(2) \times U(1)$ is discussed in the two generations (topless) mode. The possibilities of neutron oscillations and neutrino masses are indicated in both topless and standard (t-quark containing) versions.

1. INTRODUCTION

I shall discuss the possibilities of the exceptional group E_6 as the gauge group of a grand unified theory. The talk will be based on previous[1,2] as well as ongoing work[3] in which I was involved, with some reference to publications by other authors.[4,5]

E_6 belongs to the exceptional series of simple Lie groups that comprises G_2, F_4, E_6, E_7, and E_8 as its members. All these groups are connected with octonions,[6] the way that SO(n), SU(n), and Sp(n) are respectively connected with real, complex and quaternionic numbers. G_2 is the automorphism group of the algebra of seven octonionic imaginary units. Six of those units obey an algebra (the Malcev algebra[7]) that admits SU(3) as an automorphism group. Hence this $SU(3)_c$ which we can identify with the color group[8] occurs naturally as a subgroup of all exceptional groups. Another SU(3) subgroup of F_4 and the E series arises from the triality property of octonions.[6,8] Hence, theories with 3n colored quarks are naturally embedded in the exceptional series. Also a global (B - L) group involving the baryon number minus the lepton number can be defined for the E series. The E_7 and E_8 groups, although the former is pseudoreal (i.e., admits complex representations that

---

* Research supported in part by the U.S. Department of Energy under Contract No. EY-76-C-02-3075.

are equivalent to their conjugates) and the latter is real (has no complex representations) admit as subgroup E_6 which is a complex group (with complex representations that are not equivalent to their conjugates). Thus the complex representations of E_6 can be classified by means of a global U(1) (that changes the phase of the 27-plet) and they will occur in pairs in the irreducible representations of E_7 and E_8. Such pairs will be split if the global U(1) is violated by interaction terms. There need not be such doubling in E_6, which is therefore the simplest grand unified exceptional group.

Interesting subgroups of E_6 that also have complex representations are $SO(10)\times U(1)$, $SU(6)\times SU(2)$, $SU(5)\times SU(2)\times U(1)$, and $SU(3)\times SU(3)\times SU(3)_c$. The color group is then a subgroup of SO(10), SU(6), or SU(5). All these groups have already been proposed as grand unified groups: SU(5) by Georgi and Glashow,[9] SO(10) by Fritzsch and Minkowski[10] and also by Georgi,[11] SU(6) by Segré and Weldon.[12]

In the E_6 theory, left handed fermion (lepton-quark) families are described by the complex 27-plet, which is the generic fundamental representation of the group. In E_7 and E_8 $\overline{27}$ multiplets will occur as well. E_6 being anomaly free,[1] one (27) per family does not conflict with renormalizability which, on the other hand, requires the combinations $(\bar{5}+10)$ for SU(5) and $(\bar{6}+\bar{6}+15)$ for SU(6). Those combinations fit naturally in E_6, since

$$27 = (\bar{5}+10) + (\bar{5}+5) + 1 + 1 \qquad \text{under } SU(5) \tag{1.1}$$

and

$$27 = \bar{6} + \bar{6} + 15 \qquad \text{under } SU(6) \; . \tag{1.2}$$

The fact that an SO(10) family is described by the 16 dimensional spinor representation of the group also finds a natural explanation in E_6, since

$$27 = 16 + 10 + 1 \qquad \text{under } SO(10) \; . \tag{1.3}$$

Finally, the possible $(3,\bar{3})$ structure of leptons[1] under a flavor $SU(3)\times SU(3)$ group follows from the decomposition

$$27 = (\bar{3},3,1) + (3,1,3) + (1,\bar{3},\bar{3}) \tag{1.4}$$

$$\text{under } SU(3)\times SU(3)\times SU(3)_c \; .$$

It is remarkable that the next exceptional group E_7 occurs in a non-compact version as a global symmetry group of the SO(8) supergravity theory.[13] Its SU(8) subgroup acts as a gauge group.[14]

Finally, E_8 has been proposed as a group that unifies eight SO(10) generations, two of them superheavy, by Bars and Günaydin.[15]

Eventually the E_6 model could be embedded in an E_8 theory that can be made supersymmetric because, in that scheme, both the fundamental fermions and the gauge bosons belong to the adjoint representations. Then, it is natural to expect that in the resulting E_6 gauge model not only leptoquarks and some Higgs scalars will be superheavy, but also some neutral (right handed) fermions. The natural smallness of left handed neutrino masses would then follow from the diagonalization of the mass matrix.[16]

In the following I shall discuss quark and lepton assignments in the topless and standard (t-containing) versions of the E_6 theory, the natural breaking directions of the group, the structure of the Higgs sector, and the possible distributions of the Higgs vev's (vacuum expectation values), the structure of the electroweak group and its possible extension to $SU(2)_W \times U(1) \times SU'(2)$ with the addition of three neutral currents.[17] I will conclude with remarks on neutrino masses[16,18] and the possibility of neutron oscillations.[19]

2. PARTICLE ASSIGNMENTS

A. Standard and Topless Fermion Assignments

Using the decomposition (1.4) for the 27-plet Ψ, we group leptons into a 3×3 matrix, the quarks and antiquarks into two triplets

$$L = \begin{pmatrix} \hat{N}^{\theta} - \hat{\theta}_R & \hat{\ell}_R \\ \theta^-_L - \nu^{\theta}_L & \beta^{\ell}_L \\ \| \quad\;\; \| & \| \\ \ell^-_L - \nu^{\ell}_L & \alpha^{\ell}_L \end{pmatrix} , \qquad q^i_L = \begin{pmatrix} U^i_L \\ | \\ D^i_L \\ \\ B^i_L \end{pmatrix} ,$$

$$\hat{q}^i_R = \begin{pmatrix} \hat{U}^i_R \\ \hat{D}^i_R \\ \| \\ \hat{B}^i_R \end{pmatrix} \tag{2.1}$$

where $(i = 1,2,3)$ is the quark color index. We have also indicated the doublet and singlet structure under the $SU(2)_W \times U(1) \times SU'(2)$ subgroup of the $SU(3) \times SU(3)$ flavor group: A bar joins the members of a doublet under $SU(2)_W$ that belongs to the Weinberg-Salam-Glashow group, while a double bar joins the members of a doublet under the neutral $SU'(2)$ group considered in Ref. 3 and also by Georgi and Glashow.[17] We use the notation

$$\hat{\psi}_R = i\sigma_2 \psi^*_R = (\psi^c)_L \quad .$$

The charges are as follows:

$$Q = -1 : \theta^-, \ell^- ; \qquad Q = \frac{2}{3} : U , \qquad Q = -\frac{1}{3} : D, B , \tag{2.2}$$

with the rest being neutral.

We have now two possible assignments.

1. The three generation (standard) assignment. The three 27-plets are denoted $\psi^{(e)}$, $\psi^{(\mu)}$, and $\psi^{(\tau)}$ by taking successively $\ell = e,\mu,\tau$; $U = u,c,t$; $D = d,s,b$. There are three new charged leptons $\theta^{(e)}$, $\theta^{(\mu)}$, and $\theta^{(\tau)}$ and three new quarks with charge $-1/3$: $B^{(e)}$, $B^{(\mu)}$, and $B^{(\tau)}$, plus a number of additional neutral leptons. All these additional fermions can be either superheavy (to within a few orders of magnitude of the Planck mass) or heavy (beyond the range of present accelerators). This is a 9-quark model containing the present standard 6-quark model including the top quark t.

2. The two generation (topless) assignment. There are only two 27-plets $\psi^{(1)}$ and $\psi^{(2)}$. For those we have $\ell = e,\mu$; $\theta = \tau,M$; $U = u,c$; $D = d,s$; $B = b,h$. The model contains the additional charged lepton M and the 6th quark h with charge $-1/3$. This gives the topless 6-quark model.

Under $SU(6)\times SU'(2)$ we have

$$27 \;=\; (\bar{6},2) + (15,1) \tag{2.3}$$

so that we can regroup the fermions in each generation into $\psi_{\bar{6}}$, $\psi'_{\bar{6}}$ and one 6×6 antisymmetrical matrix ψ_{15}, using the same notation as before for $SU(2)_W$ and $SU'(2)$ doublets.

$$\psi_{\bar{6}},\psi_{\bar{6}} : \begin{pmatrix} \hat{D}^1_R \\ \hat{D}^2_R \\ \hat{D}^3_R \\ \nu^\ell_L \\ \ell^-_L \\ \hline \alpha^\ell_L \end{pmatrix} = \begin{pmatrix} \hat{B}^1_R \\ \hat{B}^2_R \\ \hat{B}^3_R \\ \nu^\theta_L \\ \theta^-_L \\ \hline \beta^\theta_L \end{pmatrix}$$

$$\psi_{15} = \left(\begin{array}{ccccc|c} 0 & \hat{U}^3_R & -\hat{U}^2_R & -D^1_L & -U^1_L & B^1_L \\ -\hat{U}^3_R & 0 & \hat{U}^1_R & -D^2_L & -U^2_L & B^2_L \\ \hat{U}^2_R & -\hat{U}^1_R & 0 & -D^3_L & -U^3_L & B^3_L \\ D^1_L & D^2_L & D^3_L & 0 & \hat{\ell}_R & \hat{N}_R \\ -U^1_L & -U^2_L & -U^3_L & -\hat{\ell}_R & 0 & \hat{\theta}_R \\ \hline -B^1_L & -B^2_L & -B^3_L & -\hat{N}_R & -\hat{\theta}_R & 0 \end{array}\right)$$

It is seen that $\psi_{\bar{6}}$ and ψ_{15} contain respectively the $\bar{5}$ and 10 of Georgi and Glashow's SU(5) theory in the standard assignment. The additional fermions are separated by dotted lines. In this case, $\psi_{\bar{6}}^{\perp}$ and the additional fermions in $\psi_{\bar{6}}$ and ψ_{15} are heavy or superheavy. In the topless assignment a generation contains two $\bar{5}$ and a 5 of comparable masses besides the usual 10 of SU(5). Only α_L^{ℓ} and β_L^{θ} can be heavy or superheavy. They can be identified with the conjugates of the right handed neutrinos, namely

$$\alpha_L^{\ell} = \hat{\nu}_R^{\ell} \quad , \qquad \beta_L^{\theta} = \hat{\nu}_R^{\theta} \quad . \tag{2.5}$$

b. Gauge Bosons

We now turn to the structure of gauge bosons that belong to the adjoint representation (78) of E_6 and have the following decompositions

$$SU(3) \times SU(3) \times SU(3)_c : 78 = (8,1,1) + (1,8,1) + (1,1,8_c) + (3,3,\bar{3}) + (\bar{3},\bar{3},3) \tag{2.6}$$

$$SO(10): 78 = 45 + 1 + 16 + 16' \tag{2.7}$$

$$SU(6) \times SU'(2): 78 = (35,1) + (1,3) + (20,2) \tag{2.8}$$

$$SU(5) \times SU'(2): 78 = (24,1) + (5,1) + (\bar{5},1) + (10,2) + (\overline{10},2) + (1,3) + (1,1) \quad . \tag{2.9}$$

The SU(5) adjoint representation (24) corresponds to the electroweak bosons ($W_\mu^{\pm}$, Z_μ, γ_μ), the eight color gluons g_μ^a, the colored heavy gauge bosons X_μ and Y_μ which are color triplets and have respective charges 1/3 and 4/3. The (1,3) gauge bosons in (2.9) are the neutral SU(2)' gauge bosons U_μ^i (i = 1,2,3). Other additional bosons in 5, $\bar{5}$, 10 and $\overline{10}$ representations have the same charge and color structure as the fermion and antifermion families in the same representations, so that in E_6 there are new color triplet gauge bosons X'_μ and V_μ with charges ± 1/3 and ± 2/3. These additional gauge bosons give new modes in baryon decay.

c. Higgs Scalars (and Pseudoscalars)

Spontaneous symmetry breaking is induced by s = 0 Higgs particles that may be elementary or composite (with binding due to new gauge bosons). These effective Higgsions must have Yukawa coupling to fermions in order to give them masses in the tree approximation. On the other hand, they are always coupled to gauge bosons through the covariant deriviatives in the Higgs Lagrangian. Because of the Fermi-Dirac statistics of fermions they must occur in the symmetric part of (27×27):

$$(27\times 27)_s = \overline{27} + 351' \quad . \tag{2.10}$$

Thus, the products of the elements of $\psi_{27}^T i\sigma_2$ with the elements of ψ_{27} will belong to a linear combination of the representations $\overline{27}$ and $351'$ and transform like scalars and pseudoscalars associated respectively with the real and imaginary parts of these representations. They belong to the (10), (16), and (126) dimensional representations of SO(10). If any of the fermions in ψ_{27} are superheavy, then the Higgs fields $\phi_{\overline{27}}$, $\eta_{351'}$ develop superheavy vacuum expectation values (vev). If none of the 27 basic fermions are superheavy, then the only superheavy vev's are in representations that have no scalar Yukawa coupling with ψ_{27}, namely (351) in $(27\times 27)_a$ or (78) and (650) in $(27\times \overline{27})$. Starting with only $\phi_{\overline{27}}$ we can generate effective Higgs particles in higher orders through $\phi_{\overline{27}}\times\phi_{\overline{27}}$. Elementary or effective Higgs scalars χ_{78} and H_{650} will give masses not to the fermions, but to some vector gauge bosons.

d. Conservation and Violation of the Quark-Antilepton Number

Let us consider the hypercharge Y_F associated with the $SU(3)_F$ subgroup of the $SU(3)\times SU(3)$ flavor group in E_6. This $SU(3)_F$ leaves invariant the trace of the $(3,\bar{3})$ color singlet representation of $SU(3)\times SU(3)$. Under this $U(1)_F$ group we have the following Y_F quantum numbers for basic fermions

	U	D	B	U^c	D^c	B^c	ℓ	ν^ℓ	ℓ^c	$(\nu^\ell)^c$	θ	θ^c
Y_F:	$\frac{1}{3}$	$\frac{1}{3}$	$-\frac{2}{3}$	$-\frac{1}{3}$	$-\frac{1}{3}$	$\frac{2}{3}$	-1	-1	1	1	0	0

(2.11)

In the standard assignments, $U = u,c,t$; $D = d,s,b$; $\ell = e,\mu,\tau$; we see that the F hypercharge is given by

$$Y_F = B - L \quad , \tag{2.12}$$

where B is the baryon number and L the lepton number. This is not so in the topless assignment, since in that case $Y_F = -2/3$ for $B = b$ and vanishes for $\theta = \tau$. In processes where not only the electric charge U(1) subgroup of $SU(3)_F$, but also $U_Y(1)_F$ is conserved, we have the conservation of Q and $B-L$. An example is baryon decay due to the exchange of X and Y gauge bosons between two quarks. However, the exchange of gauge bosons X' and V with charges $\pm 1/3$ and $\pm 2/3$ will lead to baryon decays which violate Y_F, hence $B-L$, in the standard assignment.

3. SOME BREAKING PATTERNS FOR E_6

In this section I will show how the minimization of a Higgs potential leads to special Higgs fields[2] that satisfy definite algebraic relations which generalize those found by Michel and Radicati[20] for SU(3) and SU(3) × SU(3).

a. Algebraic Preliminaries

Let us first consider a scalar ϕ_{27} in the generic fundamental representation. From ϕ we can construct a quadratic invariant I_2, a quartic invariant I_4, and two cubic invariants I_3 and J_3 that combine in a complex C_3. We first write ϕ as a 3×3 complex octonionic matrix that is hermitian with respect to octonionic conjugation. In terms of the seven octonionic imaginary units e_A $(A = 1,\ldots,7)$, the four complex units u_α $(\alpha = 0,1,2,3,4)$ can be written

$$u_0 = \frac{1}{2}(1 + ie_7) \;, \qquad u_j = \frac{1}{2}(e_j + ie_{j+3}) \;, \qquad (j = 1,2,3) \tag{3.1}$$

together with their complex conjugates u^*_α. The multiplication rules are given by

$$u_0^2 = u_0 \;, \qquad u_0^* u_0 = 0 \;, \qquad u_0 u_i = u_i u_0^* = u_i \;, \tag{3.2a}$$

$$u_0^* u_i = u_i u_0 = 0 \;, \qquad u_i u_j^* = - u_0 \delta_{ij} \;, \qquad u_i u_j = \varepsilon_{ijk} u_k^* \;. \tag{3.2b}$$

Then, denoting octonionic conjugation by a bar, so that

$$\bar{e}_A = - e_A \;, \qquad \bar{u}_0 = u_0^* \;, \qquad \bar{u}_i = - u_i \tag{3.3}$$

we can write

$$\phi = \bar{\phi}^T = u_0 M + u_0^* M^T + u_i R^i + u_i^* S^i \;, \tag{3.4}$$

where M is a 3×3 complex matrix and R and S are complex and antisymmetrical. This corresponds to the algebraic decomposition (1.4). It can be rewritten in the form

$$\phi = \begin{pmatrix} \alpha & c & \bar{b} \\ \bar{c} & \beta & a \\ b & \bar{a} & \gamma \end{pmatrix} \;, \qquad \phi^\dagger = \bar{\phi}^{*T} = \phi^* \;, \tag{3.5}$$

where α, β, γ are complex numbers and a, b, c are complex octonions. Let us define the Freudenthal product $\phi \times \phi$ by

$$\phi \times \phi = \phi^{-1} \text{ Det } \phi$$

$$= \phi^2 - \phi \text{ Tr } \phi - \frac{1}{2} I \, [\text{Tr } \phi^2 - (\text{Tr } \phi)^2] \quad , \tag{3.6}$$

where

$$\text{Det } \phi = \alpha\beta\gamma - \alpha a\bar{a} - \beta b\bar{b} - \gamma c\bar{c} + (ab)c + \bar{c}(\bar{b}\bar{a}) \quad . \tag{3.7}$$

If ϕ transforms like (27), then $\phi \times \phi$ transforms like $\phi^*(\overline{27})$. Denoting the Jordan product (half the anticommutator) by a dot, we have

$$I_2(\phi) = \frac{1}{2} \text{Tr}(\phi^\dagger \cdot \phi) \;, \qquad I_4(\phi) = \frac{1}{2} \text{Tr}[(\phi \times \phi)^\dagger \cdot (\phi \times \phi)] \tag{3.8}$$

$$C_3(\phi) = I_3 + iJ_3$$

$$= \text{Det } \phi \tag{3.9}$$

$$= \frac{1}{3} \text{Tr}[(\phi \times \phi) \cdot \phi] \quad .$$

b. Generalized Michel-Radicati Relations

The most general quartic potential $V(\phi)$ that can be used in a renormalizable theory with an elementary ϕ_{27} reads

$$V(\phi) = -\mu^2 I_2 + \kappa_1 I_3 + \kappa_2 J_3 + \lambda_1 I_2^2 + \lambda_2 I_4 \quad . \tag{3.10}$$

Note that if ϕ is not elementary, we can write a potential of higher degree (say 6) in ϕ.

Because there are four invariants constructed out of ϕ, we can always diagonalize ϕ by a E_6 transformation and bring it to the canonical form

$$\phi_0 = e^{i\delta} \begin{pmatrix} \rho_0 & 0 & 0 \\ 0 & \sigma_0 & 0 \\ 0 & 0 & \tau_0 \end{pmatrix} \tag{3.11}$$

where ρ_0, σ_0, τ_0 and the phase δ are functions of the invariants I_2, I_3, J_3, and I_4. The relations are given by

$$I_2 = \frac{1}{2}\,\mathrm{Tr}(\phi^\dagger\cdot\phi)$$

$$= \frac{1}{2}\,\mathrm{Tr}(\phi_0^\dagger\cdot\phi_0)$$

$$= \frac{1}{2}\,(\rho_0^2+\sigma_0^2+\tau_0^2)\quad, \tag{3.12}$$

$$I_3 + iJ_3 = \mathrm{Det}\,\phi_0$$

$$= e^{3i\delta}\rho_0\sigma_0\tau_0\quad,$$

$$I_4 = \frac{1}{2}\,(\sigma_0^2\tau_0^2+\rho_0^2\tau_0^2+\rho_0^2\sigma_0^2)\quad. \tag{3.13}$$

At this point it is useful to introduce the triple Jordan product

$$\{ABC\} = (A\cdot B)\cdot C + A\cdot(B\cdot C) - (A\cdot C)\cdot B\quad. \tag{3.14}$$

The identity

$$\frac{1}{2}\,\{\phi\phi^\dagger\phi\} = -\,(\phi\times\phi)\times\phi^\dagger + \frac{1}{2}\,\phi\,\mathrm{Tr}(\phi\cdot\phi^\dagger) \tag{3.15}$$

can be proved by going to the canonical form.

Minimization of the potential gives

$$\phi^\dagger\times(\phi\times\phi) - \frac{1}{2}\,\phi\,\mathrm{Tr}(\phi^\dagger\cdot\phi) = -\frac{1}{2}\,c^2\phi\quad, \tag{3.16}$$

where c is some function of μ, κ_1, κ_2, λ_1 and λ_2. This relation can be rewritten in triple product form as

$$\{\phi\phi^\dagger\phi\} = c^2\phi\quad. \tag{3.17}$$

If $\mathrm{Det}\,\phi \neq 0$, we find

$$\phi\times\phi = k\phi^\dagger \tag{3.18}$$

as a solution. The other solution is the trivial one $\phi = 0$.

From Springer's identity

$$(\phi\times\phi)\times(\phi\times\phi) = \phi\,\mathrm{Det}\,\phi \tag{3.19}$$

that is easily proved by going to the canonical form, one obtains

$$\text{Det}\,\phi = \frac{2}{3}k\,I_2 \quad , \qquad k^{-1}\,\text{Det}\,\phi - I_2 = -\frac{1}{2}c^2 \quad , \tag{3.20}$$

so that

$$c^2 = \frac{2}{3}I_2 \quad . \tag{3.21}$$

Going to the canonical form, the relation

$$\phi_0 \times \phi_0 = k\phi_0^\dagger \tag{3.22}$$

gives

$$\rho_0^2 = \sigma_0^2 = \tau_0^2 \quad . \tag{3.23}$$

This leads to

$$2I_2^2 - 3I_4 = 0 \quad , \qquad k = \frac{3}{2}I_2^{-1}(I_3 + iJ_3) \quad . \tag{3.24}$$

If ρ_0, σ_0, τ_0 all have the same sign, we find

$$\phi_0 = \rho_0 I \quad , \tag{3.25}$$

which is invariant under the F_4 subgroup of E_6. Thus, in this case E_6 is broken spontaneously down to F_4.

If two of ρ_0, σ_0, τ_0 have opposite sign to the remaining one, ϕ_0 is invariant under $SO(9)$.

Now, consider the case

$$\text{Det}\,\phi = I\phi \cdot (\phi \times \phi) = 0 \quad . \tag{3.26}$$

Excluding the trivial solution $\phi = 0$, we have the possibility

$$\phi \times \phi = 0 \quad , \qquad k = 0 \quad , \tag{3.27}$$

where we have put this solution in the more general relation (3.18). In terms of ϕ_0 we obtain

$$\sigma_0\tau_0 = \rho_0\tau_0 = \rho_0\sigma_0 = 0 \tag{3.28}$$

The solution which corresponds to $\rho_0 \neq 0$ is

$$\phi_0 = e^{i\delta}\rho_0 \begin{pmatrix} 1 & 0 & 0 \\ 0 & 0 & 0 \\ 0 & 0 & 0 \end{pmatrix} \quad , \tag{3.29}$$

which is left invariant by $SO(10) \times O(2)$. In terms of the original invariants this case corresponds to

$$I_3 = J_3 = I_4 = 0 \quad , \qquad I_2 \neq 0 \quad . \tag{3.30}$$

Equation (3.26) can also be solved by a ϕ such that

$$\phi \times \phi \neq 0 \quad . \tag{3.31}$$

In this case,

$$\phi_0 = e^{i\delta} \begin{pmatrix} 0 & 0 & 0 \\ 0 & \sigma_0 & 0 \\ 0 & 0 & \tau_0 \end{pmatrix} \quad , \tag{3.32}$$

and we have

$$I_3 = J_3 = 0 \, , \qquad I_2 \neq 0 \, , \quad I_4 \neq 0 \quad , \tag{3.33}$$

$$c^2 = \mathrm{Tr}\, \phi^\dagger \phi = \sigma_0^2 + \tau_0^2 = 2I_2 \, , \qquad \{\phi\phi^\dagger\phi\} = 2I_2\phi \tag{3.34}$$

or,

$$\sigma_0^3 = \tau_0^3 \, , \qquad \sigma_0 = \tau_0 \quad . \tag{3.35}$$

The solution takes the form

$$\phi_0 = e^{i\delta}\sigma_0 \begin{pmatrix} 0 & 0 & 0 \\ 0 & 1 & 0 \\ 0 & 0 & 1 \end{pmatrix} \quad , \tag{3.36}$$

which is left invariant by $SO(9) \times O(2)$.

To summarize, ϕ_{27} gives a spontaneous breaking of E_6 with natural directions left invariant by F_4 (Det $\phi_0 \neq 0$) or $SO(10) \times O(2)$ (with Det $\phi = 0$, $\phi \times \phi = 0$), $SO(9) \times SO(2)$ (Det $\phi = 0$, $\phi \times \phi \neq 0$). In all cases ϕ_0 obeys the cubic relation

$$\{\phi_0 \phi_0^\dagger \phi_0\} = c^2 \phi_0 \quad . \tag{3.37}$$

This exhibits the breaking to $SO(10)$ as a natural one. As a particular case, ρ_0 can be superheavy or very heavy (beyond the range of present machines).

Finally, if ϕ_{27} is an effective field, $V(\phi)$ is no longer restricted to be of degree four and its minimization can be achieved by different but definite values of the four invariants δ, ρ_0, σ_0, and τ_0. In that case, this set is left invariant by SO(8), which is the norm group of a real octonion.

Let us also remark that at a two-loop level an effective fermion-fermion-Higgs-Higgs coupling can be induced, resulting into a Yukawa coupling of the fermions to the symmetric product of two ϕ_0's which act like an effective 351' vev. With respect to the SO(10) subgroup these vev's will belong to the 10 and 126 dimensional representations of SO(10).

c. <u>Breaking by</u> χ_{78}

We now turn to a Higgs field χ_{78} that belongs to the adjoint representation. From χ we can form six invariants K_n of degree n in χ. The allowed values of n are obtained by adding one to the Betti numbers of the group manifold. We find $n = 2,5,6,8,9$ and 12. If χ is elementary, the Higgs potential $W(\chi)$ must be quartic, so that

$$W(\chi) = -M^2 K_2(\chi) + \lambda'[K_2(\chi)]^2 \quad , \tag{3.38}$$

χ_0 which minimizes W will only give masses to some gauge bosons.

Let us consider the color singlet part of χ only. It transforms like the (1,8) + (8,1) representations of the SU(3) × SU(3) flavor subgroup. Both octets can be diagonalized separately so that we can write

$$K_2(\chi) = \chi_3^{(1)^2} + \chi_8^{(1)^2} + \chi_3^{(2)^2} + \chi_8^{(2)^2} \quad . \tag{3.39}$$

$W(\chi)$ is minimized either by the trivial solution, or

$$K_2(\chi_0) = M^2/2\lambda' \quad . \tag{3.40}$$

For instance, if $\chi_3^{(1)} = \chi_3^{(2)} = 0$, SU(3) × SU(3) is broken spontaneously to SU(2) × U(1) × SU(2) × U(1). If all four vev's are different from zero, the remaining group is U(1) × U(1) × U(1) × U(1). A coupling between ϕ and χ will break the group further.

Such combined vev's will give masses to the fermions and make the gauge bosons in E_6/G massive if E_6 is broken down spontaneously to G, which will always include $SU(3)_c \times U(1)_{el}$.

A coupling between ϕ_{27} and χ_{78} will add three new cross terms to $V(\phi)$ and $W(\chi)$, i.e.,

$$H(\chi,\phi) = -\alpha m(\phi_{27},\phi_{\overline{27}})_{78}\cdot\chi_{78} + \beta(\phi_{27},\phi_{\overline{27}})_{1}\cdot(\chi_{78},\chi_{78})_{1} + \gamma(\phi_{27},\phi_{\overline{27}})_{650}\cdot(\chi_{78},\chi_{78})_{650} \tag{3.41}$$

where α, β, γ are dimensionless and m is some mass scale. In addition, there will be another effective Higgs field $\eta(351')$ generated by higher order interactions involving both ϕ and χ and having a Yukawa coupling with the fundamental fermions.

d. Combined Breaking Due to ϕ_{27} and χ_{78}

In order to illustrate the possible mechanisms for the generation of masses through the minimization of the total effective Higgs potential involving χ, ϕ, η and their couplings, we shall treat a simple case. We shall assume that the couplings of ϕ to χ and to η are negligible, but the coupling of η to χ will be taken into account. In this case, the vev's of η will involve the superheavy mass M in (3.38). Let us define the scale of this vev by κM, where κ is a dimensionless factor. Thus, the scales of the vev's of ϕ_{27}, $\eta_{351'}$, and χ_{78} will be taken respectively as μ, κM, and M.

In the $SU(3)\times SU(3)$ frame in which the color singlets $\chi^{(1)}$ and $\chi^{(2)}$ that belong respectively to the $(1,8)$ and $(8,1)$ representations are diagonal, the vacuum expectation values of ϕ will belong to the neutral components of the $(3,\bar{3})$ representation. Let us seek a solution of the following form:

$$\chi^{(1)} = \frac{1}{\sqrt{3}}\begin{pmatrix}\alpha & 0 & 0\\ 0 & \alpha & 0\\ 0 & 0 & -2\alpha\end{pmatrix}, \qquad \chi^{(2)} = \frac{1}{\sqrt{3}}\begin{pmatrix}-2\beta & 0 & 0\\ 0 & \beta & 0\\ 0 & 0 & \beta\end{pmatrix}$$

$$\phi = \begin{pmatrix}\rho & 0 & 0\\ 0 & 0 & 0\\ 0 & 0 & \tau\end{pmatrix} \tag{3.42}$$

which for ϕ_{27} is of the type (3.36). This set of vev's is left invariant by the electric charge and the 3d component of the U-spin.

With the neglect of the term H in (3.41), the part of the Higgs potential involving ϕ and χ takes the value

$$V_0 = -M^2(\alpha^2+\beta^2) + \lambda'(\alpha^2+\beta^2)^2 - \mu^2(\rho^2+\tau^2) + \lambda_1(\rho^2+\tau^2) + \lambda_2\rho^2\tau^2 \tag{3.43}$$

Its minimization gives

$$\alpha = \frac{M}{\sqrt{\lambda'}} \sin\gamma \ , \qquad \beta = \frac{M}{\sqrt{\lambda'}} \cos\gamma \ ,$$

$$\rho = \tau = \mu(\lambda_2 + 4\lambda_1)^{-1/2} \ , \tag{3.44}$$

where γ is an arbitrary angle. Let ν be the Yukawa coupling constant of the fundamental fermions with the Higgs field ϕ. Through the vev's ρ and τ the following mass terms and transition mass terms are generated for the fermions in Eq. (2.1) when the topless assignment is used with $\alpha^{\ell}_{L} = \hat{\nu}^{\tau}_{R}$ and $\beta^{\ell}_{L} = -\hat{\nu}^{\ell}_{R}$:

$$\nu\rho\{(\nu^{\tau\dagger}_{R}\nu^{\tau}_{L} + \nu^{e\dagger}_{R}\nu^{e}_{L}) + (N^{\dagger}_{R}\nu^{\tau}_{L} - \tau^{\dagger}_{R}\tau_{L}) + (u^{\dagger}_{R}u_{L} + b^{\dagger}_{R}b_{L})\} + \text{h.c.} \tag{3.45}$$

e. Further Breaking Due to $\eta_{351'}$

The vev's α and β in (3.44) give superheavy masses to leptoquarks and leave the gauge bosons of $SU(2)_W$ and the neutral gauge bosons of $SU'(2)$ massless. Those in turn will gain masses of order $g\rho$, where g is the universal E_6 gauge coupling constant. On the other hand, the (351') Higgs field η has two color singlet parts: $\eta^{(1)}$ and $\eta^{(2)}$, that belong respectively to the representations $(\bar{3},3)$ and $(6,\bar{6})$ of the flavor group $SU(3) \times SU(3)$. We shall here assume that the large vev κM belongs to $\eta^{(2)}$, while $\eta^{(1)}$ has a vev of order μ like ϕ. Such a set guarantees that none of the quarks is superheavy, this condition being essential for the topless assignment. On the other hand $\eta^{(2)}$ couples to leptons and can give very large Majorana masses to some neutral leptons. The vev's of $\eta^{(1)}$ will split the degeneracy between ν^{τ} and ν^{e} and also between τ, u, and b and break the invariance with respect to the 3d component of the U-spin, leaving only $SU(3)_c \times U(1)_{el}$ invariance.

The Majorana mass terms $N^{\dagger}_{R}\hat{N}_{R}$, $\nu^{\tau\dagger}_{R}\hat{\nu}_{R}$, and $\nu^{e\dagger}_{R}\hat{\nu}_{R}$ all belong to the $(6,\bar{6})$ color singlet part $\eta^{(2)}$ of $\eta(351')$ and they are of order κM. Diagonalization of the mass matrix now gives

$$m(\hat{N}_R) \sim \kappa M \ , \qquad m(\nu^{\tau}_{R}) \sim m(\nu^{e}_{R}) \sim \kappa M + 2\frac{\nu^2\rho^2}{\kappa M} \tag{3.46}$$

$$m(\nu^{e}_{L}) \sim m(\nu^{\tau}_{R}) \sim 2\frac{\nu^2\rho^2}{\kappa M} \ . \tag{3.47}$$

With $\nu\rho$ in the GeV region, M in the 10^{15} GeV region and $\kappa \sim 10^{-4}$, the left handed neutrinos in the topless version can fall in the eV region, while the quarks are in the GeV region and both the weak gauge bosons and the SU'(2) neutral gauge bosons are in the 100 GeV region. As shown by D'Jakanov[21] who applied the SU(5) treatment of Georgi, Quinn, and Weinberg[22] to E_6, such a value of M leads to a proton lifetime around 10^{31} years and to a renormalized Weinberg parameter $\sin^2\theta = 0.22$, a value in reasonable agreement with recent experiments.

If we use the standard assignment with three generations, the situation is very similar to the SO(10) case discussed by Ramond,[16] Harvey, Ramond, and Reiss,[23] and also by Witten[18] who takes the (126) as generated in higher order. In their work the mass scale μ is assigned to (16) of SO(10) contained in (27) of E_6, κM to (126) of SO(10) in (351') of E_6, while the superheavy scale M belongs to (45) of SO(10) contained in (78) of E_6. In this case both the $(\bar{3},3)$ and $(6,\bar{6})$ parts of $\eta_{351'}$ can have vev's of order κM, sending the mass of the quark B out of the observable region alongside ν_R^e. The same large $\eta^{(1)}$ masses will also raise the masses of the SU'(2) gauge bosons, leaving the Weinberg-Salam group $SU(2)_W \times U(1)$ as the effective electroweak group at low energies.

4. CONCLUDING REMARKS ON PHENOMENOLOGY

We have seen that in both the topless and standard versions of E_6 the use of effective Higgs fields belonging to the representations 27, 351', and 78 can lead to an acceptable phenomenological model with small left handed neutrino masses, very heavy right handed neutrinos, charged leptons and quarks in the GeV region, weak bosons of the Weinberg-Salam-Glashow kind and possibly additional neutral SU'(2) weak bosons in the 100 GeV region and superheavy colored leptoquarks causing quark-lepton and quark-antiquark transitions.

Furthermore, in the standard version an approximate B - L number conservation is expected. However, from the remarks of part (d) of Section 2, we have seen that the F hypercharge does not give B - L for the observed leptons and quarks in the topless version. Hence there is no reason to expect any approximate (B - L) conservation in this case and exchange of leptoquarks X, Y, X', and V between quarks will lead not only to the reaction

$$n \rightarrow \pi^- + e^+ \quad , \tag{4.1}$$

but also to

$$n \rightarrow \pi^+ + e^- \quad , \tag{4.2}$$

which is forbidden if only exchange of X, Y of the SU(5) subgroup is allowed. As a result, we can have neutron oscillations

$$n \to \pi^- + e^+ \to \bar{n} \tag{4.3}$$

discussed by Mohapatra and Marshak.[19]

As far as weak interactions are concerned, the standard mode of E_6 does not differ from the three-generation SU(5) or SO(10) models. The b quark belongs to an $SU(2)_W$ doublet with the top quark t as its other member. The upsilon decay modes are similar to those of the meson $\phi(s\bar{s})$ with a smaller Cabibbo-type mixing angle. On the other hand, in the topless mode of E_6 we may expect the decay through V + A SU'(2) currents

$$\gamma(b\bar{b}) \to M_0(d\bar{b}) + U \to M_0(d\bar{b}) + \begin{cases} \tau^- + e^+ \quad (\text{or } \tau^+ + e^-) \\ \text{or} \\ \nu_L^e + \bar{\nu}_L^\tau \quad (\text{or } \bar{\nu}_L^e + \nu_L^e) \end{cases} \tag{4.4}$$

where M_0 is a bound state of the quarks d and $\bar{b}$. In turn, M decays to ω or ρ^0 and another neutral combination of τ and e, so that in the end the upsilon can decay into π^0's and four charged leptons, as the τe pair can get replaced by the $M^-\mu$ pair of the second generation, or two charged leptons and two neutrinos.

To sum up, if the E_6 grand unified theory is valid, the topless and standard modes can be distinguished by the charge of the sixth quark, the existence of a fourth charged lepton M^-, the variety and masses of neutral leptons, the details of neutron oscillations, the existence of new neutral weak gauge bosons and the V + A decay of the upsilon into four leptons and π^0's (gamma pairs). Both versions of E_6 lead to acceptable values for the proton lifetime and the Weinberg parameter.

Acknowledgments. I am grateful to I. Bars, M. Serdaroglu, and P. Ramond for many illuminating discussions. I would like to thank A. Yildiz and P. Frampton for hospitality at the site of the New Hampshire workshop, which provided such a stimulating scientific atmosphere.

REFERENCES

1. F. Gürsey, P. Ramond, and P. Sikivie, *Phys. Lett.* 60B, 177 (1976).

2. F. Gürsey, in *Group Theoretical Methods in Physics, 5th Colloquium*, R.T. Sharp (ed.), Academic Press, 1977, p. 213; F. Gürsey and M. Serdaroglu, *Lett. al Nuovo Cimento* 21, 28 (1978); F. Gürsey, in *Second Workshop on Current Problems in High Energy Particle Theory*, G. Domokos and S. Kövesi-Domokos (eds.), Johns Hopkins Univ., 1978, p. 3; F. Gürsey, in *The Why's of Subnuclear Physics*, A. Zichichi (ed.), Plenum Press, 1979, pp. 1059-1164.

3. F. Gürsey and M. Serdaroglu, "E_6 gauge field theory model revisited", to be published.

4. Y. Achiman and B. Stech, *Phys. Lett.* 77B, 389 (1978); Q. Shafi, *Phys. Lett.* 79B, 301 (1978); P. Ramond, "The family group in grand unified theories", Cal Tech Preprint 68-709 (1979); H. Ruegg and T. Schücker, *Nucl. Phys.* B161, 388 (1979).

5. R. Barbieri and D.V. Nanopoulos, "An exceptional model for grand unification", CERN Preprint TH. 2810 (1980); O.K. Kalashnikov, S.E. Konshtein, and E.S. Fradkin, *Sov. J. Nucl. Phys.* 29(6), 852 (1979); N. Ogievietsky and V. Tzeitlin, Dubna Preprint JINR-E2-11136.

6. For a review, see H. Freudenthal, *Advances in Math* 1, 145 (1965).

7. A.I. Malcev, *Izv. Akad. Nauk SSSR, Ser. Mat.* 8, 775 (1967).

8. M. Günaydin and F. Gürsey, *Phys. Rev.* D9, 3387 (1974).

9. H. Georgi and S.L. Glashow, *Phys. Rev. Lett.* 32, 438 (1974).

10. H. Fritzsch and P. Minkowski, *Ann. of Phys.* 93, 193 (1975).

11. H. Georgi, in *Particles and Fields*, C.E. Carlson (ed.), AIP, New York, 1974, p. 575.

12. G. Segre and H.A. Weldon, *Hadronic Physics*, 1, 424 (1978).

13. E. Cremmer and B. Julia, Ecole Normale Superieure Preprint, 1979.

14. J. Ellis, M. Gaillard, and B. Zumino, "A grand unified theory obtained from broken supergravity", CERN Preprint TH2842 (1980). See J. Ellis in these Proceedings.

15. I. Barx and M. Günaydin, Yale Preprint YTP80-09 (1980).

16. M. Gell-Mann, P. Ramond, and R. Slansky, unpublished; H. Fritzch, M. Gell-Mann, and P. Minkowski, *Phys. Lett.* 59B, 256 (1975); P. Ramond, Corol Gables talk on SO(10), 1980.

17. See Refs. 2 and 3 and H. Georgi and S.L. Glashow, Harvard Preprint HUTP-79/A073 (1980).

18. E. Witten, Harvard Preprint HUTP-79/A076 (1980).

19. R.N. Mohapatra and R.E. Marshak, Virginia Polytechnic Preprint VPI-HEP-80/1 (1980).

20. L. Michel and L.A. Radicati, *Ann. of Phys.* 66, 758 (1971).

21. D.I. D'Jakanov, Leningrad Institute on Nuclear Physics Preprint, No. 303 (1977).

22. H. Georgi, H.R. Quinn, and S. Weinberg, *Phys. Rev. Lett.* 33, 451 (1974).

23. J.A. Harvey, P. Ramond, and D.B. Reiss, Cal Tech Preprint CALT-68-758 (1980).

CHARGE CONJUGATION AND ITS VIOLATION IN UNIFIED MODELS

R. Slansky

Theoretical Division
Los Alamos Scientific Laboratory*
University of California
Los Alamos, NM 87545

Abstract

Yang-Mills theories admitting a charge conjugation C, which reflects the representation of left-handed fermions $\underline{f}_L$ onto itself, are reviewed with particular attention to flavor chiral theories, where $\underline{f}_L$ is non-self-conjugate. Simple cases of the fermion mass matrices in SO_{10} and E_6 are studied, and it is observed that the weak isospin $\underline{I}^W$ conserving part of the mass can be classified into its C conserving and C violating pieces. If the left-handed fermions are assigned to families of $\underline{16}$'s of SO_{10} or $\underline{27}$'s of E_6, then the hypothesis of the $\underline{I}^W$ invariant mass violating C maximally, with the C conserving part put to zero, gives a simple explanation of the low-mass "$\underline{\bar{5}}+\underline{10}$" structure of the families.

This talk begins with a sketch of the solution of the problem of finding the possible charge conjugation operators C, which reflect the left-handed fermions to their left-handed antiparticle states, in Yang-Mills theories where C can be defined. The general analysis can be found in a paper with M. Gell-Mann;[1] besides reviewing those results, this report provides an explicit construction of the C operator for SO_{10} and E_6 theories that unify electromagnetic, weak, and strong interactions. The uniqueness of the construction follows from the general analysis; although it adds little in principle, it does ease the analysis of the fermion mass matrix. For example, the $\underline{27}$ of E_6 has five neutral lepton states, but setting up the mass matrix and identifying the properties of the elements under C is not complicated, and is carried out here explicitly. The main reason for showing these examples is to demonstrate that the mass matrix may have special properties under C. Specifically, it is found that the hypothesis of maximal C violation of the $\underline{I}^W$ (weak isospin) invariant part of the fermion mass matrix can provide large $I^W = 0$ masses in some models to just those states that have not been observed. Of course, confirmation of the hypothesis is not possible until the assignment

*Work supported by the U.S. Department of Energy.

of the left-handed fermions to the correct $\underline{f}_L$ has been discovered. However, the hypothesis may be a helpful guide in searching for the gauge group G and the representation $\underline{f}_L$ to which the left-handed fermions should be assigned.

Let us consider first the problem of defining C for a "family" of left-handed fermions consisting of the u, $\bar{u}$, d', $\bar{d}'$, e^-, e^+, and ν_e. In the SU_5 model[2] it is not possible to define C, since although u and $\bar{u}$ are both in the same $\underline{10}$, d' and e^+, also in $\underline{10}$, are not in the same irreducible representation (irrep) as their antiparticles $\bar{d}'$ and e^-, which are in the $\bar{\underline{5}}$. Thus a reflection taking the SU_5 quantum numbers of the u onto those of the $\bar{u}$ in the $\underline{10}$ will not reflect the $\bar{d}'$ and e^+ onto the correct quantum numbers in the $\bar{\underline{5}}$; C cannot be defined.

In the standard SO_{10} model[3] the state of affairs is much nicer. Each family is assigned to a 16-dimensional spinor, and the u, d', e^-, and their antiparticles are in the same irrep. In addition, the ν_L is matched up by the C operation defined below with a neutral, weak singlet state $(\bar{\nu})_L$ to form the two halves of a Dirac spinor. It is possible to define C in a unified model if a particle and its antiparticle image are always assigned to the same irrep of the gauge group G.

The generators of the unifying gauge group G must include the electric charge operator Q^{em} and the eight color generators of SU_3^c of the strong interactions. Charge conjugation must flip the sign of Q^{em}, that is, C must anticommute with Q^{em}. Similarly, C must invert the SU_3^c root diagram through the origin, exchanging I_+^c for I_-^c, U_+^c for U_-^c, and V_+^c for V_-^c. Thus the color generators F_1^c, F_3^c, F_4^c, F_6^c, and F_8^c anticommute with C and F_2^c, F_5^c, and F_7^c commute with C. Note that F_2^c, F_5^c, and F_7^c, which are left invariant by C, form an SO_3 subgroup of SU_3^c; it is a symmetric subgroup, as is now discussed.

The Yang-Mills Lagrangian must be invariant under C, which implies that C must be an automorphism of the Lie algebra of G that reverses the signs of some generators A of G, while leaving the remaining generators S of G invariant:

$$C(S) = S \;, \qquad C(A) = -A \;. \tag{1}$$

In order for C to be an automorphism of G, S must define a symmetric subgroup:

$$[S,S] \subseteq S \;; \qquad [S,A] \subseteq A \;; \qquad [A,A] \subseteq S \;. \tag{2}$$

In the cases where the action of C on any irrep of G is to reflect its weight system onto itself, C is an inner automorphism; but if C reflects a complex irrep onto its conjugate, or a self-conjugate irrep onto an inequivalent irrep of the same dimension, then it is an outer automorphism. The mathematical analysis is found in Ref. 1.

Let us examine the simplest example of such a reflection. The CP reflection, which takes $\underline{f}_L$ to $\underline{f}_R$, must invert the root diagram of G through the origin of root space; without it, there can be no gauge invariant kinetic energy term. (Root space is an Euclidean space of dimension equal to the rank of G; the root vectors describe the shift in quantum numbers due to the action of the generators, or currents, on the states of an irrep. A weight vector is a list of rank(G) quantum numbers carried by a Hilbert-space vector in the representation. This language is reviewed in Ref. 4. The inversion of the roots and weights implies that a non-self-conjugate irrep is carried onto its conjugate; for example, CP must reflect a $\underline{3}^c$ weight onto minus itself, which is in the $\underline{\bar{3}}^c$ weight system. Thus, there is no member of the Cartan subalgebra of G in the symmetric subgroup associated with CP. (The Cartan subalgebra is the maximal set of diagonalizable generators of G, of which there are rank(G) in number.) The reflected representation $\underline{f}_R$ must be such that $\underline{f}_R \times \underline{f}_L$ contains the identity and the adjoint, which is the group theoretical restatement of the requirement that the kinetic energy be gauge invariant. The symmetric subgroups that are left invariant by CP are

$$\begin{array}{llll} SU_n \supset SO_n & & G_2 \supset SU_2 \times SU_2 \\ SO_{2n+1} \supset SO_{n+1} \times SO_n & & F_4 \supset SU_2 \times Sp_6 \\ Sp_{2n} \supset SU_n \times U_1 & & E_6 \supset Sp_8 \\ SO_{2n} \supset SO_n \times SO_n & & E_7 \supset SU_8 \\ & & E_8 \supset SO_{16} \end{array} \tag{3}$$

Notice that in every case, the dimension of the symmetric subgroup is $\frac{1}{2}(\dim(G) - \mathrm{rank}(G))$, which is due to the fact that the symmetric subgroup is generated by

$$\frac{1}{i\sqrt{2}} \left(E_\alpha - E_{-\alpha}\right) \quad ,$$

where α is a root and E_α is the corresponding ladder operator (or generator) of G. This is an obvious generalization of the discussion above of SU_3^c.

In the case of a flavor chiral theory, which is a theory where $\underline{f}_L$ is not self conjugate, the reflection by C of $\underline{f}_L$ onto itself cannot coincide in its group structure with CP, since CP reflects $\underline{f}_L$ onto $\underline{\bar{f}}_L$. Moreover, since $\underline{f}_L \times \underline{f}_L$ does not contain a gauge singlet, any fermion mass violates the gauge symmetry. Only SU_n, SO_{4n+2}, and E_6 have complex representations, so they are the only candidate simple groups that can lead to a flavor chiral theory. The emphasis on flavor chiral theories is, of course, due to the economical way that they incorporate the standard model of the weak interactions.

The symmetric subgroups of SU_{p+q}, where the associated C reflects a complex irrep onto itself, are $SU_p \times SU_q \times U_1$. The number of Cartan subalgebra generators outside the symmetric subgroup is min(p,q); the remaining members of the Cartan subalgebra are invariant under C. Similarly, SO_{p+q} contains the symmetric subgroup $SO_p \times SO_q$, and flavor chiral theories are defined by the constraint that $p + q = 4n + 2$. Then, only if p and q are even does C reflect a complex irrep onto itself. The number of Cartan subalgebra generators that are flipped in sign by C is min(p,q). In addition, SO_{2n} contains $SU_n \times U_1$ as a symmetric subgroup; the integer part of n/2 diagonal generators are changed in sign by C. Finally, in E_6, the C associated with $SU_2 \times SU_6$ changes the signs of four diagonal generators, and that associated with $SO_{10} \times U_1$ changes only two. Thus, the C associated with $SU_2 \times SU_6$ is the only suitable candidate.

In Ref. 1 we carried out in a coordinate independent language the analysis of several models. We carry out the same discussion here using a definite coordinatization of root space. There are practical advantages of each formulation, but they are, of course, physically equivalent.

The discussion of applications begins with the SO_{10} model: After selecting C, we show in detail what it does to the SO_{10} generators. Then, the action of C on the weights in the $\underline{16}$ can be studied, and finally a classification of the neutral lepton mass matrix is possible. We do not study the charged particles in this example, because they have, trivially, just C-conserving,

$$|\underline{I}^W| = \frac{1}{2} \text{ masses} \quad .$$

There are six symmetric subgroups of SO_{10}: for $SO_5 \times SO_5$, the reflection flips the sign of five diagonal generators and $\underline{16}$ onto $\overline{\underline{16}}$, so the reflection is suitable for CP as it simply reverses the sign of each root and weight; the reflection associated with $SO_4 \times SO_6$ takes $\underline{16}$ onto $\underline{16}$ and flips the signs of four diagonal generators, and turns out to be the only candidate for C; the reflection associated with $SO_3 \times SO_7$ takes $\underline{16}$ onto $\overline{\underline{16}}$; the reflection associated with $SO_2 \times SO_8$ takes $\underline{16}$ onto $\underline{16}$, but flips only two quantum numbers; the reflection associated with SO_9 takes $\underline{16}$ onto $\overline{\underline{16}}$; and the reflection associated with $SU_5 \times U_1$ flips the sign of only two diagonal generators. This exhausts the list of symmetric subgroups and the action of the associated reflection on complex irreps. We conclude that there is only one candidate for C, and it leaves one quantum number in SO_{16} invariant.

SO_{10} contains color and flavor in a well-known way; for example, we may follow the maximal subgroup chain

$$SO_{10} \supset SU_5 \times U_1^r$$

with

$$SU_5 \supset SU_2^w \times U_1^w \times SU_3^c \quad ,$$

where Y^w generates the U_1^w and Q^r generates the U_1^r. This embedding can be specified uniquely (up to a Weyl reflection) in terms of the root diagram. If it is required that the highest weight of an SO_{10} irrep is projected onto the highest weights of the SU_5 irreps contained in its branching rule, then the embedding in root space is specified by the matrix,[5]

$$P(SO_{10} \supset SU_5) = \begin{pmatrix} 1 & 1 & 0 & 0 & 0 \\ 0 & 0 & 1 & 0 & 1 \\ 0 & 0 & 0 & 1 & 0 \\ 0 & 1 & 1 & 0 & 0 \end{pmatrix} \tag{4}$$

where this matrix acts on an SO_{10} weight, written in the integer basis of Dynkin, as a column vector, to give the Dynkin labels of the SU_5 weight. The axis defined by the Q^r generator, which is in the Cartan subalgebra, is (-1 1 -1 0 1). (The weights and axes in root space are always written here in the Dynkin integer basis, which is dual to the weight written as a linear combination of simple roots. The Dynkin basis is not orthonormal, so the computation of scalar products requires knowledge of the metric tensor, which is essentially the inverse of the Cartan matrix. The reader who wants a more detailed resume of these points might enjoy looking at Ref. 4.) The $SU_2^w \times SU_3^c$ can be embedded in SU_5 with the projection matrix,[5]

$$P(SU_5 \supset SU_2 \times SU_3) = \begin{pmatrix} 0 & 1 & 1 & 0 \\ 1 & 1 & 0 & 0 \\ 0 & 0 & 1 & 1 \end{pmatrix} \tag{5}$$

Now that color and flavor are embedded explicitly into SO_{10}, we can identify the physical significance of each of the 45 SO_{10} roots. It is easy to find that the nonzero color roots are (0 1 0 0 0), (1 0 0 -1 1) and (-1 1 0 1 -1), and their negatives, and the electric charge axis, properly normalized, is

$$\frac{1}{3}(-2\ -2\ 3\ -1\ 1) \quad .$$

The action of C on the generators is to flip the signs of these roots and axes. The remaining equations can be gotten from the generators, but it is slightly simpler to study the weights in the <u>10</u>. The procedure is to write out the weights of the 10, compute their flavor and color content according to Eqs. (4) and (5), and then require that the action of C on the weights do what it must to color and electric charge. It follows that the action of C on the SO_{10} weights is

$$C(SO_{10}) = \begin{pmatrix} -1 & 0 & 0 & 0 & 0 \\ 0 & -1 & 0 & 0 & 0 \\ 0 & 0 & -1 & -1 & -1 \\ 0 & 0 & 0 & 0 & 1 \\ 0 & 0 & 0 & 1 & 0 \end{pmatrix} \qquad (6)$$

Thus C leaves invariant the axis with Dynkin labels (0 0 1 -1 -1), which corresponds to the diagonal generator

$$3Y^W - 4Q^r - 10\, I_3^W \quad ;$$

C inverts the SU_3^c roots, electric charge, and $2Q^r + Y^W$, where the Y^W axis is

$$\frac{1}{3}\,(-4\ -1\ 6\ -5\ -1) \quad .$$

We now study the action of C on the weights in the 16. The u quark weight

(0 0 0 0 1), (-1 0 0 1 0), and (0 -1 0 0 1)

are refelcted to the $\bar{u}$ weights,

(0 0 -1 1 0), (1 0 -1 0 1), and (0 1 -1 1 0) ,

respectively; the d quark weights

(0 1 0 -1 0), (-1 1 0 0 -1), and (0 0 0 -1 0)

are reflected to the $\bar{d}$ weights

(0 -1 1 0 -1), (1 -1 1 -1 0), and (0 0 1 0 -1) ,

respectively; and the e^- (1 0 0 0 -1) is reflected to the e^+ weight (-1 0 1 -1 0). Finally, the ν_L with weight (1 -1 0 1 0) is reflected to (-1 1 -1 0 1), which is the SU_5 singlet and is called the $(\bar{\nu})_L$.

The weights of the neutral lepton mass matrix are the sums of the weights of the corresponding states. Thus, the ν_L mass matrix element $\langle \nu_L | M | \nu_L \rangle$ has weight (2 -2 0 2 0) with

$$|\Delta \underline{I}^W| = 1 \quad ;$$

certainly we expect it to be less than about 1 eV. It is reflected by C onto $\langle(\bar{\nu})_L|M|(\bar{\nu})_L\rangle$, which has weight (-2 2 -2 0 2) and is a weak isospin singlet. The off-diagonal element $\langle\nu_L|M|\bar{\nu}_L\rangle$ and its transpose have weight (0 0 -1 1 1),

$$|\Delta \underline{I}^W| = \frac{1}{2} \quad ,$$

and are invariant under C. The mass matrix can be written in the useful form

$$\begin{array}{cc} & \begin{array}{cc} \nu_L & \bar{\nu}_L \\ (1\ -1\ 0\ 1\ 0) & (-1\ 1\ -1\ 0\ 1) \end{array} \\ \begin{array}{c} \nu_L \\ (1\ -1\ 0\ 1\ 0) \\ \\ \\ \bar{\nu}_L \\ (-1\ 1\ -1\ 0\ 1) \end{array} & \left(\begin{array}{ccc} \begin{array}{c} (2\ -2\ 0\ 2\ 0) \\ |\Delta \underline{I}^W| = 1 \end{array} & & \left[\begin{array}{c} (0\ 0\ -1\ 1\ 1) \\ |\Delta \underline{I}^W| = \frac{1}{2} \end{array} \right] \\ & \nwarrow\ C\ \searrow & \\ \left[\begin{array}{c} (0\ 0\ -1\ 1\ 1) \\ |\Delta \underline{I}^W| = \frac{1}{2} \end{array} \right] & & \begin{array}{c} (-2\ 2\ -2\ 0\ 2) \\ |\Delta \underline{I}^W| = 0 \end{array} \end{array} \right) \end{array} \tag{7}$$

where the [...] signify that the mass matrix element is reflected onto itself by C.

The $|\Delta \underline{I}^W| = \frac{1}{2}$ mass has the same weight as the u quark, and is expected to have a value of a few MeV. In order for the small eigenvalue of (7) to be a few eV or less, the $|\Delta \underline{I}^W| = 0$ term must be huge, and if we ignore the $|\Delta \underline{I}^W| = 1$ term, the mass matrix has the form,[6]

$$\begin{pmatrix} 0 & m \\ m & M \end{pmatrix} \tag{8}$$

which has small eigenvalue m^2/M, approximately. Note that Eq. (8) can be restated as: *the weak isospin conserving mass violates* C *maximally*, while the $|\Delta \underline{I}^W| = \frac{1}{2}$ mass conserves C.

The second example is less trivial: The unifying group is E_6 and a single family is assigned to a $\underline{27}$.[7] The $\underline{27}$ has two charge -1/3 quarks and their antiparticles, so there is an opportunity to study the C properties of the quark masses in this example.

The symmetric subgroups of E_6 are Sp_8, $SU_2 \times SU_6$, $SO_{10} \times U_1$, and F_4. Of these, the reflections associated with Sp_8 and F_4 reflect $\underline{27}$ to $\underline{\overline{27}}$; CP is associated with Sp_8. We have already argued that C must be associated with $SU_2 \times SU_6$, because the reflection associated with $SO_{10} \times U_1$ flips the signs of only two diagonal generators. Thus C leaves invariant two of the six quantum numbers in E_6.

The embedding of color and flavor in E_6 can be described by the subgroup chain

$$E_6 \supset SO_{10} \times U_1^t \supset SU_5 \times U_1^r \times U_1^t \supset SU_2 \times U_1^w \times SU_3^c \times U_1^r \times U_1^t \quad ,$$

with the projection of the E_6 to SO_{10} weights given by,[5]

$$P(E_6 \supset SO_{10}) = \begin{pmatrix} 0 & 1 & 1 & 1 & 0 & 0 \\ 0 & 0 & 0 & 0 & 0 & 1 \\ 0 & 0 & 1 & 0 & 0 & 0 \\ 0 & 0 & 0 & 1 & 1 & 0 \\ 1 & 1 & 0 & 0 & 0 & 0 \end{pmatrix} \tag{9}$$

the remaining projections are given by Eqs. (4) and (5).

The C reflection is constructed in the same fashion as Eq. (6) for SO_{10}. It is

$$C(E_6) = \begin{pmatrix} 0 & 0 & 0 & 1 & 0 & 0 \\ 0 & 0 & 0 & 0 & 1 & 0 \\ -1 & -1 & -1 & -1 & -1 & 0 \\ 1 & 0 & 0 & 0 & 0 & 0 \\ 0 & 1 & 0 & 0 & 0 & 0 \\ 0 & 0 & 0 & 0 & 0 & -1 \end{pmatrix} \tag{10}$$

It inverts color roots and reverses the signs of electric charge and $2Q^r + Y^w$, while leaving invariant $3Y^w - 4Q^r - 10\,I_3^w$ and Q^t.

The weight diagram for the $\underline{27}$ is derived from the highest weight (1 0 0 0 0 0) in the usual way.[4] Three of the neutral lepton weights are eigenvectors of $C(E_6)$ with eigenvalues $+1$; (-1 0 1 -1 0 0), (0 1 -1 0 1 0), and (1 -1 0 1 -1 0). The other two neutral weights (0 0 0 1 0 -1) and (1 0 -1 0 0 1) are transformed into one another by C.[8] The remaining weights carry electric charge and transform under C as expected.

The charge 2/3 u quark has weights

(1 0 0 0 0 0), (1 -1 0 0 1 0), and (1 0 0 0 0 -1) ,

which are reflected by C in Eq. (10) to the $\bar{u}$ weights

(0 0 -1 1 0 0), (0 1 -1 1 -1 0), and (0 0 -1 1 0 1) ,

respectively. The u quark mass carries weight (1 0 -1 1 0 0), which is a C conserving $|\Delta \underline{I}^w| = \frac{1}{2}$ mass.

The mass matrix of the charge $-1/3$ quarks and the charged lepton mass matrix are similar; they have precisely the same weight structure. The charge $-1/3$ quarks in the SU_5 $\underline{10}$ of the SO_{10} $\underline{16}$, to be denoted $\underline{10}(\underline{16})$, have the weights

$$(0\ 0\ 0\ 0\ {-1}\ 1), \quad (0\ {-1}\ 0\ 0\ 0\ 1), \quad \text{and} \quad (0\ 0\ 0\ 0\ {-1}\ 0) \ ;$$

the C partners

$$(0\ {-1}\ 1\ 0\ 0\ {-1}), \quad (0\ 0\ 1\ 0\ {-1}\ {-1}), \quad \text{and} \quad (0\ {-1}\ 1\ 0\ 0\ 0),$$

respectively, are in $\underline{\bar{5}}(\underline{16})$. The other charge $-1/3$ quark is in $\underline{5}(\underline{10})$, with weights

$$({-1}\ 1\ 0\ 0\ 0\ 0), \quad ({-1}\ 0\ 0\ 0\ 1\ 0), \quad \text{and} \quad ({-1}\ 1\ 0\ 0\ 0\ {-1}) \ ,$$

with C partners

$$(0\ 0\ 0\ {-1}\ 1\ 0), \quad (0\ 1\ 0\ {-1}\ 0\ 0), \quad \text{and} \quad (0\ 0\ 0\ {-1}\ 1\ 1) \ ,$$

respectively, in $\underline{\bar{5}}(\underline{10})$. Let us write out the mass matrix for one color state, (1 0) for quarks, (-1 0) for antiquarks as

$$\begin{array}{c|cccc}
 & D\ \underline{5}(\underline{10}) & d\ \underline{10}(\underline{16}) & \bar{d}\ \underline{\bar{5}}(\underline{10}) & \bar{D}\ \underline{\bar{5}}(\underline{16}) \\
 & ({-1}\ 1\ 0\ 0\ 0\ 0) & (0\ 0\ 0\ 0\ {-1}\ 1) & (0\ 0\ 0\ {-1}\ 1\ 0) & (0\ {-1}\ 1\ 0\ 0\ {-1}) \\
\hline
D\ \underline{5}(\underline{10}) \ ({-1}\ 1\ 0\ 0\ 0\ 0) & 0 & 0 & \left[\begin{array}{c} ({-1}\ 1\ 0\ {-1}\ 1\ 0) \\ |\Delta \underline{I}^W| = 0 \end{array}\right] & \begin{array}{c} ({-1}\ 0\ 1\ 0\ 0\ {-1}) \\ |\Delta \underline{I}^W| = 0 \end{array} \\
 & & & \overset{C}{\longleftrightarrow} & \\
d\ \underline{10}(\underline{16}) \ (0\ 0\ 0\ 0\ {-1}\ 1) & 0 & 0 & \begin{array}{c} (0\ 0\ 0\ {-1}\ 0\ 1) \\ |\Delta \underline{I}^W| = \tfrac{1}{2} \end{array} & \left[\begin{array}{c} (0\ {-1}\ 1\ 0\ {-1}\ 0) \\ |\Delta \underline{I}^W| = \tfrac{1}{2} \end{array}\right] \\
\bar{d}\ \underline{\bar{5}}(\underline{10}) \ (0\ 0\ 0\ {-1}\ 1\ 0) & [({-1}\ 1\ 0\ {-1}\ 1\ 0)] & (0\ 0\ 0\ {-1}\ 0\ 1) & 0 & 0 \\
 & \overset{C}{\longleftrightarrow} & & & \\
\bar{D}\ \underline{\bar{5}}(\underline{16}) \ (0\ {-1}\ 1\ 0\ 0\ {-1}) & ({-1}\ 0\ 1\ 0\ 0\ {-1}) & [(0\ {-1}\ 1\ 0\ {-1}\ 0)] & 0 & 0
\end{array} \tag{11}$$

(The charged lepton mass matrix has the same weight structure if the weight of D is replaced by the charge 1 lepton weight, (0 0 1 -1 1 -1), the weight of d by (1 -1 1 -1 0 0) of charge 1, the weight of $\bar{d}$ by (-1 1 -1 0 0 1) of charge -1, and the weight of $\bar{D}$ by (-1 0 0 1 -1 0) of charge -1.) There are two candidate assignments for the weak isospin conserving mass: either the (-1 0 1 0 0 -1) mass is nonzero, the d state is left massless (before the weak breaking), and C is maximally violated; or the (-1 1 0 -1 1 0) mass is nonzero, the D is massless, and the mass is C conserving. For the purposes of studying the charged particle masses, these situations appear interchangeable, although the $\underline{\bar{5}}+\underline{10}$ left massless in the limit of no weak breaking differs in the two cases. Recall that

$$\underline{27} = \underline{16} + \underline{10} + \underline{1}$$

of SO_{10}. In the first case (d massless), the $\underline{\bar{5}}$ belongs to the SO_{10} $\underline{10}$; in the second case (D massless), the $\underline{\bar{5}}$ comes from the SO_{10} $\underline{16}$. The same considerations also apply to the two charged leptons in the $\underline{27}$.

In order to decide which assignment is more attractive, we turn to a study of the neutral lepton mass matrix, which can be written as a matrix of weights where the labels on the rows and columns should, by now, be obvious (just divide the diagonal entries by 2):

$$\begin{pmatrix} (0~0~0~2~0~{-2}) & [(1~0~{-1}~1~0~0)] & (1~{-1}~0~2~{-1}~{-1}) & (\underline{-1~0~1~0~0~{-1}}) & (0~1~{-1}~1~1~{-1}) \\ & \nwarrow\!\searrow & \updownarrow & \updownarrow & \updownarrow \\ [(1~0~{-1}~1~0~0)] & \underline{(2~0~{-2}~0~0~2)} & \underline{(2~{-1}~{-1}~1~{-1}~1)} & (0~0~0~{-1}~0~1) & (1~1~{-2}~0~1~1) \\ (1~{-1}~0~2~{-1}~1) \leftrightarrow & \underline{(2~{-1}~{-1}~1~{-1}~1)} & [\underline{(2~{-2}~0~2~{-2}~0)}] & [(0~{-1}~1~0~{-1}~0)] & [(1~0~{-1}~1~0~0)] \\ (\underline{-1~0~1~0~0~{-1}}) \leftrightarrow & (0~0~0~{-1}~0~1) & [(0~{-1}~1~0~{-1}~0)] & [({-2}~0~2~{-2}~0~0)] & [\underline{({-1}~1~0~{-1}~1~0)}] \\ (0~1~{-1}~1~1~{-1}) \leftrightarrow & (1~1~{-2}~0~1~1) & [(1~0~{-1}~1~0~0)] & [(\underline{-1~1~0~{-1}~1~0})] & [(0~2~{-2}~0~2~0)] \end{pmatrix} \tag{12}$$

where the I_3^w value of the mass matrix element is one-half the sum of the first five Dynkin labels. Let us first assume that the weak isospin conserving part of Eq. (12) is maximally C violating, so that only the entries with weights (2 0 -2 0 0 2), (2 -1 -1 1 -1 1), and (-1 0 1 0 0 -1) are nonzero. For a general choice of parameters, Eq. (12) has four nonzero eigenvalues and one zero eigenvalue; the massless fermion has weight (0 1 -1 0 1 0), which is in $\underline{\bar{5}}(\underline{10})$. Thus, with maximal C violation, the massless fermions at the weak isospin conserving level are classified by $\underline{\bar{5}}+\underline{10}$.

In the case of C conservation, the elements with weights (2 -2 0 2 -2 0) and (-1 1 0 -1 1 0) are nonzero, and the neutrals in the SO_{10} $\underline{1}+\underline{10}$ get masses. Both neutral states in the $\underline{16}$ remain massless, at least until some C violation is introduced at the SO_{10} level. Thus, the C conservation hypothesis leaves a $\underline{1}+\underline{\bar{5}}+\underline{10}$ of SU_5 to get masses from other sources, such as the weak interactions. If the four component ν mass comes from the weak interactions, then its mass is of order the u mass, not in accord with experience.

States in a slightly different way, all the C conserving weak isosinglet masses leave SO_{10} invariant, so the fermions occur in SO_{10} irreps, $\underline{16}$'s in this case, but the C violating masses leave just SU_5 invariant, while violating SO_{10}, and the low mass fermions in the $\underline{27}$ occur in a $\bar{5}(\underline{10}) + \underline{10}(\underline{16})$ pattern.

In summary, we find that the hypothesis of maximal C violation of the weak isospin invariant masses leads to a satisfactory fermion spectrum in several flavor chiral models. Of course, this selection rule must be tested on the "correct" representation before it can be confirmed or rejected. However, the general structure of the mass matrices is such that the hypothesis may provide a helpful guideline in searching for satisfactory theories.

Acknowledgments. Most of the work reported in this talk was done in collaboration with M. Gell-Mann. P. Ramond has provided us with much encouragement and many helpful comments.

REFERENCES

1. M. Gell-Mann and R. Slanksy, Preprint (1980).

2. H. Georgi and S.L. Glashow, *Phys. Rev. Lett.* 32, 438 (1974).

3. H. Georgi, *Particles and Fields* (1974), APS/DPF Williamsburg, C.E. Carlson (ed.) AIP, New York, 1975, p. 575; H. Fritzsch and P. Minkowski, *Ann. Phys.* 93, 193 (1975).

4. R. Slansky, Coral Gables Talk (1980), Los Alamos Preprint LA-UR-80-591; G. Shaw and R. Slansky, Los Alamos Preprint LA-UR-80-1001. Further references are given there.

5. W. McKay, J. Patera, and D. Sankoff, *Computers in Non Associative Rings and Algebras*, R. Beck and B. Kolman (eds.), Academic Press, New York, 1977, p. 235.

6. P. Ramond, Sanibel Lectures, Caltech Preprint CALT-68-709; M. Gell-Mann, P. Ramond, and R. Slansky, *Supergravity*, P. Van Nieuwenhuizen and D. Freedman (eds.), North Holland, Amsterdam, 1979, p. 315; E. Witten, Harvard Preprint HUTP-79/A076.

7. F. Gürsey, P. Ramond, and P. Sikivie, *Phys. Lett.* B60, 177 (1975).

8. For a pedagogical discussion of the distinction of Dirac and Majorana masses, see E. Witten, contribution to this conference.

BARYON AND LEPTON NON-CONSERVATION, MAJORANA NEUTRINOS AND NEUTRON ($N \leftrightarrow \bar{N}$) OSCILLATIONS

Rabindra N. Mohapatra*

Department of Physics
City College, CUNY
New York, N.Y. 10031

Abstract

Baryon and lepton number violating processes are discussed as a tool to study the existence of intermediate mass scales below 10^{15} GeV. We emphasize that particularly interesting in this connection are $\Delta B = 2$ processes that cause $N \leftrightarrow \bar{N}$ oscillations and $N_1 + N_2 \to$ pions. Detailed phenomenology of $N \leftrightarrow \bar{N}$ oscillation is presented. A partial unification model where B- and L- breakdown arises from spontaneous breakdown is reviewed and its implications are presented. Implications of grand unified models based on SU(5) and SO(10) for such processes are noted.

1. INTRODUCTION

The subject of baryon (B) and lepton (L) non-conservation is a topic of great current interest to particle physicists. The theoretical motivation for their study comes from the following reasons: It is now generally believed[1] that all interactions among elementary particles owe their origin to a fundamental gauge principle. Since baryon number appears to be conserved to a high degree,[2] it would, therefore, be logical to try to associate a gauge degree of freedom with baryon number and attempt a dynamical understanding of baryon conservation in a manner similar to the electric charge conservation. Such a program, however, runs into problems with the experimental constraints[3] coming from Eötvos experiments. It is therefore reasonable to entertain the possibility that baryon number may not be absolutely conserved. Furthermore, the currently prevalent philosophy of grand unification[4,5] of all elementary particle forces in a single gauge framework, almost invariably leads to violation of baryon number and hence to the interesting possibility of proton decay. The necessity to postulate baryon non-conservation, from a purely phenomenological point of view, to understand the observed baryon asymmetry in the universe was, of course, recognized by Sakharov[6] as early as 1967.

*Work supported by National Science Foundation Grant No. Phys78-24888 and CUNY-PSC-BHE research award No. RF 13096.

Once we accept the possibility of baryon non-conservation, it is important to study the various complexions (or selection rules) of B-violating processes and try to extract their implications for physics. In the context of particular grand unification models such as SU(5), it is well known that the strength of B-violating decay is related to the unification mass of weak, electromagnetic and strong couplings.[7] Therefore, experimental detection of proton decay[8] and a knowledge of its lifetime would reveal the existence of new thresholds in physics beyond 10^2 GeV, the characteristic mass scale of weak interactions. An important first step in studying the selection rules operating in B-non-conservation and its possible connection with mass scales in particle physics was the work of Weinberg[9] and Wilczek and Zee.[9] Their observation was that baryon non-conserving operators of lowest scale dimensions that respect the low energy $SU(3)_C \times SU(2)_L \times U(1)$ symmetry conserve the quantum number B-L. Therefore, they allow only $\Delta B = \Delta L$ decays of the type $p \to e^+\pi^0$, $n \to e^+\pi^-, \ldots$ and forbid in lowest order the processes such as $n \to e^-\pi^+$, $n \leftrightarrow \bar{n}$, etc., which do not respect this selection rule. Therefore, detection of the latter kind of processes at a level of strength comparable to that expected for $\Delta B = \Delta L$ processes would signal the existence of intermediate mass scales between 10^2 GeV and 10^{15} GeV. A systematic study of this question has been conducted by Marshak and the author[10] as well as others.[11,12] In particular, it has been shown in Ref. 10 that there exist interesting gauge models based on the left-right symmetric gauge groups,[13] where the signature of the intermediate mass scale corresponding to the right-nanded W_R-boson is the existence of $\Delta B = 2$ transitions such as $N \leftrightarrow \bar{N}$ oscillation and $N_1 + N_2 \to$ pions. Other signals characteristic of intermediate mass scales may arise in different grand unified models.

In this paper I plan to present a brief review of baryon and lepton non-conserving processes that throw light on the question of intermediate mass scales and their implication for grandunification. The paper is organized as follows. In Section 2 some of the relevant higher dimensional operators[14] consistent with $SU(3)_C \times SU(2)_L \times U(1)$ will be listed along with their characteristic experimental signature. In Section 3, I focus on $\Delta B = 2$ transitions such as $N \leftrightarrow \bar{N}$ oscillation and $N_1 + N_2 \to$ pions and discuss the phenomenon of coherent regeneration of antineutrons starting with a beam of neutrons under various laboratory conditions. In Section 4, a left-right symmetric model based on the gauge group $SU(2)_L \times SU(2)_R \times SU(4)_C$ is presented, which naturally generates a large $\Delta B = 2$ amplitude while suppressing proton decay. In Section 5, I comment on an extended SU(5) model[15] and other grand unification models with reference to the possible existence of such effects.

2. INTERMEDIATE MASS SCALES AND $\Delta B \neq 0$ AND $\Delta L \neq 0$ OPERATORS

In this section we present an extended catalog of baryon and lepton number non-conserving operators which can be used in the probe of intermediate mass scales. For this purpose, we fix our notation for fermion fields of both helicities and their $SU(3)_C \times SU(2)_L \times U(1)$ quantum numbers:

$$\begin{aligned} q^p_{L,i} &: \quad (3,\ 1/2,\ 1/3) \\ q^c_{L,i} & \quad\ \ (3^*,\ 0,\ 2t_{3R} - 1/3) \\ \ell^p_L & \quad\ \ (1,\ 1/2,\ -1) \\ \ell^c_L & \quad\ \ (1,\ 0,\ 2t_{3R} + 1) \end{aligned} \tag{2.1}$$

where i is the color index $(i = 1,2,3)$ and p is the $SU(2)$ index; superscript c stands for charge-conjugate field: $\psi^c = C\bar{\psi}^T$ where C is the Dirac charge conjugation matrix. t_{3R} takes the following values:

$$t_{3R}(u^c_L) = -\frac{1}{2} \; ; \qquad t_{3R}(d^c_L) = +\frac{i}{2} \; ; \qquad t_{3R}(e^+_L) = +\frac{1}{2} \; .$$

An important point to note is that in terms of the newly defined quantum number, we can write electric charge as

$$Q = I_{3L} + t_{3R} + \frac{B-L}{2} \; . \tag{2.2}$$

Our object is to write down $\Delta B \neq 0$ or $\Delta L \neq 0$ Hamiltonians of various scale dimensions involving the fields in Eq. (1).

i) Four Fermi Operators

If we do not include the Higgs bosons in the theory, this is the lowest dimension operator (dim = 6) of interest. Then, it follows from inspection of all available operators made out of u, d, e, ν (for one generation) that respect the low energy symmetry given earlier that $\Delta Q = 0$ implies $\Delta t_{3R} = 0$. Therefore, from (2.2) it follows that $\Delta(B-L) = 0$, a result first obtained in Ref. 9. This will allow decays of the types $p \to e^+\pi^0$, $n \to e^+\pi^-$, ... The strength for these processes, from dimensional considerations, can be given as α/M_0^2 (where the subscript 0 stands for the fact that the operator has lowest dimension). From the present lower bounds on proton lifetime, therefore, one deduces that $M \simeq 10^{15}$ GeV. Since $\Delta(B-L) \neq 0$ processes must be higher dimension, their experimental detection with a lifetime $\simeq 10^{30}$ yrs. (or a strength $\approx 10^{-29}$-10^{-30} $(\text{GeV})^{-m}$, m = integer) would provide clear evidence for the existence of intermediate mass scales. We now discuss such operators.

ii) D = 7 Operators

The $\Delta B \neq 0$ operators with dimension 7 that do not involve Higgs bosons were noted in Ref. 10 to be of the following type:

$$0^{(7)} = \bar{q}^p_{Li} \gamma_\mu \ell^p_L q^{c^T}_{Lj} C^{-1} D_\mu q^c_{Lk} \varepsilon^{ijk} + \text{h.c.} \tag{2.3}$$

where D_μ is the color gauge invariant operator. This operator will give rise to $\Delta(B+L) = 0$ processes like $p \to e^- \pi^+ \pi^+$, $n \to e^- \pi^+$, etc. From dimensional considerations, the strength of this operator is h/M_1^3, where h is expected to be $\approx 10^{-4}$ to 10^{-5}. The existence of this process at the level mentioned earlier would therefore indicate the existence of an intermediate mass scale $\approx 10^{10}$ GeV. If one allows for Higgs mesons $(\phi_i \equiv (0,1/2,1))$ to exist, other operators can exist.[14] An example is

$$q^{c^T}_{Li} C^{-1} q^c_{Lj} \bar{q}^p_{Lk} \ell_R \phi_p \varepsilon^{ijk} \tag{2.4}$$

This will also give rise to processes of $\Delta B = -\Delta L$ type. This will correspond to a strength for $n \to e^- \pi^+$

$$\sim \frac{h m_W}{g M_1^3}$$

and is therefore enhanced compared to the operator in Eq. (2.3). This also brings out an interesting point, that caution has to be taken in deducing precise values for the mass scale from the "observed" strength of $\Delta B \neq 0$ operators. While this will not affect the general conclusion about the existence of intermediate mass scales very much, more precise conclusions will have to depend on particular models.

iii) Operators with D = 9

It was pointed out in Refs. 10 and 12 that an interesting $\Delta B \neq 0$ operator is the one with $D = 9$, since it gives rise to the interesting phenomenon of $N \leftrightarrow \bar{N}$ transitions as well as $\Delta B = 2$ nuclear transitions: The operator is

$$0^{(9)} = d^{c^T}_{Li_1} C^{-1} d^c_{Li_2} u^{c^T}_{Lj_1} C^{-1} u^c_{Lj_2} d^{c^T}_{Lk_1} C^{-1} d^c_{Lk_2} \varepsilon^{i_1 j_1 k_1} \varepsilon^{i_2 j_2 k_2} \tag{2.5}$$

$$+ \text{ permutations}$$

Its strength is given by h_9/M_3^5; therefore, even though h_9 will be a model dependent quantity, the detection of this process with strength 10^{-30} (GeV)$^{-6}$

will indicate the existence of a very low intermediate mass scale indeed ($M_3 \simeq 10^4$ to 10^6 depending on where $h_9 \simeq 10^{-10}$ to 1). Our major concern in the following sections will be to analyze the question of experimental accessibility of this process and its implications for unified gauge theories. Before we proceed to that discussion, we note another operator of the same dimension where $\Delta B = 0$ but $\Delta L = 2$, $O_L^{(9)}$.

$$O_L^{(9)} = \bar{q}^c_{Li} \gamma_\mu q^c_{Li} \bar{q}^c_{Lj} \gamma_\mu q^c_{Lj} \ell^{c^T} C^{-1} \ell^c_L \quad . \tag{2.6}$$

Operators of this kind are responsible for double β-decay. The search for other dim = 9 operators has been carried out in Ref. 14 and an interesting process of this kind is $p \to 3\nu\pi^+$.

iv) D = 12 Operators

Finally, we also note that dim 12 operators may also be of future phenomenological interest. A typical operator of this type is

$$\varepsilon^{i_1 j_1 k_1} \varepsilon^{i_2 j_2 k_2} d^{c^T}_{Li_1} C^{-1} d^c_{Li_2} u^{c^T}_{Lj_1} C^{-1} u^c_{Lj_2} d^{c^T}_{Lk_1} C^{-1} d^c_{Lk_2} \ell^{c^T}_L C^{-1} \ell^c_L \tag{2.7}$$

This will give rise to processes[15] like $N + p \to e^+ \nu^c$ or hydrogen to anti-hydrogen transitions. The existing astrophysical limits on the strength of such processes is $\lesssim G_F \times 10^{-21}$. Since the expected strength of these transitions $\sim h_{12}/M_4^8$, if we assume $h_{12} \simeq 10^{-2}$, the above upper bound on the strength implies $M_4 > 10^3$ GeV. Such processes will therefore in the future yield important information on the mass scale. We wish to point out parenthetically that the higher the dimension of the operator, the less model dependent the value of M_4 will be.

3. PHENOMENOLOGY OF $N \leftrightarrow \bar{N}$ OSCILLATIONS

As we saw in the last section, detection of $\Delta B = 2$ transitions will be very important from a theoretical point of view. In this section we present a phenomenology of $N \leftrightarrow \bar{N}$ oscillation and other $\Delta B = 2$ transitions of type $N_1 + N_2 \to \pi$'s leading to nuclear instability and indicate level of the $\Delta B = 2$ transition amplitude that may be detectable with presently available experimental facilities. For this purpose, we define an effective Hamiltonian for nucleons that we can abstract out of Eq. (2.5):

$$\mathcal{L}_m = \delta m N^T C^{-1} N + \text{h.c.} \tag{3.1}$$

A characteristic oscillation time for $N \leftrightarrow \bar{N}$ can be defined in terms of δm

as follows:

$$\tau_{N-\bar{N}} \simeq \frac{\hbar}{\delta m} \simeq \left(\frac{10^{-32}}{\delta m \text{ in GeV}}\right) \text{ yrs.} \tag{3.2}$$

Next we look for experimentally acceptable ranges of values for $\tau_{N-\bar{N}}$ from upper limits on nuclear stability. To do that, observe that the same Hamiltonian in Eq. (2.5) that causes $N \leftrightarrow \bar{N}$ oscillation also drives two nucleons to pions, i.e., $N_1 + N_2 \to \pi$'s; this can cause nuclear instability. One can therefore roughly write:[12]

$$\delta m \approx \sqrt{\Gamma_{N_1 + N_2 \to \pi's} M} \tag{3.3}$$

From the knowledge that

$$\left(\Gamma_{N_1 + N_2 \to \pi's}\right)^{-1} \gtrsim 10^{30} \text{ yrs} \quad ,$$

and choosing $M \approx 1$ GeV, we obtain from Eq. (3.3) that

$$\delta m \lesssim 10^{-21} \text{ eV} \tag{3.4}$$

This corresponds to $\tau_{N-\bar{N}} \gtrsim 10^6$ sec. Below we demonstrate that a characteristic transition time of this order should be well within present experimental capabilities. We consider the phenomenon of coherent regeneration of anti-neutrons starting with a beam of neutrons. First, we will discuss the idealized case of free neutron oscillation and subsequently extend these considerations to the realistic case of $N \leftrightarrow \bar{N}$ oscillations in the presence of external fields. The analysis changes drastically provided the external field acts differently on N and $\bar{N}$, such as the case with external magnetic field or the nuclear forces.

i) Free Neutron Oscillations

The phenomenology of neutral kaon oscillations is well-known and this $\Delta S = 2$ weak transition has played an important role in the theory of weak interactions. The $\Delta B = 2$ $N - \bar{N}$ transition can be treated in a similar fashion in the absence of an external field. The starting point is the $N - \bar{N}$ mass matrix exactly as in the case of the $K^0 - \bar{K}^0$ system:

$$L_{mass} = \bar{\psi} M \psi \tag{3.5}$$

where

$$\psi = \begin{pmatrix} N \\ \bar{N} \end{pmatrix}$$

and

$$M = \begin{pmatrix} A & \delta m \\ \delta m & A \end{pmatrix} \tag{3.6}$$

In Eq. (3.6), δm is the $\Delta B = 2$ transition mass between N and $\bar{N}$ states (see Eq. (3.1)). The equality of the diagonal elements follows from CPT invariance and that of the off-diagonal elements from CP invariance. The eigenstates of M can be written as

$$N_{1,2} = \frac{N \pm \bar{N}}{\sqrt{2}} \tag{3.7}$$

with masses:

$$m_{1,2} = A \pm \delta m \quad . \tag{3.8}$$

Next, the amplitude for finding an $\bar{N}$ at time t starting with a beam of neutrons at $t = 0$ is:

$$|N(t)> \simeq e^{\frac{-\gamma t}{2}} \left\{ \frac{|N_1(0)> e^{-im_1 t} + |N_2(0)> e^{-im_2 t}}{\sqrt{2}} \right\} \tag{3.9}$$

where we have assumed that the decay widths of N_1 and N_2 are equal and denoted by γ. The probability for finding an $\bar{N}$ at time t, i.e., $P_{\bar{N}}(t)$, follows from Eq. (3.9)

$$P_{\bar{N}}(t) \simeq \frac{e^{-\gamma t}}{2} (1 - \cos 2\,\delta m t) \quad . \tag{3.10}$$

If $t \ll 1/\delta m$, then we obtain

$$P_{\bar{N}}(t) \simeq e^{-\gamma t} (\delta m)^2 \quad . \tag{3.11}$$

It follows that if a beam of N free neutrons is allowed to travel for a time T $(T \ll \tau_{N-\bar{N}})$ before hitting a target, the number of antineutrons, $\bar{N}$, at the target after T should be (assuming $\gamma T \ll 1$):

$$\bar{N} \lesssim N \left(\frac{T}{\tau_{N-\bar{N}}} \right)^2 \quad . \tag{3.12}$$

With reasonable values of N and T, Eq. (9) would be rather promising for a "free" neutron oscillation experiment. However, there is a complicating factor resulting from the presence of the earth's magnetic field which shifts the energy levels of N and $\bar{N}$ by opposite amounts $\mu B \sim 10^{-11}$ eV, which is much larger than δm. We deal with this situation below.

ii) $N \leftrightarrow \bar{N}$ Oscillation in an External Field

If the neutron oscillation takes place in an external field (such as the earth's magnetic field or any other field) so that the CPT theorem need not be respected, the picture outlined in the previous section changes and we have:

$$M \simeq \begin{pmatrix} A_1 & \delta m \\ \delta m & A_2 \end{pmatrix} \tag{3.13}$$

The neutron mass eigenstates in this case are:

$$\begin{aligned} |N_1\rangle &\simeq |N\rangle + \theta|N\rangle \\ |N_2\rangle &\simeq -\theta|N\rangle + |\bar{N}\rangle \quad , \end{aligned} \tag{3.14}$$

where

$$\theta \simeq \frac{\delta m}{\Delta M} \tag{3.15}$$

where

$$\Delta M \equiv M_1 - M_2 \approx A_1 - A_2$$

if

$$A_1 - A_2 \gg \delta m \quad .$$

The analog of Eq. (3.10) for this case is

$$P_{\bar{N}}(t) \simeq \frac{\theta^2}{2}\,[1 - \cos \Delta M t] \quad . \tag{3.16}$$

Equation (3.16) is interesting in two limiting cases: a) $\Delta M t \gg 1$ and (b) $\Delta M t \ll 1$. In case (a), the second term oscillates rapidly and the average probability for finding an $\bar{N}$, $\langle P_{\bar{N}} \rangle_{av}$, becomes:

$$\langle P_{\bar{N}} \rangle_{av} \simeq \left(\frac{\delta m}{\Delta M}\right)^2 \quad . \tag{3.17}$$

In case (b) we get:

$$P_{\bar{N}}(t) \simeq (\delta m t)^2 \quad , \tag{3.18}$$

which recaptures the field-free result (3.11) (as long as $\gamma t \ll 1$).

iii) Experimental Detection of Neutron Oscillations

Equations (3.17) and (3.18) can now be used to suggest optimal conditions for a proposed reactor experiment to detect neutron oscillations.[16] In such an experiment with thermal neutrons, $t \sim 10^{-2}$ sec, and if the magentic field of the earth is not shielded, $\Delta Mt \gg 1$ and Eq. (3.17) applies. On the other hand, if the earth's magnetic field is "degaussed" by a factor of 10^3 or more, Eq. (3.18) applies and the experiment becomes quite favorable. The sensitivity of the experiment depends, of course, on the detailed design, but $N \simeq 6 \times 10^{13}$ neutrons/sec seems possible[16] and with the numbers of Eq. (3.18) corresponding to a "degaussing" factor of 10^3, values of $\bar{N}$ as large as 10^6/yr. are not incompatible with the present lower limit on the lifetime for nuclear stability.

Whether $\Delta B = 2$ nucleon transitions take place or not, the above considerations do indicate that the search for "free" neutron oscillations $(N \to \bar{N})$ shielded from earth's magnetic field can, in principle, yield a sensitivity many orders of magnitude greater than the search for "bound" $\Delta B = 2$ nucleon transitions, e.g., through the processes $N + P \to \pi$'s in a nucleus. We would also like to add here that $\tau_{N-\bar{N}} \simeq 10^6$ sec corresponds to a

$$\Gamma_{N_1 + N_2 \to \pi's} \gtrsim 10^{30} \text{ yrs.}$$

It should therefore be possible to search for $\Delta B = 2$ transitions in the experimental setups designed to detect proton decay.[8] The pions resulting in this manner would have average energies $\gtrsim 500$ MeV making it easier to detect them.

4. $\Delta B = \Delta L = 2$ TRANSITIONS IN A UNIFICATION MODEL WITH LOCAL B - L SYMMETRY

In the present section we discuss a gauge model in which $B - L$ is a dynamical quantum number locally conserved in the symmetry limit. Thus, its spontaneous breakdown leads naturally to $\Delta B = \Delta L = 2$ transitions of the type discussed earlier. The strength of these transitions is characterized by a mass scale which is linked to the existence of $V + A$ weak currents. Therefore, $\Delta B = \Delta L = 2$ transitions are suppressed to the extent that $V + A$ weak currents are suppressed. The model is based on the left-right symmetric gauge group $SU(2)_L \times SU(2)_R \times U(1) \times SU(3)_c$ with full quark lepton symmetry or the extended local symmetry $SU(2)_L \times SU(2)_R \times SU(4)_c$ that unifies quarks and leptons in a single framework.

The first important observation is that, unlike the case of the $SU(2)_L \times U(1)$ model, the vector $U(1)$ generator in the left-right symmetric theories[13] can be identified[17] with the $(B - L)$ symmetry. One implication of this observation is that the mass scales associated with spontaneous breakdown of parity could be associated with the breakdown of local $(B-L)$ electroweak symmetry. To see this explicitly, note that in these theories the electric charge is given by

$$Q = I_{3L} + I_{3R} + \frac{B-L}{2} \tag{4.1}$$

where $\vec{I}_{L,R}$ are generators of $SU(2)_{L,R}$ groups. In the energy regime above the m_{W_L} mass, this implies that (since $\Delta I_{3L} \simeq 0$)

$$\Delta I_{3R} \simeq -\frac{1}{2}\Delta(B-L) \quad . \tag{4.2}$$

Therefore, breakdown of parity and that of $(B-L)$ symmetry are linked. The bounds on the masses of $\vec{W}_R$ gauge bosons associated with spontaneous breakdown of parity can be gotten from the data on neutral currents[18] to be

$$m_{W_R} \gtrsim 5m_{W_L} \quad .$$

We thus expect the mass associated with $\Delta B = 2$ transitions to be of order ≈ 500 GeV to 10^3 GeV.

To realize these ideas explicitly, we work with an $SU(2)_L \times SU(2)_R \times SU(4)_c$ model. The quarks and leptons belong to the same multiplet of $SU(4)_c$ as follows:

$$\psi \equiv \begin{pmatrix} u_1 & u_2 & u_3 & \nu \\ d_1 & d_2 & d_3 & e^- \end{pmatrix} \tag{4.3}$$

ψ_L and ψ_R transform as $(\frac{1}{2}, 0, 4)$ and $(0, \frac{1}{2}, 4)$ representations of the group. The $U(1)$ generator identified with $(B-L)$ is given by

$$B - L = \begin{pmatrix} 1/3 & & & \\ & 1/3 & & \\ & & 1/3 & \\ & & & -1 \end{pmatrix} \tag{4.4}$$

There are two gauge couplings: $g_L = g_R = g$ and f, the $SU(4')$ coupling. We envision a mass scale m_X such that this local group breaks down to the level of $SU(3)_c \times SU(2)_L \times SU(2)_R \times U(1)_{B-L}$, which breaks down in stages to $SU(3)_c \times SU(2)_L \times U(1)$ and subsequently to $SU(3)_c \times U(1)_{em}$ with m_{W_R} and m_{W_L} being the corresponding mass scales, respectively. Since we expect $m_X \gtrsim 10^5$ GeV due to the absence of $K_L \to \mu\bar{e}$ decay, the mass hierarchy would be

$$m_X > m_{W_R} \gg m_{W_L} \gg 1 \quad \text{GeV} \quad .$$

To implement the Higgs mechanism that will lead to this pattern of symmetry breaking, we choose only such Higgs multiplets that can be expressed as bilinears in the basic fermion multiplet. We choose the Higgs multiplets $\phi \equiv (\frac{1}{2}, \frac{1}{2}, 0)$ (transforming as $\bar{\psi}\psi$) and $\Delta_L \equiv (1, 0, 10)$ and $\Delta_R \equiv (0, 1, 10)$ which transform as $\psi_i^T C^{-1} \tau_2 \tau_a \psi_j$. This pattern of Higgs multiplets allows only two stages of symmetry breakdown:

$$SU(2)_L \times SU(2)_R \times SU(4)_c$$

$$\Big\downarrow \quad \langle \Delta_{R,44} \rangle = v \neq 0 \; , \quad \langle \Delta_L \rangle = 0$$

$$SU(2)_L \times U(1) \times SU(3)_c$$

$$\Big\downarrow \quad \langle \phi \rangle \neq 0$$

$$U(1)_{em} \times SU(3)_c$$

The masses of m_X and m_{W_R} are related:

$$m_X \approx \frac{f}{g} m_{W_R} = fv \quad .$$

We choose[13]

$$\langle \phi \rangle = \begin{pmatrix} \kappa & 0 \\ 0 & \kappa' \end{pmatrix}$$

which gives mass both to the fermions and the W_L-bosons.

i) Majorana Neutrinos

To study the implications of $(B - L)$ breakdown in detail, we write down the gauge-invariant Yukawa couplings:

$$L_Y = ih \, [\, \psi^T_{L,i} \tau_2 \tau_a C^{-1} \psi_{L,j} \Delta^a_{L,ij} + \Psi^T_{R,i} \tau_2 \tau_a C^{-1} \Psi_{R,j} \Delta^a_{R,ij}]$$
$$+ h_1 \tilde{\Psi}_L \Phi \Psi_R + h_1' \Psi_L \tilde{\Phi} \Psi_R + h.c. \tag{4.5}$$

This pattern of symmetry breaking was recently discussed in the context of electroweak models by Senjanovic and the author[19] and it was noted that

$$\langle \Delta^{1+i2}_{R,44} \rangle \neq 0 \quad ,$$

along with the contribution of $\langle\phi\rangle$, led to the following mass matrices[20] for the neutrinos (call $\nu \equiv \nu_L$ and $N \equiv \nu_R$):

$$\begin{array}{c} \\ \nu \\ N \end{array}\begin{array}{c} \begin{array}{cc} \nu & N \end{array} \\ \begin{pmatrix} 0 & am_e \\ m_e a & m_N \end{pmatrix} \end{array} \tag{4.6}$$

where $m_e \approx h_1\kappa$ and $m_N \approx hv \approx gm_{W_R}$ and a is a numerical constant that denotes the ratio of two typical Yukawa couplings. From this it follows that the usual neutrinos must be Majorana particles with

$$m_\nu = \frac{a^2 m_e^2}{gm_{W_R}} \quad .$$

We thus see that the smallness of the neutrino mass is related to the dominant (V - A) nature of the weak interactions at low energy.[19] On the other hand, since $g \approx e$, $m_N \gtrsim 100$ GeV to 300 GeV for $m_{W_R} \simeq 300$ to 1000 GeV reflecting the breaking of I_{3R}. This implies the breaking of (B - L) which is consistent with a Majorana neutrino (for which $\Delta L = 2$). Furthermore, we note that if we choose $a \approx 1$, the above formula gives $m_\nu \lesssim 1$ to 2 eV. The important point about this model is a smooth connection between the massless neutrino with pure left-handed weak currents and a massive neutrino with mass being linked to smallness of V + A currents at low energies [or $m_\nu \to 0$ as $m_{W_R} \to \infty$]. This model therefore explains the Parity puzzle in terms of the small neutrino mass.

The Majorana character of ν_e predicts the existence of double β decay coming from the exchange of two W_R bosons with the heavy Majorana neutral Lepton N_R as the intermediate state. The strength of this amplitude is

$$\approx G_F^2 \left(\frac{m_{W_L}^2}{m_{W_R}^2}\right)^2 \frac{1}{m_N} \lesssim G_F^2 \times 10^{-4} \quad ,$$

$$\text{for} \quad \frac{m_{W_L}}{m_{W_R}} \lesssim \frac{1}{5} \quad \text{and} \quad m_N \gtrsim 100 \text{ GeV} \quad .$$

It is easy to convince oneself that the contribution from the graph in which the contracted intermediate lepton line corresponds to the ν_e is very small (i.e., $\lesssim G_F \times 10^{-6}$, if $m_{\nu_e} \simeq 1$ eV). The present experimental situation has been analyzed[21] to yield upper bounds on the double β-decay amplitude of order $\lesssim G_F^2 \times 10^{-4}$. It will therefore be of great interest to redouble the efforts to search for double β-decay.[22]

So far, we have concentrated only on the first generation (i.e., ν_e, e^-) to study the question of neutrino mass. Similar considerations can be extended to higher generations. In general, of course, the situation is quite complicated due to the presence of mixings among the various generations. However, if we ignore the mixings, we get similar formulae:

$$\begin{aligned} m_{\nu_e} &\simeq a_e \cdot \frac{m_e^2}{g m_{W_R}} \\ m_{\nu_\mu} &\simeq a_\mu \cdot \frac{m_\mu^2}{g m_{W_R}} \qquad (4.7) \\ m_{\nu_\tau} &\simeq a_\tau \cdot \frac{m_\tau^2}{g m_{W_R}} \end{aligned}$$

The parameters $a_{e,\mu,\tau}$ are arbitrary. Therefore, apart from displaying a connection between the small neutrino mass and the maximality of parity violation, any further conclusion to be drawn from Eq. (4.6), such as exact values of m_{ν_e}, etc., will require fixing of $a_{e,\mu,\tau}$. In obtaining $m_{\nu_e} \simeq 1$ eV, we choose $a_e \simeq 1$. If we postulate generation independence of a, i.e., $a_e = a_\mu = a_\tau$, and choose it to be 1, then, one would predict $m_{\nu_\mu} \lesssim 40$ to 80 KeV and $m_{\nu_\tau} \lesssim 12$ to 24 MeV. Needless to say, these values can be lowered by either giving up the requirement of generation independence or changing the values of $a_{e,\mu,\tau}$. One point worth noting here is that if the masses of the various neutrinos are of the order shown above, they are unlikely to be changed very much due to mixing, in which case one would expect ν_μ and ν_τ to be unstable. The predominant decay mode is $\nu_\tau \to \nu_\mu\gamma$, $\nu_e\gamma$ and $\nu_\mu \to \nu_e\gamma$. They give the following lifetimes for ν_τ and ν_μ:

$$\tau_{\nu_\tau} \simeq \frac{\pi}{\alpha}\left(\frac{m_\mu}{m_{\nu_\tau}}\right)^5 \cdot \frac{1}{\delta^2} \cdot \tau_\mu$$

where

$$\delta \simeq \sin 2\phi \cdot \left(\frac{m_L^2}{m_{W_L}^2}\right) \quad . \qquad (4.8)$$

ϕ is a mixing angle and m_L is a charged lepton mass, and a similar expression for τ_{ν_μ}. Assuming maximal mixing and $m_L \simeq m_\tau$, one obtains $\tau_{\nu_\tau} \approx 10^7$ sec and $\tau_{\nu_\mu} \approx 10^{19}$ sec. On the other hand, if there existed a fourth generation of leptons with the mass of the next charged lepton $\simeq 100$ GeV, then one would expect $\tau_{\nu_\tau} \simeq$ few sec. and $\tau_{\nu_\mu} \simeq 10^{13}$ sec. We might note that if the results from astrophysical analysis of the neutrino mass are to be taken seriously,[23] then while the ν_τ and ν_e would be consistent, ν_μ would not be. Another consequence of the Majorana neutrino is the production of "wrong"-type charged leptons in inverse β-decay; here, the model prediction is much lower than the experimental limit[24] due to the

$$\left(\frac{m_\nu}{E_\nu}\right)$$

supression factor (E_ν is the neutrino energy).

ii) "Neutron Oscillations"

We now discuss the implications of $(B-L)$ violation for the hadronic sector. First we note that there exists a self-coupling of the scalar multiplet $\Delta_{L,R}$ as follows:

$$L_S = \lambda \left[\varepsilon^{ikmp} \varepsilon^{jlnq} \Delta^a_{L,ij} \Delta^a_{L,kl} \Delta^b_{L,mn} \Delta^b_{L,pq} + L \leftrightarrow R + \text{h.c.} \right] \tag{4.9}$$

From Eqs. (4.5) and (4.6), it follows that there exists a six-fermion vertex of type (see Fig. 2)

$$L_{6f} = h_{\text{eff}}\, d^T_{R,i_1} C^{-1} d_{R,i_2}\, d^T_{R,j_1} C^{-1} d_{R,j_2}\, u^T_{R,k_1} C^{-1} u_{R,j_2}\, \varepsilon^{i_1 j_1 k_1} \varepsilon^{i_2 j_2 k_2} \tag{4.10}$$

$$+ \text{ similar terms}$$

As noted in Section 2, this Lagrangian causes the transition $N \leftrightarrow \bar{N}$ or neutron oscillation. The important point, which becomes obvious looking at Fig. 2, is that the $\Delta B = 2$ $N \leftrightarrow \bar{N}$ transition is caused by the same spontaneous breaking mechanism (i.e., $\langle \Delta_{R,44} \rangle \neq 0$) that causes $\Delta L \neq 0$. We now estimate the strength of the $N \leftrightarrow \bar{N}$ transition in our model.

$$h_{\text{eff}} \approx \frac{\lambda \cdot h^3 \langle \Delta_{R,44} \rangle}{m^6_{\Delta_R}}$$

We see that as $\langle \Delta_{R,44} \rangle \to 0$ (i.e., restoration of parity as well as $B-L$ symmetry), $N-\bar{N}$ oscillations disappear. We may visualize that

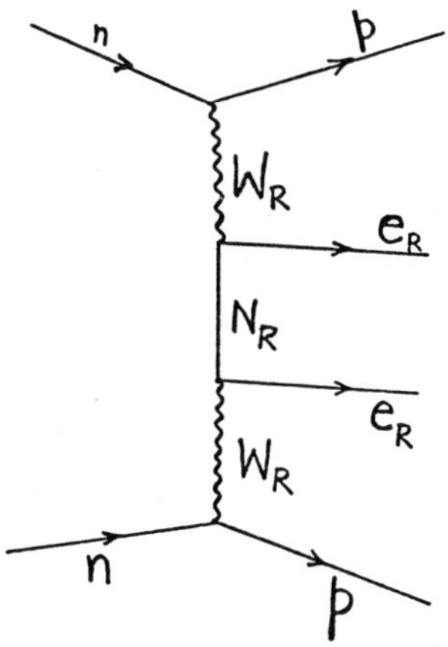

Figure 1. The Feynman diagram in the left-right symmetric model that gives the dominant contribution to neutrinoless double β-decay.

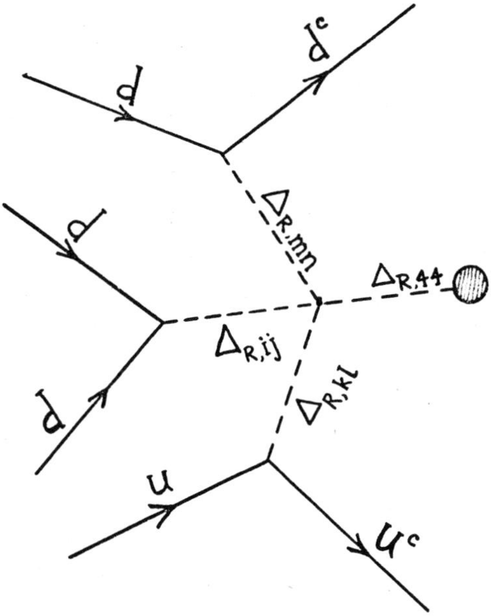

Figure 2. A typical Higgs-boson mediated graph responsible for $\Delta B = 2$ transitions in the $SU(2)_L \times SU(2)_R \times SU(4)_c$ model.

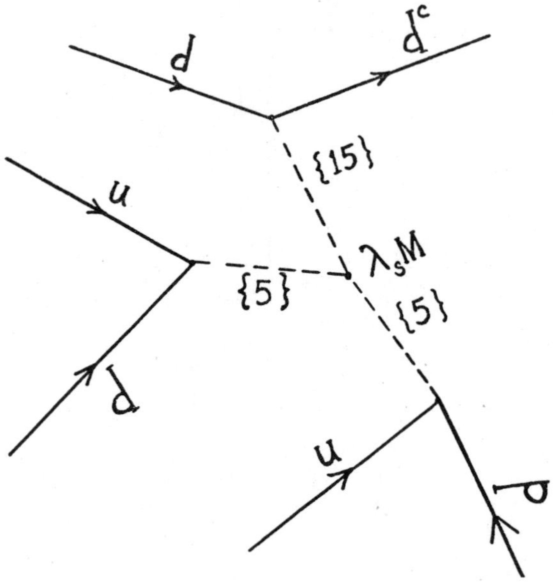

Figure 3. The Feynman diagram responsible for $\Delta B = 2$ transitions in the extended SU(5) model.

$$m_{\Delta_R} \approx m_{W_R} \approx 10^4 \text{ GeV} \qquad \text{and} \qquad h \approx 10^{-2}$$

(since this is responsible for the mass of m_N); putting $\lambda \approx \alpha^2$, we get $h_{eff} \approx 10^{-30}$ $(\text{GeV})^{-5}$; this would correspond to a characteristic time $\tau_{N-\bar{N}}$ for the $N-\bar{N}$ oscillation of about 10^5 to 10^6 sec. The strength of other $\Delta B = 2$ processes such as $N_1 + N_2 \to \pi$'s is also characterized by the same strength leading to a double neucleon decay lifetime for nuclei of about 10^{30} years. This could be observable in the experiments presently under way.[8]

iii) Other Exotic $\Delta B \neq 0$ Processes

This model allows for the $\Delta(B-L) = 0$ process $p + e^- \to \bar{p} + e^+$ (hydrogen-antihydrogen transition) with strength of

$$\frac{\lambda h^4}{m_H^8} \approx 10^{-44} \ (\text{GeV})^{-8} \quad ,$$

which is some 18 orders of magnitude smaller than present astrophysical limits.[15]

iv) Absence of Proton Decay

An interesting feature of this model is the stability of proton against any semileptonic decay such as $p \to e^+\pi^0$, etc. The reason for this is the existence of an exact discrete symmetry of the theory:

$$q_i \to e^{i\pi/3} q_i \ , \qquad \Delta_{ij} \to e^{-2\pi i/3} \Delta_{ij} \ , \qquad \Delta_{i4} \to e^{-i\pi/3} \Delta_{i4}$$

with everything else unchanged. This hidden discrete symmetry would, of course, be broken by the addition of extra Higgs multiplets into the theory.

5. ΔB = 2 TRANSITIONS IN GRANDUNIFICATION MODELS

Having observed[10] that partial unification models of the type $SU(2)_L \times SU(2)_R \times SU(4)_C$ lead naturally to $\Delta B = 2$ transitions due to $(B - L)$ being a local symmetry, an immediate question arises as to whether this phenomenon persists in any grand unified model that describes weak, electromagnetic, as well as strong interactions. Here we report on investigations of this question in the context of two grand unified models: i) an extended SU(5) model,[12,25] ii) a model based on the $[SU(4)]^4$ grand unification group.[26] Barring certain unnatural possibilities, in extended SU(5) models, a large $\Delta B = 2$ transition amplitude appears to conflict with lower bounds on the proton lifetime. On the other hand, the $[SU(4)]^4$ model which incorporates the partial unification group can yield a significant $\Delta B = 2$ transition amplitude. Thus in the very least, observation of $\Delta B = 2$ transitions would have profound impact on the future direction of grand unification.

i) Extended SU(5) Model

As is well-known, the minimal set of Higgs multiplets required for the desired breakdown of the local SU(5) symmetry is {5} and {24}-plet Higgs multiplets.[5] Several other phenomenological requirements often need the inclusion of several {45}-dim Higgs multiplets in the model.[27] It is easy to check that in the presence of {5} or {45} and {24}-plet of Higgs multiplets, the $B - L$ quantum number remains an exact symmetry of the model subsequent to spontaneous breakdown of SU(5) symmetry. Without further modification of the Higgs structure of the model, therefore, a $B - L$ violating transition such as $N \leftrightarrow \bar{N}$ and $N_1 + N_2 \to \pi$'s will be forbidden.

It has been suggested[12,25] that inclusion of an SU(5) symmetric {15}-plet Higgs multiplet causes a breakdown of the $\Delta(B - L) = 0$ selection rule and can thereby lead to $\Delta B = 2$ transitions. In this section, it is argued that phenomenological requirements on this extended SU(5) model are such that the $\Delta B = 2$ transitions are extremely suppressed. To see this, let us denote by ϕ_p, S_{pq}, H^p_q the {5}, {15}, and {24} dimensional Higgs multiplets. They satisfy the following conditions:

$$S_{pq} = S_{qp}$$

and

$$\sum_p H_p^p = 0 \quad .$$

The superrenormalizable scalar coupling and the Yukawa couplings that cause (B - L)-symmetry breakdown are:[10]

$$L_s = \lambda_5 M S^\dagger_{pq} \phi_p \phi_q$$
$$L_Y = h_s \psi_p^T C^{-1} \psi_q S^\dagger_{pq} \tag{5.1}$$

It is clear from Eq. (4.1) that $\langle\phi_5\rangle \neq 0$ leads to $\langle S_{55}\rangle \neq 0$; as a result of L_Y, the neutrino becomes a Majorana particle[10] with mass $h_s \langle S_{55}\rangle$. The indirectly observed equality of $m_{W_L} = m_Z \cos\Theta_W$ to within a few percent leads to the constraint that

$$\frac{\langle S_{55}\rangle}{\langle\phi_5\rangle} \leq .1 \tag{5.2}$$

since S_{pq} contains a Higgs triplet under $SU(2)_L \times U(1)$ group.

The Feynman graph that leads to $\Delta B = 2$ transitions in this model is given in Fig. 3. The resulting strength of $\Delta B = 2$ transition amplitudes can be given as

$$A_{\Delta B=2} \approx \frac{\lambda_s M \cdot h_s \cdot h_5^2}{m_5^4 m_s^2} \quad . \tag{5.3}$$

Furthermore, we note that there exist Higgs mediated $\Delta B = 1$ type proton-decay amplitudes whose strength is given by

$$A_{\Delta B=1} \approx \frac{h_5^2}{m_5^2} + \frac{h_5 h_s \cdot \lambda_s M \langle\phi_5\rangle}{m_5^2 m_s^2} \quad . \tag{5.4}$$

It follows from Eq. (5.4) and present lower limits on the proton lifetime that

$$\frac{h_5^2}{m_5^2} \lesssim 10^{-30}\ (\text{GeV})^{-2}$$

and

$$\frac{h_s}{h_5} \cdot \frac{\lambda_s M \langle \phi_5 \rangle}{m_s^2} < 1 \quad . \tag{5.5}$$

Equations (5.3) and (5.5) imply that

$$A_{\Delta B=2} \lesssim 10^{-30} \cdot \frac{h_5}{\langle \phi_5 \rangle m_5^2} \ (\mathrm{GeV})^{-5} \quad . \tag{5.6}$$

Using typical values for h_5, $\langle \phi_5 \rangle$ and m_5, one obtains a characteristic transition time for $N \leftrightarrow \bar{N}$ oscillation of about 10^{27} years or an unobservably long lifetime for $\Delta B = 2$ transitions.[28]

ii) An $[SU(4)]^4$ -Model for $\Delta B = 2$ Transitions

The purpose of this subsection is to describe a grand unification model based on the gauge group $[SU(4)]^4$, where, by straightforward extension of the analysis in Ref. 10, one shows that a sizable $\Delta B = 2$ transition amplitude may exist.

We start by giving the fermion representations under this group: If we denote[29]

$$\psi = \begin{pmatrix} u_{i=1,2,3} & \nu \\ d_i & e^- \\ s_i & \mu^- \\ c_i & \nu' \end{pmatrix} \tag{5.7}$$

then $\psi_L \equiv (4,1,\bar{4},1)$ and $\psi_R = (1,4,1,\bar{4})$ under the $[SU(4)]^4$ group (decomposed as $SU(4)_L \times SU(4)_R \times SU(4')_L \times SU(4')_R$). The Higgs bosons relevant for our purpose will be chosen to belong to $\Delta_L \equiv (10, 1, \overline{10}, 1)$ and $\Delta_R \equiv (1, 10, 1, \overline{10})$ representations of the group. They will be denoted by

$$\Delta_{L,ab,pq} \ , \ \Delta_{R,ab,pq} \quad ,$$

where $a,b = 1,\ldots,4$ are flavor indices and $p,q = 1,\ldots,4$ are color indices. The invariant Yukawa and Higgs couplings which are straightforward generalizations of Eqs. (4.5) and (4.9) are

$$L_Y = h\left[\psi^T_{La,p} C^{-1} \psi_{Lb,q} \Delta^L_{ab,pq} + L \leftrightarrow R\right] + \mathrm{h.c.} \quad ,$$

$$L_S = \lambda \Big[\varepsilon^{a_1a_2a_3a_4} \varepsilon^{b_1b_2b_3b_4} \varepsilon^{p_1p_2p_3p_4} \varepsilon^{q_1q_2q_3q_4} \cdot \Delta^L_{a_1b_1,p_1q_1} \Delta^L_{a_2b_2,p_2q_2} \Delta^L_{a_3b_3,p_3q_3} \Delta^L_{a_4b_4,p_4q_4} + L \leftrightarrow R \Big] + \text{h.c.} \tag{5.8}$$

As in the $SU(2)_L \times SU(2)_R \times SU(4')$ case discussed in Ref. 10, breakdown of parity and $B-L$ is achieved via the following vacuum expectation value:

$$\langle \Delta^R_{44,44} \rangle = V \neq 0 \quad . \tag{5.9}$$

Thus, this gives rise to $\Delta B = 2$ transition amplitudes of order

$$A_{\Delta B=2} \approx \frac{\lambda h^3 v \sin 2\Theta_C}{m_\Delta^6} \tag{5.10}$$

Note the presence of the Cabibbo suppression factor $\sin 2\Theta_C$ in this case. An appropriate choice of m_Δ, etc., can lead to $A_{\Delta B=2} \approx 10^{-30}\ (\text{GeV})^{-5}$ as in Section 4. We will not discuss the rest of the symmetry breakdown in this case, and refer the reader to Ref. 26 for further details and other consequences of such a model.

Before concluding this section, we would like to remark that observation of $\Delta B = 2$ transition amplitudes may have profound implications for the future of grand unification.

6. CONCLUSION

In conclusion, we would like to reiterate that a reasonably large strength for $\Delta B = 2$ and $\Delta L = 2$ processes appears to be characteristic of a class of models where spontaneous breakdown of local $B-L$ symmetry and parity symmetry are linked. In such models, there can be an intermediate mass scale as low as 10^4 GeV. Such large strengths, however, appear to be precluded by other phenomenological constraints in grand unification models. based on $SU(5)$ local symmetry and could therefore distinguish between these two classes of theories.

As far as observability of these processes goes, we show that a characteristic $N \leftrightarrow \bar{N}$ transition time of $\tau \simeq 10^6$ sec allowed in the former class of models can be observable with experiments using the existing reactors. Improvements on the double β-decay experiments will also be very useful in deciding the future of unified gauge theories.

The author would like to thank R.E. Marshak and G. Senjanovic for collaborations which resulted in the bulk of this work. Thanks also are due to N.P. Chang, J.C. Pati, B. Sakita, and R. Wilson for comments and criticism at various stages of this work.

REFERENCES

1. For a review of the recent developments on the subject, see J.C. Pati, *Topics in Quantum Field Theory*, J.A. de Azcárraga (ed.), pp. 221-334. See also J.C. Taylor, *Gauge Theories of Weak Interactions*, Academic Press, New York, 1976.

2. For a review of the subject, see M. Goldhaber, *Unification of Elementary Forces and Gauge Theories*, D. Cline and F. Mills (eds.), Academic Press, New York, 1977, p. 531; H.S. Gürr, W.R. Kropp, F. Reines, and B. Meyer, *Phys. Rev.* 158, 1321 (1967); J. Learned, F. Reines, and A. Soni, *Phys. Rev. Lett.* 43, 907 (1979).

3. T.D. Lee and C.N. Yang, *Phys. Rev.* 98, 1501 (1955). See also A. Pais, Rockefeller Preprint COO-2232B-18 (1973).

4. J.C. Pati and A. Salam, *Phys. Rev.* D8, 1240 (1973); *Phys. Rev. Lett.* 31, 661 (1973); *Phys. Rev.* D10, 275 (1974).

5. H. Georgi and S.L. Glashow, *Phys. Rev. Lett.* 32, 438 (1974).

6. A. Sakharov, *Zh. Exp. Theor. Fiz. Pisma Red.* 5, 32 (1967).

7. H. Georgi, H. Quinn, and S. Weinberg, *Phys. Rev. Lett.* 33, 451 (1974).

8. The present status of various experimental searches for proton decay has been reviewed in reports at this workshop by L. Sulak, D. Winn, and R. Steinberg.

9. S. Weinberg, *Phys. Rev. Lett.* 43, 1566 (1979); F. Wilczek and A. Zee, *Phys. Rev. Lett.* 43, 1571 (1979).

10. R.N. Mohapatra and R.E. Marshak, *Phys. Rev. Lett.* 44, 1316 (1980); VPI-HEP-80/2, to be published, *Proc. Orbis Scientiae*, 1980.

11. F. Wilczek and A. Zee, *Phys. Lett.* B (to appear).

12. S.L. Glashow, *Future of Elementary Particle Physics*, HUTP-79/A029 and HUTP-79/A059.

13. J.C. Pati and A. Salam, *Phys. Rev.* D10, 275 (1974); R.N. Mohapatra and J.C. Pati, *Phys. Rev.* D11, 566, 2558 (1975); G. Senjanovic and R.N. Mohapatra, *Phys. Rev.* D12, 1522 (1975). For a review and other references to the complete literature in the field to 1977, see R.N. Mohapatra, *New Frontiers in High Energy Physics*, A. Perlmutter and L. Scott (eds.), Plenum, 1978.

14. I was informed at the workshop of two papers, which generalize the analysis of Refs. 10, 11, and 12 to catalogue a whole list of such operators. S. Weinberg, HUTP-80/A023 (1980); A. Weldon and A. Zee, University of Penn. Preprint (1980).

15. G. Feinberg, M. Goldhaber, and G. Steigman, *Phys. Rev.* D18, 1602 (1979).

16. R. Wilson, MIT Proposal (1979).

17. R.E. Marshak and R.N. Mohapatra, *Phys. Lett.* B (to appear). See also A. Davidson, *Phys. Rev.* D20, 776 (1979).

18. H. Williams, talk at the Weak Interaction Workshop, Virginia Polytechnic Institute, 1979.

19. R.N. Mohapatra and G. Senjanović, *Phys. Rev. Lett.* 44, 912 (1980).

20. Such mass matrices for neutrinos were first discussed by M. Gell-Mann, P. Ramond, and R. Slansky in connection with the SO(10) group (unpublished). However, in their model the scenario of breakdown of SO(10) is as follows:

$$SO(10) \xrightarrow{\{45\}} SU(4)\times SU(2)\times U(1) \xrightarrow{\{126\}} SU(3)_c \times SU(2)\times U(1) \quad .$$

Therefore, the physically appealing relation between m_{ν_e} and $1/m_{W_R}$ that we obtain does not appear in their case. See P. Ramond, Coral Gables talk, 1980.

21. A. Halperin, P. Minkowski, H. Primakoff, and S.P. Rosen, *Phys. Rev.* D13, 2567 (1976).

22. The experimental situation has been surveyed by E. Fiorini, *Rev. Nuovo Cim.* 2, 1 (1971) and Neutrino '77, M. Markov et al. (ed.), p. 315.

23. D. Dicus, E. Kolb, V. Teplitz, and R.V. Wagoner, *Phys. Rev.* D17, 1529 (1978); D18, 1829 (1978).

24. R. Davis, *Proc. Int. Conf. on Radio Isotopes*, Pergamon, 1958.

25. L.N. Chang and N.P. Chang, *Phys. Lett.* B (to appear).

26. A detailed investigation of this question will be the subject of a forthcoming publication by R.N. Mohapatra and J.C. Pati (to appear).

27. See, for example, P. Frampton, S. Nandi, and J. Scanio, *Phys. Lett.* 85B, 225 (1979); H. Georgi and C. Jarlskog, HUTP-79/A026 (1979); R.N. Mohapatra and D. Wyler, *Phys. Lett.* 89B,181 (1980); S. Nandi and K. Tanaka, *Phys. Lett.* B (to appear).

28. G. Senjanović, private communication.

29. J.C. Pati, see Ref. 1.

POSSIBLE ENERGY SCALES IN THE DESERT REGION*

T. Goldman

California Institute of Technology
Pasadena, CA 91125

Abstract

Combining current suggestions for new interactions with the constraints of grand unification of all of the interactions, predictions are made for the mass scales of new (and sometimes, additionally required) interactions. The corresponding vector bosons are found to have masses in the range $10^7 - 10^{11}$ GeV. The conclusions drawn are that unless there are very many new unexpected degrees of freedom discovered below these scales: 1) It will be very difficult to distinguish among possible grand unified theories by direct tests and 2) The proton lifetime is likely to be near the presently predicted range.

With a modicum of exaggeration, I will begin with the premise that everything we know about quantum dynamics may be summarized by stating that the nongravitational interactions have the form of a non-Abelian gauge theory based on the group $SU(3)\otimes SU(2)\otimes U(1)$, and that the variation with momentum scale Q of the associated coupling constants is given by the renormalization group equations[1] (RGE's):

$$\alpha_S^{-1}(Q^2) = \alpha_S^{-1}(Q_0^2) + \frac{b_S}{4\pi}\ln\frac{Q^2}{Q_0^2} \tag{1a}$$

$$\alpha_W^{-1}(Q^2) = \sin^2\theta_W\,\alpha_{em}^{-1}(4M_W^2) + \frac{b_W}{4\pi}\ln\frac{Q^2}{4M_W^2} \tag{1b}$$

$$\alpha_B^{-1}(Q^2) = \cos^2\theta_W\,\alpha_{em}^{-1}(4M_W^2) + \frac{b_B}{4\pi}\ln\frac{Q^2}{4M_W^2}\quad . \tag{1c}$$

In Eqs. (1a), (1b), and (1c), Q_0 is an arbitrary reference scale, explicitly chosen as $2M_W$ in the latter two equations. The free parameters of the theory are the mass of the intermediate vector boson (IVB) of the weak

*Work supported in part by the U.S. Department of Energy under Contract No. DE-AC-03-79ER0068.

**Permanent address: T-Division, LASL, PO Box 1663, Los Alamos, NM 87545.

interactions M_W, the mixing angle θ_W between the neutral IVB and the B-vector bosons of the Glashow-Weinberg-Salam (GWS) $SU(2)\otimes U(1)$ theory of weak and electromagnetic interactions,[2] and the coupling constant of the strong interactions α_S, and the electromagnetic coupling constant α_{em}. We also know that

$$b_S = 11 - \frac{2}{3} n_f$$

$$b_W = \frac{22}{3} - \frac{2}{3} n_f \qquad (2)$$

$$b_B = - \frac{2}{3} n_f \frac{5}{3}$$

where n_f is the number of fermion flavors, currently believed to be six (for u, d, s, c, b and t quarks). The variation of α_{em} with scale obeys a similar equation to Eq. (1) with

$$b_{em} = - \frac{2}{3} n_f \frac{8}{3} \quad ,$$

with its initial value of 1/137 defined, on the scale Q_0 = mass of the electron, by the Thompson cross-section.

This information can be summarized by the short solid lines in Fig. 1. In this form of presentation, it is natural to ask what happens if we continue these lines, as indicated by the dashed lines. As is apparent from the figure, for $\sin^2\theta_W$ close to its experimental value[3] of .23 and for a properly SU(5) normalized Abelian coupling of $(3/5)\alpha_B^{-1}$, *all* the lines meet at one point corresponding to a scale of $M_x = (4 \pm 2) \times 10^{14}$ GeV (when calculated very carefully[4]). This raises the exciting possibility that all of these interactions derive from a single non-Abelian SU(5) grand unified theory (GUT), which undergoes spontaneous symmetry breaking at that scale.[5]

But does this crossing have to mean anything? The answer is no; it could be a fortuitious crossing (FC). As illustrated in Fig. 2, if one insists on grand unification, this may always be accomplished by enlarging the SU(2) group at some $Q > M_x$ to a group $G \supseteq SU(4)$ so that the couplings still come to a common value. Since the overall unifying group is always much larger than SU(5) and involves scales even greater than M_x, we characterize these theories as Bigger Speculations, which are not of interest here. Instead, we ask ourselves what may happen as we trek through this vast desert (so called because in SU(5) it contains no new physics) between the scales of M_W and M_x. Surely, this vast extrapolation is too bold?

To emphasize how vast this extrapolation is, consider that the strong and weak couplings at M_W differ by about an order of magnitude, while M_x

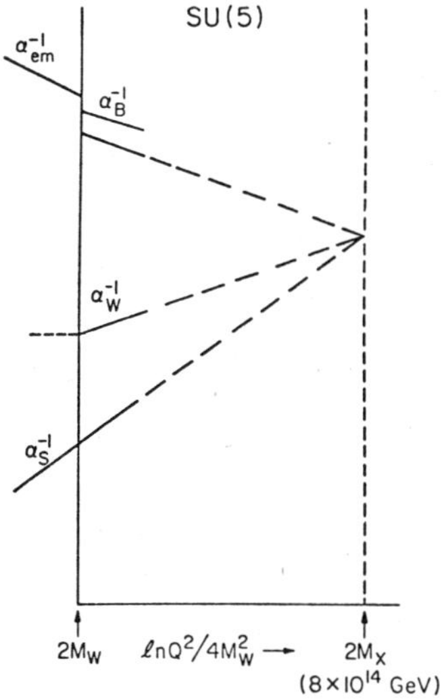

Figure 1. Variation of the inverse coupling constants versus momentum scale of definition for the known interactions. Dashed lines are SU(5)-model projection from the "experimentally tested" region (solid lines).

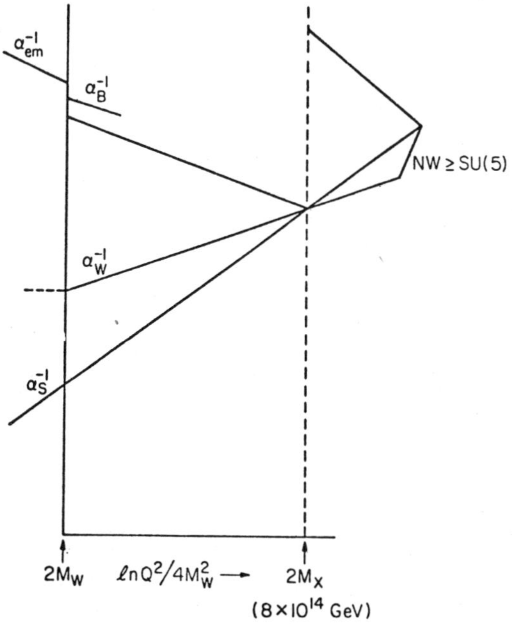

Figure 2. An example of a fortuitous crossing (FC) of coupling constant trajectories and the effect of requiring eventual grand unification.

is some thirteen orders larger than M_W. Bringing these couplings to the same value as in Fig. 1 corresponds, on a linear scale, to a triangle with a base (separation between α_S^{-1} and α_W^{-1} at M_W in Fig. 1) chosen to be the length of, say, a hockey rink here in New Hampshire, and with an apex (unification point in Fig. 1) in the region of Pluto's orbit!

We will be looking for oases with tall vector boson resonances growing around them;[6] that is, for mass scales characteristic of massive vector bosons that are the force carriers for new, as yet unknown, interactions. New fermions may possibly occur at any scale, and we have little to say about fundamental scalars (especially as not one has yet been experimentally observed).

However, we will still have very little predictive power if we allow ourselves to invent arbitrary new interactions as factor groups, and/or allow FC's as shown in Figs. 3 and 4. We will see that FC's try to arise in some interesting cases, even though we try to eliminate them.[7] As for arbitrariness, we will instead take a conservative approach, adding only a little in the way of new interactions and observing to what scales they are restricted. In fact, we shall even only consider adding those new interactions which are already motivated by other considerations (than GUTs).

The first alterantive we consider may be termed chiral-Pati-Salam theories (CPS). This is suggested by noting that the U(1)-quantum numbers of the G-S-W theory can be decomposed into the sum of the quantum numbers of a U(1) associated with the fifteenth generator of an SU(4) of color with lepton number as the fourth color;[8] and into the third component of a weak isospin analog of $SU(2)_L$, but one associated with right-chiral fields: $SU(2)_R$. Thus the "known" interactions taken as a starting point for unification here are described by the group

$$[SU(2)_R \otimes SU(2)_L] \otimes [SU(3)_c \otimes U(1)_c] \quad , \tag{3}$$

where the factor in first square brackets is "chiral" and that in the second square brackets can unify to form the $SU(4)_c$ suggested by Pati and Salam.[8] It is interesting that this system fits neatly into O(10) which also includes $SU(5) \otimes U(1)$.

Let me remark parenthetically that the motivation for this approach can be seen graphically in Fig. 5. In Fig. 5b the quark set from 5a has been shifted to the left by 1/3 unit, and the lepton set to the right by 1 unit. This removes the (1, - 1/3, - 1/3, - 1/3) $U(1)_c$ quantum numbers leaving only T_{3R} (unnormalized). A right chiral neutrino must be added to complete the set.

The new coupling constant RGE's are

$$\alpha_R^{-1}(Q^2) = \sin^2\phi \; \alpha_B^{-1}(4R^2) + \frac{b_W}{4} \ln \frac{Q^2}{4R^2} \tag{4a}$$

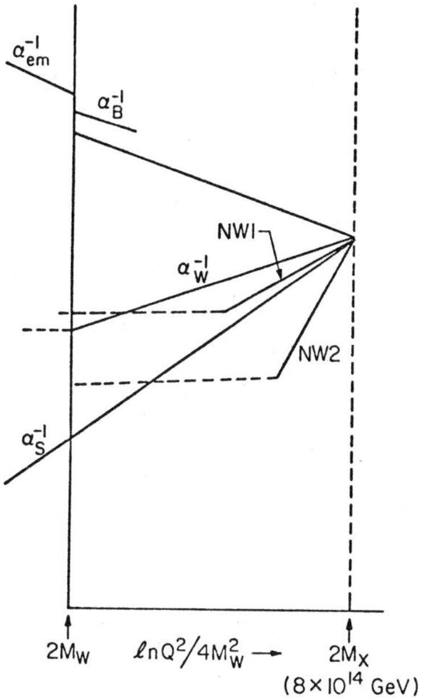

Figure 3. Coupling constant trajectories for arbitrary additional factor groups.

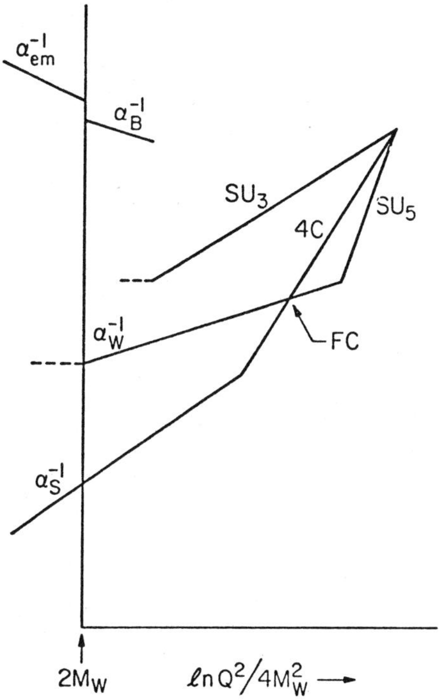

Figure 4. A possible set of trajectories if arbitrary FC's are allowed.

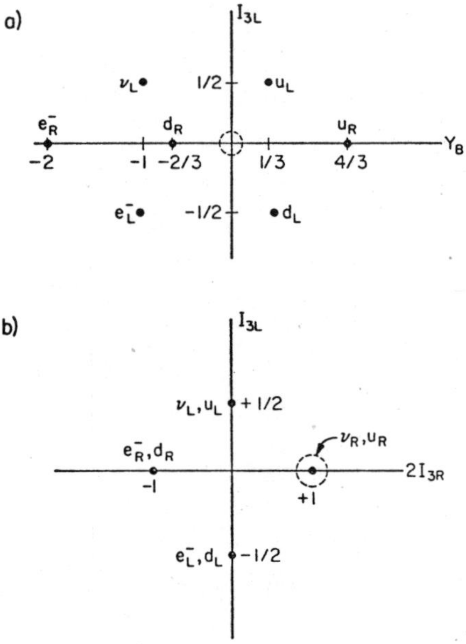

Figure 5. a) Weak isospin and hypercharge of quarks and leptons in the Glashow-Weinberg-Salam model

b) Effect of removing the Pati-Salam $U(1)_C$ quantum number, leaving an (unnormalized) right weak isospin.

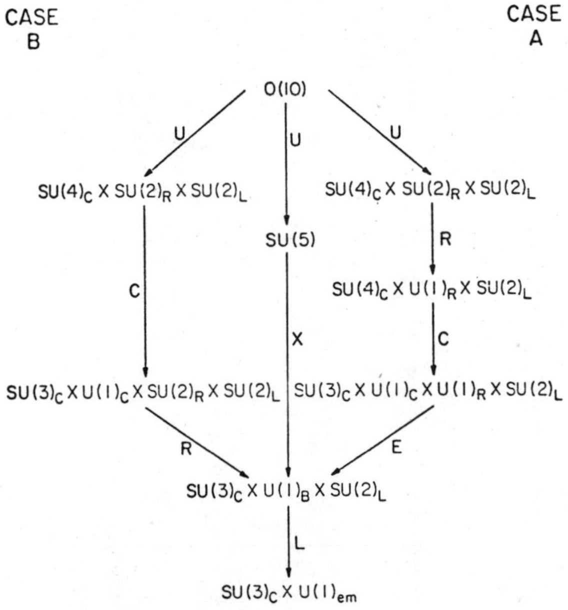

Figure 6. Structure and labelling of cases of O(10) decomposition in Reference 9.

$$\alpha_C^{-1}(Q^2) = \frac{3}{2}\cos^2\phi\, \alpha_B^{-1}(4R^2) - \frac{n_f}{6\pi}\,\ell n\,\frac{Q^2}{4R^2} \quad , \tag{4b}$$

where ϕ is the analog of θ_W and describes the mixing of the $U(1)_C$ and neutral $SU(2)_R$ vector bosons to form the B of G-W-S, and the 3/2 factor comes from $SU(4)_C$ normalization and is the analog of the 3/5 in the SU(5) analysis. R is the mass scale of the W_R-vector bosons of $SU(2)_R$. We also define C to be the scale for

$$SU(4)_C \rightarrow SU(3)_C \times U(1)_C$$

breaking, so that

$$\alpha_S(4C^2) = \alpha_C(4C^2) \quad . \tag{5}$$

Using Eqs. (5), (4), and (1), plus the chiral symmetry restoration constraint

$$\alpha_W(4R^2) = \alpha_R(4R^2) \quad , \tag{6}$$

we can eliminate ϕ and n_f. Some algebra plus addition of known parameters yields the result[6,9]

$$RC \sim 10^{30 \pm 2}\ \text{GeV}^2 \quad , \tag{7}$$

and this value decreases as $\sin^2\theta_W$ increases. We have actually assumed $R < C$ in the above; this is case B of the O(10) decomposition discussed by Shafi, Sondermann, and Wetterich[10] (SSW). Note that Eq. (7) therefore requires $C > X$. Since gravity has been neglected, the analysis is not valid unless $C < 10^{19}$ GeV (Planck mass), and so $R \gtrsim 10^{11\pm2}$ GeV is required.

The coupling constant trajectories then appear either as in Fig. 7 or Fig. 8. In the former, there is an FC at X *and* the weak interaction group must grow to $G \supseteq SU(5)$ above C. In the latter, the FC at X is avoided by having the weak group grow to $G \supseteq SU(4)$, but a new FC appears. Ignoring this, $SU(4)_C$ must also grow above C. Note that the overall unifying group cannot be as small as O(10) in either case, unless $R \sim C \sim X$.

If $C < R$, we are in SSW case A. Now if $X < C$, X is again an FC, so let us require $C < X$ and again forbid FC's. The result appears in Fig. 9: Either the O(10) unification scale U is less than X, which implies that the proton lifetime must be less than the present experimental limit,[11] or U is an FC *and* the overall unifying group is larger than O(10).

We conclude that CPS schemes cannot fit nicely into O(10) unless $R \sim C \sim X$, so that there is no hierarchy of hierarchies.[12]

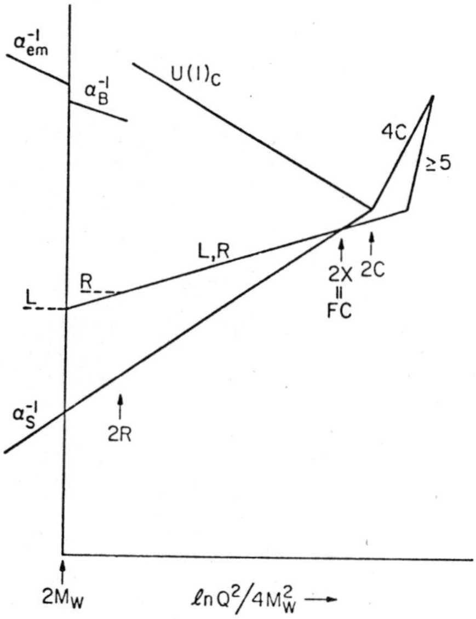

Figure 7. Coupling constant trajectories in a chiral-Pati-Salam (CPS) theory with an FC.

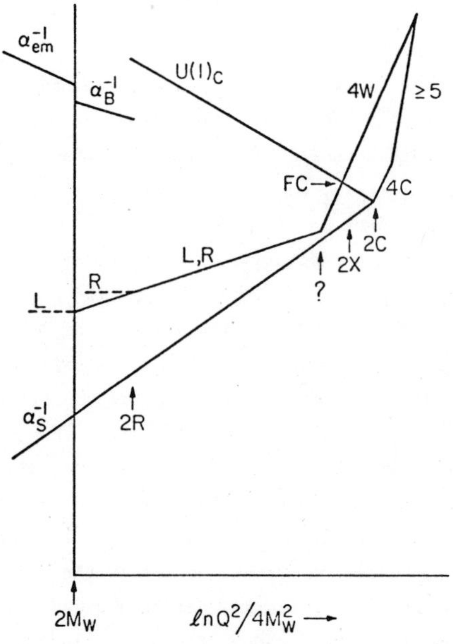

Figure 8. Effect on trajectories of avoiding the FC in Fig. 7.

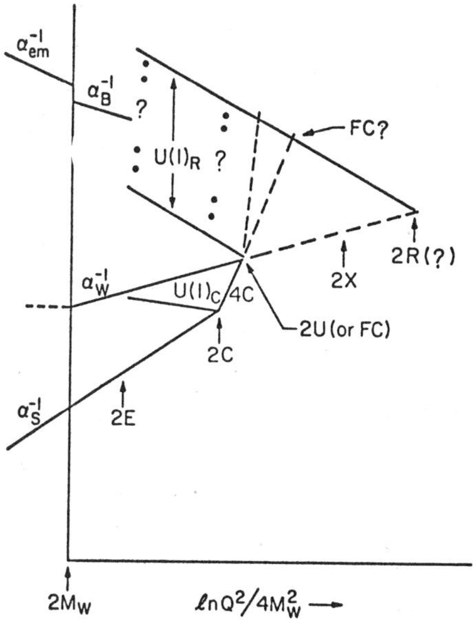

Figure 9. Coupling constant trajectories in CPS theories for which the $SU(4)_c$ group unifies before any other.

The second class of alternatives that we consider involves "technicolor" theories.[13] The crucial feature of such theories that is of interest here[14] is that the technicolor interactions are non-Abelian and have a coupling strength $\alpha_{TC} \sim 1$ for $Q \sim 1$ TeV. The results depend, however, on what technicolor fermions are introduced.

In the original form of the idea,[13] an $SU(4)_{TC}$ group of technicolor interactions was introduced. The simplest possibility for the associated fermions is to add a quartet of TC-quarks for every color triplet of ordinary quarks. This already ruins the asymptotic freedom (AF) of the weak interaction group, as shown in Fig. 10. Following the appropriate trajectories, we find that the color and technicolor groups unify to $G \supseteq SU(7)$ at $Q7 \sim 10^{10\pm1}$ GeV. To avoid FC's, the weak interaction group must grow larger before this scale, but not larger than G so that ultimately all forces may unify. We find that if the new weak group G_{NW} is $SU(6)$ or $SU(6)_L$, then the scale Q_{NW} for the vector bosons in G_{NW} that are outside the GWS theory must be $\gtrsim 10^7$ GeV.

If $G_{NW} = SU(4)_L$, with the six "known" flavors in a six-plet of $SU(4)$, then

$$Q_{NW} < 10^5 \text{ GeV} \quad .$$

This occurs for $U \sim 10^{14}$ GeV in Fig. 10, as limited by the lower bound on the

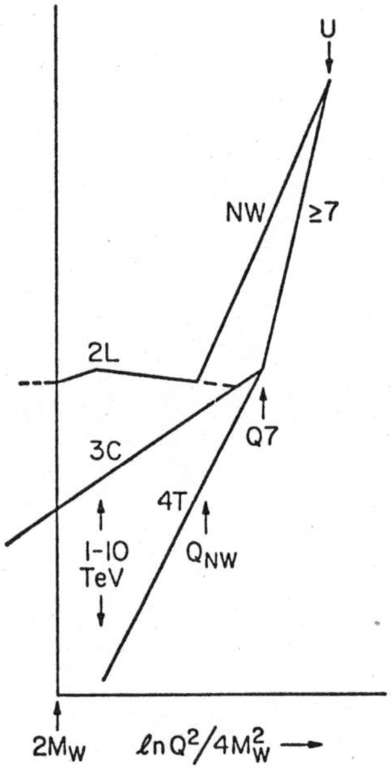

Figure 10. Coupling constant trajectories for an SU(4)-technicolor theory.

proton lifetime.[11] As U rises to $\sim 10^{17}$ GeV, Q_{NW} falls to ~ 1 TeV, which is of the order of the experimental limit for the strength of new weak interactions involving the known fermions.[16] The upper limit on Q_{NW} occurs at minimum U because of the fixed slope for α_{NW}^{-1} in Fig. 10. This is the only case we consider which allows a low (read "experimentally accessible") mass scale for new vector bosons. It may have problems, however, in connection with fermion masses which will be mentioned later.

Still within the framework of $SU(4)_{TC}$, one may, instead of introducing TC-quartets for both left and right chiral TC-quarks, use the antisymmetric product of fundamental representations, namely, a TC-six-plet, for the right chiral fields[17] following the example[5] of SU(5). If one imagines these as arising from the corresponding antisymmetric product of an encompassing SU(7) or larger group, there would also appear a set of quarks that are both color triplets and technicolor quartets. These latter would be a disaster because they would destroy the AF of the TC group. They must somehow be made too heavy to affect the RGE's; this illustrates the difficulty of going on to even higher dimension representations.

With this structure, we find

$$Q7 \sim 10^{17 \pm 2} \text{ GeV}$$

and again a G_{NW} must appear (below 10^9 GeV) to avoid FC's. The only half-way plausible system that we have found here is $G_{NW} = SU(3)_L$ with the six flavors in two triplets,[18] for which

$$Q_{NW} \sim 10^8 \text{ GeV} \quad .$$

There are further problems with anomaly cancellations in these systems that have been pointed out by Frampton;[19] requiring such cancellations may always cost TC the AF property.

A new kind of TC has been described by Farhi and Susskind.[14,20] As shown in Fig. 11, the TC group is only $SU(2)_{TC}$ above 1 TeV, but grows to an SU(4) above the threshold for extended technicolor bosons Q_{ETC}. For this system, we find

$$Q_{ETC} \gtrsim 10^{10} \text{ GeV} \quad .$$

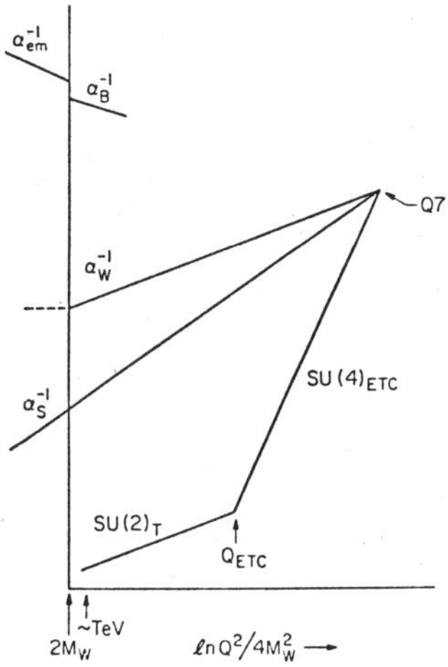

Figure 11. Coupling constant trajectories for an SU(2)-technicolor theory which grows to SU(4)-extended technicolor at the mass scale Q_{ETC}.

In connection with lower mass scales, we must also mention the $[SU(4)]^4$ of Elias, Pati, and Salam.[21] In this scheme, the non-Abelian couplings may also attain a factor of 1/2 normalization correction relative to each other similar to the U(1) factor of 3/5 in SU(5). This reduces the unification scale to $U \sim 10^6$ GeV, but no conflict with the proton lifetime appears because it occurs as a triple beta-decay. In this particular model,

$$\sin^2 \theta_W > .25 \quad ,$$

and so it is probably ruled out. Recent work[22] on $[SU(6)]^4$ and $[SU(2n)]^4$ shows that this defect can be remedied with minimal effects on the value of U.

Quark masses may also be used to constrain allowed interaction scales. In particular, the relation $m_b = m_\tau$ for the bottom quark and tau-lepton masses at the unification scale yields $m_b \sim 5$ GeV at that scale in the SU(5) model, consistent with the observed mass of the upsilon $b\bar{b}$ composite state.[23] This result has been used to limit the number of flavors possible.[24] For CPS theories with $C < R$, the m_b/m_τ ratio does not vary from one until $Q < C$, because quarks and leptons are not distinct when the $SU(4)_c$ symmetry is unbroken. This reduces the predicted value of m_b and so of the upsilon mass. To avoid conflict with experiment, C must be $\gtrsim 10^{11}$ GeV. In the TC cases with $Q7 \sim 10^{10}$ GeV, the opposite effect occurs: Above Q7 the quark-lepton mass ratio changes with scale at a faster rate than in SU(5). This leads to $m_b > 6.5$ GeV, which suggests that too massive an upsilon is also predicted.

These quark mass arguments are weaker than our previous ones for four reasons: 1) The scale-dependent masses are gauge dependent, albeit weakly so; 2) The relation to lepton masses is not required in TC theories; 3) Additional flavors may compensate for a lower value of C in CPS theories; 4) The relation between the quark mass and the associated onium ($q\bar{q}$ composite vector meson) is not precise. They may be very important, however, to specific theories.

Finally, we should mention the work of Dawson and Georgi on generalized hierarchies for symmetry breaking.[25] By requiring $SU(3) \times SU(2) \times U(1)$ to devolve from a sequence of larger SU(n) product groups, they show that the intermediate scales for an arbitrary number of symmetry breaking steps may be made to cancel (approximately). One may thus reach a relation between the GWS - and grand unification mass scales that depends only on the angle θ_W and the strong and electromagnetic couplings:

$$\cos 2\theta_W \, \alpha_{em}^{-1}(W) - \frac{2}{3}\alpha_S^{-1}(W) = \frac{22}{3} \ln \frac{X}{W} \quad . \tag{8}$$

With θ_W set to its experimental value, this yields $X \sim 10^{14}$ GeV, which corresponds to a proton lifetime of less than 10^{31} years.

We may summarize the conclusions of these peregrinations as follows: The scale for O(10) unification must be very close to that determined for SU(5) unification. Most new mass scales motivated by other considerations (than just having new scales) lie in the range $10^7 - 10^{11}$ GeV. There are a few special cases below this, but virtually none of the cases are easy to test. Precision null experiments such as proton decay, neutrino masses and mixings and rare decay modes (e.g., $\mu \to e\gamma$) offer the best hope of testing for new mass scales. If, as perhaps seems more reasonable, our theoretical expectations are confounded by the discovery of some new degrees of freedom at very modest energy scales, then the constraint of ultimate grand unification strongly suggests that there are likely to be even (many) more new degrees of freedom to be found.[26] And finally, on a hopeful note, unless there is very much new to be discovered, the value of the proton lifetime is probably not very much greater than can be determined by present detectors.

REFERENCES

1. H.D. Politzer, *Phys. Rev. Lett.* 26, 1346 (1973); D. Gross and F. Wilczek, *Phys. Rev. Lett.* 26, 1343 (1973).

2. S.L. Glashow, *Nucl. Phys.* 22, 579 (1961); S. Weinberg, *Phys. Rev. Lett.* 19, 1264 (1967); A. Salam, *Proc. Eighth Nobel Symp. on Elementary Particle Theory, Relativistic Groups, and Analyticity*, N. Svartholm (ed.), Wiley, New York, 1969.

3. P. Langacker et al., The weak neutral current, University of Pennsylvania/ BNL Preprint COO-3071-243 (1979).

4. T. Goldman and D.A. Ross, *Phys. Lett.* 84B, 208 (1979); T. Goldman and D.A. Ross, How accurately can we estimate the proton lifetime in an SU(5) grand unified model?, Caltech Preprint, CALT-68-759; W.J. Marciano, *Phys. Rev.* D20, 274 (1979).

5. SU(5) was first suggested by H. Georgi and S.L. Glashow, *Phys. Rev. Lett.* 32, 438 (1974). The RGE's were first applied to it by H. Georgi, H. Quinn, and S. Weinberg, *Phys. Rev. Lett.* 32, 451 (1974).

6. This talk is based mostly on work reported in T. Goldman and D.A. Ross, *Nucl. Phys.* B162, 102 (1979), but includes some results of others, which are referenced where mentioned.

7. In the discussion following presentation of this talk, S. Weinberg summarized this aversion to FC's by stating that if you want to meet someone, it is probably better not to go past them.

8. J.C. Pati and A. Salam, *Phys. Rev.* D8, 1240 (1973); D10, 275 (1974).

9. Although not phrased in this manner, this result has also been obtained by H. Georgi and D.V. Nanopoulos, *Nucl. Phys.* B159, 16 (1979), and by Q. Shafi, M. Sondermann, and Ch. Wetterich, Fourth colour in O(10), Universität Freiburg Preprint THEP 79/13, December 1979.

10. See second paper in Ref. 9 and Fig. 6.

11. J. Learned, F. Reines, and A. Soni, *Phys. Rev. Lett.* 43, 907 (1979).

12. Shafi et al. (Ref. 9), find $C > 10^{10}$ GeV anyway. There are no bounds on the extra scale E.

13. S. Dimopoulos and L. Susskind, *Nucl. Phys.* B155, 237 (1979).

14. Unfortunately, we do not have the space to go into the attractive motivations for such theories.

15. We blandly assume that the t-quark exists and will be found at PETRA or PEP.

16. For flavor-changing neutral currents, the limit is much larger; this value obtains for models in which Cabibbo-mixing is either violated, or only apparent, with the effects due to these new interactions.

17. Such antisymmetric products have been referred to as "larks" by E. Corrigan and P. Ramond, *Phys. Lett.* 87B, 73 (1979). See also G. 't Hooft, *Nucl. Phys.* B72, 461 (1974).

18. Halfway because of the problems of producing phenomenologically satisfactory decays of b and t quarks in such a system.

19. P. Frampton, *Phys. Rev. Lett.* 43, 1912 (1979).

20. E. Farhi and L. Susskind, *Phys. Rev.* D20, 3404 (1979).

21. V. Elias, J.C. Pati, and A. Salam, *Phys. Rev. Lett.* 40, 920 (1978).

22. V. Elias and S. Rajpoot, *Phys. Rev.* D20, 2445 (1979); J.C. Pati et al., *Nature* 273, 736 (1978).

23. S.W. Herb et al., *Phys. Rev. Lett.* 39, 252 (1977).

24. D.V. Nanopoulos and D.A. Ross, *Nucl. Phys.* B157, 273 (1979).

25. S. Dawson and H. Georgi, *Phys. Rev. Lett.* 43, 821 (1979).

26. With apologies to a great poet: "There are more parameters in Heaven and Earth, Horatio, than are dreamt of in your Grand Unified Theory".

THE FATE OF GLOBAL CONSERVATION LAWS IN GRAND UNIFIED THEORIES*

P. Ramond

California Institute of Technology
Pasadena, CA 91125

Abstract

We emphasize that global conservation laws are likely to be broken in Grand Unified theories on very general grounds. We describe the mechanisms involved and then discuss the consequences of lepton number violation such as neutrino oscillations which we illustrate in a recently proposed version of the SO(10) model.

Low energy phenomena apparently respect several (global) conservation laws: 1) the impressive stability of the proton leads to baryon number (B) conservation and the absence of a long range force coupling to baryonic matter indicates the global nature of this conservation law; 2) the fact that charged leptons always appear accompanied by either their antiparticle or their associated antineutrino suggests the existence of three types of conserved quantum numbers: L_e, L_μ, L_τ, i.e., electronic, muonic and tauonic numbers. One assigns $L_e = +1$ for e^-, ν_e, and $L_e = 1$ for e^+, $\bar{\nu}_e$, etc. No long range force of detectable strength has even been observed between electrons or muons and the same is likely to be true for the tau system. Thus, it is reasonable to assume that L_e, L_μ, and L_τ correspond to global (ungauged) conservation laws.

On the other hand, the aim of grand unification is to regard all elementary particles as being part of the same entity no matter how different they may seem to be at low energy. Thus, it is no wonder that global conservation laws which label different particles stand in the way of grand unification.[1,2] The existence of very large experimental limits against the breakdown of global conservation laws suggests that unification has to occur at very large scales (or very short distances), which is, of course, the scenario envisaged by the classic paper of Georgi, Quinn, and Weinberg.[3]

Let us examine how this comes about in the simplest grand unified models. It is convenient to discuss the conservation of B, $L = L_e + L_\mu + L_\tau$ and $L_e - L_\mu$, $L_\mu - L_\tau$, separately.

---

*Work supported in part by the U.S. Department of Energy under Contract No. DE-AC-03-79ER0068.

In SU_5,[2] assuming the existence of the t-quark, there are three families of particles, each consisting of a $\underset{\sim}{\bar{5}} + \underset{\sim}{10}$. We know that families are mixed because of the Cabibbo angle and because the b quark is not stable. Since each family contains a different type of lepton, $L_e - L_\mu$ and $L_\mu - L_\tau$ are broken ab initio in the Lagrangian, leaving B and L. What happens to these is more subtle and requires some discussion. The structure of the coupling of the fermions to spinless bosons shows that the SU_5 theory of Georgi and Glashow has a global U(1) symmetry. The possible couplings are

$$\begin{array}{ccccccccccc} 5_f & \cdot\ 10_f & \cdot\ (\bar{5}_H, & \overline{45}_H) ; & \qquad & 10_f & \cdot\ 10_f & \cdot\ (5_H, & 45_H; & 50_H) & \\ -3 & 1 & 2 & 2 & & 1 & 1 & -2 & -2 & -2 & , \end{array}$$

where f(H) stands for a fermion (Higgs) and we have indicated below the representations the values of the global U(1). This U(1) is preserved by an SU(5) invariant potential built out of the shown Higgs representations. In a more complicated version, one may have extra couplings of the form

$$\begin{array}{cccccccc} (\bar{5}_f & \cdot\ \bar{5}_f)_S & \cdot\ (15_H) ; & \qquad & (\bar{5}_f & \cdot\ \bar{5}_f)_A & \cdot\ \overline{10}_H \\ -3 & -3 & +6 & & -3 & -3 & +6 \end{array}$$

but the Higgs potential involving the 15 can break this U(1) without violating SU_5 by, say, a term of the form $15_H \bar{5}_H \bar{5}_H$ which has a nonzero value for this U(1).

In a model involving only 5_H, 45_H and 50_H, where 5_H and 45_H acquire vacuum values, they of course break the global U(1) symmetry. However, they do it in such a way that they break not only this global U(1), but also the local U(1) coming from the SU_5 which is not in $SU_{2L} \times SU_3^c$ in such a way that a linear combination of these two U(1)'s is preserved. This miraculous escape from massless Goldstone bosons is known as the 't Hooft mechanism and has wide applicability. In this way, one trades one global symmetry for another. The conserved linear combination is $B - L$. Thus, in the simplest SU_5, only one global conservation law remains, and only because the minimal Higgs vacuum values respect the same linear combinations of individually broken U(1).[4] For instance, if the 15_H is introduced, its vacuum value will preserve a different linear combination, leaving no global conservation law. In this case, the Goldstone theorem demands the presence of the aforementioned cubic term in the Higgs potential to avoid massless bosons.

As is well known, $B - L$ allows for proton decay, but forbids neutrinoless double β-decay.

In the next (in order of complication) grand unified model, the SO_{10} model of Georgi[5] and Fritzsch and Minkowski,[6] the three families of fermions appear in the complex spinor representation of SO_{10} whose SU_5 content is

$$16 = \bar{5} + 10 + 1 \quad .$$

The coupling of fermions to spinless bosons appears in the form

$$16_f \cdot 16_f \; (10_H, \overline{126}_H; 120_H) \quad .$$

The $\underset{\sim}{10}$ and $\underset{\sim}{120}$ of SO(10) are real representations, so this theory contains no global U(1) unless the 10_H and 120_H are made complex by doubling. Assuming this complexification, the SO(10) model apparently displays a global symmetry in its Yukawa couplings, but the irreducibility of each fermion family now forces the Higgs to assume vacuum values that do not respect any global symmetries. Hence, the global U(1) must be broken explicitly in the Higgs potential to avoid massless Goldstone bosons.

Thus we are left with no global symmetries at the level of the simplest SO(10), which opens the way for lepton number violating processes.

In the next unifying group, E_6, of Gürsey, Ramond, and Sikivie,[7] the same situation exists. It is interesting to note that in this case, Goldstone bosons can be avoided only by cubic terms which will, in general, give rise to non-trivial discrete symmetries in the fermion sector.

Let us now examine the mechanism by which neutrinos can acquire lepton number violating masses, i.e., the so-called Majorana masses. Consider a one-family SO_{10} model with the left-handed fermion in the complex $\underset{\sim}{16}$ spinor, with the SU_5 break up

$$16_f = \bar{5}_f + 10_f + 1_f \quad .$$

Each family has one charged lepton, two quarks of charge 2/3 and - 1/3, and two neutral leptons. One neutral lepton has $I_W = 1/2$, the other $I_W = 0$, and we call them $\nu_L^{1/2}$ and ν_L^0, respectively. Here I_W stands for the weak SU_2 of the Weinberg-Salam model. Thus, the symmetric neutral lepton mass matrix will be of the Majorana form*

---

*A Majorana mass refers to the mass of a self-conjugate spin 1/2 fermion; it violates L by two units, and allows for neutrinoless double β-decay. A Dirac mass refers to the mass of a fermion that is not self-conjugate, such as a charged lepton or a quark--it does not violate L. In the Weinberg-Salam model, the neutrinos can be made to have a Majorana mass with $I_W = 1$. It is amusing to note that it is only L-conservation in the W-S model that keeps the neutrino massless since it could otherwise acquire a calculable mass without violating the $SU_2 \times U_1$ structure of the theory.

$$(\nu_L^{1/2} \quad \nu_L^0)^T \sigma_2 \left(\begin{array}{c|c} I_w = 1 & I_w = 1/2 \\ \hline I_w = 1/2 & I_w = 0 \end{array} \right) \begin{pmatrix} \nu_L^{1/2} \\ \\ \nu_L^0 \end{pmatrix}$$

where we have indicated the I_w quantum numbers. When the charge 2/3 quark is given a mass m by some mechanism, the Clebsch-Gordan structure of SO_{10} is such that the $I_w = 1/2$ sector of the neutral lepton mass matrix acquires a mass $-3m$ (because of the Pati-Salam type of SU_4 present in SO_{10}). The result is that ν_L^0 becomes the right-handed partner of $\nu_L^{1/2}$, giving rise to a lepton number conserving (Dirac) mass and an unacceptably large mass for the neutrino. Furthermore, it is known that I_w is primarily broken by $I_w = 1/2$ operators and that therefore the $I_w = 1$ part of the mass matrix should be much smaller than the $I_w = 1/2$ part. At any rate, it would not affect the existence of a large mass for the neutrino (of the order of GeV's!). It was proposed some time ago by Gell-Mann, Slansky and the author[8,4] that the $I_w = 0$ part of the mass matrix be made very large on the scale of the $I_w = 1/2$ breaking. Then, diagonalization of the mass matrix leads to a small mass for the neutrino, naturally depressed in value from a typical $I_w = 1/2$ mass by the ratio of the strengths of the $I_w = 1/2$ to $I_w = 0$ breaking

$$m_{\nu^{1/2}} = 9m_u \left(\frac{m_u}{M}\right) ,$$

where m_u is the mass of the charge 2/3 quark and M the mass of the I_w-singlet neutrino in the family. Then the phenomenologically needed suppression of nine orders of magnitude from GeV's to eV's would be provided by the ratio m_u/M, which indicates that M must be of the order of at least 10^{10} GeV. Taking the ratio of Yukawa to gauge couplings to be 10^{-4}, this would lead us to giving a mass of the order of 10^{14} GeV to the vector bosons which inhabit the coset SO_{10}/SU_5. It is amusing that this is the scale at which SU_5 must break down to $SU_3 \times SU_2 \times U_1$ in order to agree with phenomenology![9] Another advantage of this large Majorana mass has been recently pointed out by Harvey, Reiss, and the author[10] who remarked that it provides a natural mechanism for softly breaking CP at unification scales, in contradistinction with SU_5, where no such mechanism is present. This is because the $I_w = 0$ Majorana mass comes from the complex $\underset{\sim}{126}$ of SO_{10} and that its vacuum expectation value will have a non-trivial phase equal to $\pi/4$ coming from a quartic self-interaction of the form $(126)^4$.

In the trivial one-family model, the neutrino produced by the charged current will be a linear combination of the light mass eigenstate and of the heavy mass neutrino, with a tiny mixing angle of the order of m/M. Thus, neutrino oscillations[11] will take place, but will be enormously suppressed by this minuscule mixing.

This analysis can be readily generalized to three families of left-handed 16's. The only change is that the neutral lepton mass matrix is now 6×6 dimensional and each block with a given I_W is 3×3 in size.

If it weren't for mixing angles, this mechanism would give the three light neutrinos masses

$$9\,\frac{m_u^2}{M_e}\,, \qquad 9\,\frac{m_c^2}{M_\mu}\,, \qquad 9\,\frac{m_t^2}{M_\tau}\,,$$

where M_i ($i = e,\mu,\tau$) are the masses of the right-handed neutrinos of the e-, μ-, and τ-families. Thus, light neutrino mass ratios are expected to be like

$$\frac{m_{\nu_e}}{m_{\nu_\mu}} \sim \left(\frac{m_u}{m_c}\right)^2 \frac{M_\mu}{M_e}\,,$$

$$\frac{m_{\nu_\mu}}{m_{\nu_\tau}} \sim \left(\frac{m_c}{m_t}\right)^2 \frac{M_\tau}{M_\mu}$$

Since the heavier families do couple more to scalars than the lighter ones, one could expect the ratios M_μ/M_e, M_τ/M_μ, to be significantly larger than one, possibly giving large values for these ratios.[12] Now if we naively regard mixing angles to arise from mass ratios (e.g., the Oakes relation relating the Cabibbo angle and m_d/m_s), we would expect to have the $\nu_e - \nu_\mu$ mixing angle given by

$$\tan^2 \theta_{\nu_e - \nu_\mu} \sim \left(\frac{m_u}{m_c}\right)^2 \frac{M_\mu}{M_e}\,, \quad \text{etc.}$$

The role of these angles in neutrino oscillations will shortly be discussed. The main point we want to make is that in SO_{10} the neutrino mixing angles and masses depend crucially on the unknown $I_W = 0$ part of the neutral lepton mass matrix. In Ref. 10 a model was presented where the $I_W = 0$ sector was determined on the basis of "naturalness", and where $M_e = M_\mu \ll M_\tau$ with the result that ν_e- oscillations would be very hard to observe because of mixing angle suppression. It is amusing to note that a very massive t-quark gives a "very massive" τ-neutrino which can easily violate the astrophysics bound for the sum of the light (< 1 MeV) neutrino masses. At any rate, our mechanism suggests that there is a hierarchy of neutrino masses.

Let us analyze the most important consequences of this neutral lepton mass matrix: neutrino oscillations. As we have already remarked in the case of one family, it gives rise to oscillations between the two neutral leptons in the family, but the mixing angle was of the order of $I_w = \frac{1}{2}/I_w = 0$ and therefore negligible. In the case of several families, the situation becomes more interesting because the light neutrinos can mix appreciably, giving rise to the phenomenon of "flavor oscillations".

To see this, let us analyze the situation in terms of the $SU_2 \times U_1$ theory.* Label the neutral leptons by

$$N_L = \begin{pmatrix} N_L^{1/2} \\ N_L^0 \end{pmatrix} \qquad L_L^{1/2} , \qquad L_R^0 ,$$

where the superscript refers to the I_w representation. The relevant parts of the charged and neutral currents will be

$$j_+^\mu = L_L^{1/2} \sigma^\mu N_L^{1/2} + \cdots$$

$$j_0^\mu = \frac{1}{2} N_L^{1/2\dagger} \sigma^\mu N_L^{1/2} + \cdots$$

These do not refer to mass eigenstates. After spontaneous breaking, the neutral and charged lepton mass matrices will be

$$N_L^T \sigma_2 M_0 N_L + L_L^{1/2\dagger} M_1 L_R + \text{h.c.}$$

where M_0 is a symmetric matrix. For simplicity of analysis, we also take it to be real (CP conservation). We can then introduce the neutral lepton mass eigenstates

$$\nu_L = \begin{pmatrix} \nu_L^{1/2} \\ \nu_L^0 \end{pmatrix} = R^T \begin{pmatrix} N_L^{1/2} \\ N_L^0 \end{pmatrix} ,$$

$$RR^T = R^T R = 1 ,$$

where

$$M_0 = R D_0 R^T ,$$

*In there we pattern the neutral leptons after theories that are derivable from spinor representations of orthogonal groups. Our analysis would have to be slightly modified in the E_6 theory.

D_0 being a real diagonal matrix. Similarly, we can introduce the charged lepton mass eigenstates

$$E_L = \begin{pmatrix} e_L \\ \mu_L \\ \tau_L \\ \vdots \end{pmatrix} = UL_L^{1/2} \quad ;$$

$$E_R = \begin{pmatrix} e_R \\ \mu_R \\ \tau_R \\ \vdots \end{pmatrix} = VL_R \quad .$$

where U and V are hermitian matrices which diagonalize M_1,

$$M_1 = U^{\dagger} D_1 V \quad ,$$

D_1 being a diagonal matrix. We can now re-express the currents in terms of the mass eigenstates. Let us set

$$R = \begin{pmatrix} R_{11} & R_{12} \\ R_{21} & R_{22} \end{pmatrix} .$$

Then the charged current becomes

$$E_L^{\dagger} U \sigma^{\mu} [R_{11} \nu_L^{1/2} + R_{12} \nu_L^0]$$

Here $\nu_L^{1/2}$ refers to the light neutrino mass eigenstates, ν_L^0 to the heavy ones, and E_L contains the physical charged leptons, e, μ, τ,... . Similarly, the neutral current is given by

$$\frac{1}{2} \nu_L^{1/2^{\dagger}} R_{11}^T R_{11} \nu_L^{1/2} + \frac{1}{2} \nu_L^{1/2^{\dagger}} R_{11}^T R_{12} \nu_L^0 + \frac{1}{2} \nu_L^{0^{\dagger}} R_{12}^T R_{11} \nu_L^{1/2} + \frac{1}{2} \nu_L^{0^{\dagger}} R_{22}^T R_{22} \nu_L^0 \ .$$

These expressions allow us to analyze the neutrino oscillations because it gives the linear combinations of mass eigenstates produced in the weak interactions. First in the case of the charged current we note that mixing angles come from two places: U, which diagonalizes the charge-1 sector and in unified models is related to the charge $-1/3$ sector, and the R_{11} and R_{12} matrices which come from the neutral lepton mass matrix. In Ref. 10, based on the Georgi-Jarlskog mass relations,[3] the U matrix gives very little mixing between e_L and μ_L, with

$$\tan^2 \theta_{e_L - \mu_L} = \frac{m_e}{m_\mu} \sim \frac{1}{200}$$

thus much smaller than the Cabibbo mixing, and no mixing of τ_L with either μ_L or e_L. Thus, all the mixing must come from R_{11} and R_{12}. As we have previously discussed, we must take the $I_w = 0$ entries of M_0 much larger than the $I_w = 1/2$ entries. If we set

$$\varepsilon = \frac{I_w = 1/2}{I_w = 0} \ll 1 \quad ,$$

it is easy to see that R_{11} and R_{22} are of $\mathcal{O}(1)$, while R_{12} and R_{21} are of $\mathcal{O}(\varepsilon)$. In fact, set

$$R_{12} = \varepsilon \hat{R}_{12} \, , \qquad R_{21} = \varepsilon \hat{R}_{21} \, ,$$

with the careted quantities of $\mathcal{O}(1)$. Then the charged current gives

$$E_L^\dagger \sigma^\mu R_{11} \nu_L^{1/2} + \mathcal{O}\left(\frac{m_e}{m_\mu}, \varepsilon\right) \quad ,$$

showing that the charged weak current will give rise to flavor oscillations among the light neutrinos associated with the three families. Since R is a rotation matrix, it follows that

$$R_{11}^T R_{11} = 1 + \mathcal{O}(\varepsilon^2) \quad ,$$

$$R_{22}^T R_{22} = 1 + \mathcal{O}(\varepsilon^2) \quad ,$$

giving the neutral current

$$\frac{1}{2} \nu_L^{1/2^\dagger} \sigma_\mu \nu_L^{1/2} + \frac{1}{2} \nu_L^{0^\dagger} \sigma_\mu \nu_L^0 + \mathcal{O}(\varepsilon) \quad ,$$

which is seen to keep its diagonal form.

It follows from the above that the phenomenologically accessible consequences of our neutral lepton mass matrix are summarized in the (approximate) rotation matrix R_{11}, as well as in the light values of D_0.

The form of the charged current indicates that when an electron is weakly produced, a linear combination of the three light neutrinos is produced at the same time. Because they have different masses, the composition of this neutrino beam will change. At a later time it will be a linear combination of different weak eigenstates which can, in time, produce in interaction different charged leptons, thus giving rise either to the opening of new channels or the depletion of old ones. This is the phenomenon of neutrino oscillations, which is analyzed in detail elsewhere[14] for the model of Ref. 11.

Acknowledgement: The author wishes to acknowledge useful conversations with Prof. F. Boehm and P. Vogel on neutrino oscillations and with his collaborators J. Harvey and D.B. Reiss.

REFERENCES

1. J. Pati and A. Salam, *Phys. Rev.* D8, 1240 (1973).

2. H. Georgi and S.L. Glashow, *Phys. Rev. Lett.* 32, 438 (1974); A. Buras, J. Ellis, M.K. Gaillard, and D.V. Nanopoulos, *Nucl. Phys.* B135, 66 (1978).

3. H. Georgi, H. Quinn, and S. Weinberg, *Phys. Rev. Lett.* 33, 451 (1974).

4. P. Ramond, Sanibel Symposia Talk, Feb. 1979, CALT-68-709, unpublished.

5. H. Georgi, in *Particles and Fields 1975*, Carlson (ed.), AIP Press, New York.

6. H. Fritzsch and P. Minkowski, *Ann. Phys.* 93, 193 (1975).

7. F. Gürsey, P. Ramond, and P. Sikivie, *Phys. Lett.* B60, 177 (1975).

8. M. Gell-Mann, P. Ramond, and R. Slansky, unpublished.

9. T. Goldman and D.A. Ross, *Phys. Lett.* 84B, 208 (1979); W.J. Marciano, *Phys. Rev.* D20, 274 (1979).

10. J.A. Harvey, P. Ramond, and D.B. Reiss, Caltech Preprint CALT-68-758, to appear, *Phys. Lett.* B.

11. See B. Pontecorvo and S.M. Bilenky, *Phys. Reports* C41, 225 (1978) and references therein.

12. E. Witten, Harvard Preprint, 1979.

13. H. Georgi and C. Jarlskog, Harvard Preprint HUTP-79/A026.

14. J.A. Harvey, P. Ramond, and D.B. Reiss, in preparation.

QUARK-LEPTON UNIFICATION AND PROTON DECAY

Jogesh C. Pati*

and

Abdus Salam**

International Centre for Theoretical Physics
Trieste, Italy

Abstract

Complexions for proton decay arising within a maximal symmetry for quark-lepton unification, which leads to spontaneous rather than intrinsic violations of B, L, and F, are considered. Four major modes satisfying $\Delta B = -1$ and $\Delta F = 0$, -2, -4, and -6 are noted. It is stressed that some of these modes can coexist in accord with allowed solutions for renormalization group equations for coupling constants for a class of unifying symmetries. None of these remarks is dependent on the nature of quark charges. It is noted that if quarks and leptons are made of constituent preons, the preon binding is likely to be magnetic.

1. INTRODUCTION

The hypothesis of grand unification[1-3] serving to unify all basic particles--quarks and leptons--and their forces--weak, electromagnetic as well as strong--stands at present primarily on its aesthetic merits. It gives the flavor of synthesis in that it provides a rationale for the existence of quarks and leptons by assigning the two sets of particles to one multiplet of a gauge symmetry, G. It derives their forces through one principle-gauge unification.

With quarks and leptons in one multiplet of a local spontaneously broken gauge symmetry G, baryon and lepton number conservation cannot be absolute. This line of reasoning has led us to suggest in 1973 that the lightest baryon--the proton--must ultimately decay into leptons.[2] Theoretical considerations suggest a lifetime for the proton in the range of 10^{28} to 10^{33} years.[2-5] Its decay modes and corresponding branching ratios depend in general upon the details of the structure of the symmetry group and its breaking pattern. What is worth noticing at this juncture is that proton decay modes and the value of the weak angle[6] $\sin^2 \theta_W$ may be the only effective tools we would have for some time to probe into the underlying design of grand unification.

---

*Permanent address: Department of Physics, University of Maryland, College Park, Maryland. This talk was presented at the Workshop by Professor Pati.

**And Imperial College, London, England.

Experiments[7] are now underway to test proton stability to an accuracy one thousand times higher than before.[8] In view of this, we shall concentrate primarily on the question of expected proton decay modes within the general hypothesis of quark-lepton unification and on the question of intermediate mass scales filling the grand plateau between 10^2 and 10^{15} GeV, which influence proton decay. At the end we shall indicate some new features, which may arise if quarks and leptons are viewd, perhaps more legitimately, as composites of more elementary objects--the *"preons"*.

Much of what we say arises in the context of maximal quark-lepton unifying symmetries of the type proposed earlier.[9] We specify such symmetries in detail later. One characteristic feature worth noticing from the beginning is that within such symmetries a linear combination of baryon and lepton numbers as well as fermion number F are locally gauged and are therefore conserved in the gauge Lagrangian. They are violated spontaneously and unavoidably as the associated gauge particles acquire masses. The purpose of the talk would be many-fold:

1) First to restress in the light of recent developments that within the maximal symmetry framework the proton may in general decay via four major modes,* which are characterized by a change in baryon number by -1 and that in fermion number (defined below) by $0,\ -2,\ -4$ and -6 units**

$$\left.\begin{array}{ll} p \rightarrow 3\nu + \pi^+ & \\ p \rightarrow 2\nu + e^- + \pi^+\pi^+ & \text{etc.} \end{array}\right\} \left.\begin{array}{l} \Delta F = 0 \\ \Delta B_q = -3\ , \quad \Delta L = +3 \end{array}\right\} \Delta(B-L) = -4 \tag{1}$$

$$\left.\begin{array}{lll} p \rightarrow e^-\pi^+\pi^+\ , & \nu\pi^+\ , & e^-K^+\pi^+ \\ p \rightarrow e^+\nu_1\nu_2\ , & e^+e^-e^-\pi^+\pi^+ & \text{etc.} \end{array}\right\} \left.\begin{array}{l} \Delta F = -2 \\ \Delta B_q = -3\ , \quad \Delta L = +1 \end{array}\right\} \Delta(B-L) = -2 \tag{2}$$

$$\left.\begin{array}{lll} p \rightarrow e^+\pi^0\ , & \bar{\nu}_e\pi^+\ , & \mu^+K^0 \\ p \rightarrow \mu^+\pi^0\ , & \bar{\nu}_\mu\pi^+\ , & e^+e^+e^- \\ & & \text{etc.} \end{array}\right\} \left.\begin{array}{l} \Delta F = -4 \\ \Delta B_q = -3\ , \quad \Delta L = -1 \end{array}\right\} \Delta(B-L) = 0 \tag{3}$$

---

*That the proton may decay via all these four modes $(\Delta F = 0, -2, -4$ and $-6)$ was to our knowledge first observed in Ref. 9. The modes $\Delta F = 0$ and -4 occur within specific models of Refs. 2 and 3, respectively, while the questioning of baryon number conservation (Ref. 2) is based on more general considerations and is tied simply to the idea of quark-lepton gauge unification, together with spontaneous symmetry breaking.

**We are not listing decay modes satisfying $\Delta B = -1$ and $\Delta F = +2$, 4 etc. corresponding to $p \rightarrow$ (5 or 7 leptons) + π's. These appear to be suppressed compared to those listed.

$$p \to 3\bar{\nu} + \pi^+ , \quad e^+ \bar{\nu}_1 \bar{\nu}_2 , \quad e^+ e^+ e^- \nu \pi^- \ \text{etc.} \left\} \begin{array}{c} \Delta F = -6 \\ \Delta B_q = -3 , \quad \Delta L = -3 \end{array} \right\} \quad \Delta(B - L) = +2 \qquad (4)$$

(Here B_q denotes quark number which is $+1$ for all quarks, -1 for antiquarks, and 0 for leptons. The familiar baryon number* B, which is $+1$ for the proton, is one third of quark number $(B \equiv B_q/3)$. L denotes lepton number which is $+1$ for $(\nu_e, e^-, \mu^-, \nu_\mu, \ldots)_{L,R}$, -1 for their antiparticles and 0 for quarks. Fermion number F is the sum of B_q and L: $F \equiv B_q + L$. Since for proton decay quark number must change by a fixed amount $\Delta B_q = -3$, there is a *one-one relationship* between change of fermion number F and that of any other linear combination of B_q and L; for example, of $[(B_q/3) - L] = B - L$ for the proton decay modes as exhibited in (1)-(4).)

2) Our main emphasis here is that these four alternative decay modes can in general coexist.

We wish:

3) To stress that observation of any of the three decay modes satisfying $\Delta F = 0, -2$, and -6 would unquestionably signal the existence of one or several *intermediate mass scales* filling the plateau between 10^2 and 10^{15} GeV. (Existence of such intermediate mass scales is a feature which naturally rhymes with a) maximal symmetries and the consequent spontaneous rather than intrinsic nature of B, L, F violations and b) partial quark-lepton unification at moderate energies $10^4 - 10^6$ GeV.)[2,10]

4) To point out that the complexions of proton-decay selection rules alter if one introduces intrinsic left-right symmetry in the basic Lagrangian, thereby permitting the existence** of ν_R's parallel to ν_L's and

5) To stress that none of the remarks (1)-(4) is tied to the nature of quark charges. These remarks hold for integer as well as fractional quark charges. (In the latter case (fractional charges), SU(3) color symmetry is exact.)

---

*The reader may note that in our previous papers we had by *convention* chosen to call quark number B_q as baryon number B. Therefore, the 15-th generator of $SU(4)_{col}$ (Ref. 2), which was written as $(B - 3L)$, stood for $(B_q - 3L)$. With the more conventional definition of baryon number, as adopted in the present note, $B_q - 3L$ is just $3(B - L)$.

**Note that in this case one must assume that ν_R and ν_L combine to form a light 4 component Dirac particle and that

$$m_{W_R} \gg m_{W_L} .$$

(To conform with astrophysical limits, one may need

$$m_{W_R} \gtrsim 50\, m_{W_L} ,$$

To motivate these remarks, let us first specify what we mean by "maximal" symmetry. Maximal symmetry[9] corresponds to gauging all fermionic degrees of freedom with fermions consisting of quarks *and* leptons. Thus with n two component left-handed fermions F_L plus n two component right-handed fermions F_R, the symmetry G is $SU(n)_L \times SU(n)_R$. One may extend the symmetry G by putting fermions F_L and antifermions F_L^c (as a substitute for F_R) in the same multiplet. In this case, the symmetry G is SU(2n), which is truly the maximal symmetry of 2n two component fermions. As an example, for a single family of two flavors and four colors, including leptonic color, n = 8 and thus G = SU(16). One word of qualification: Such symmetries generate triangle anomalies, which are avoided, however, by postulating that there exists a conjugate mirror set of fermions $F^m_{L,R}$ supplementing the basic fermions $F_{L,R}$ with the helicity flip coupling represented by the discrete symmetry $(F_{L,R} \leftrightarrow F^m_{R,L})$. Thus, by "maximal" symmetries we shall mean symmetries which are maximal up to the discrete mirror symmetry.

Though old, it is now useful to recall the argument leading to violations of B, L, and F. If all quark-lepton degrees of freedom are gauged locally as in a maximal symmetry G specified above, then fermion number $F \equiv B_q + L = 3B + L$ as well as an independent* linear combination of baryon and lepton numbers $(B + xL)$ are among the generators of the local symmetry G. Now if all gauge particles** with the exception of the photon and (for the fractional charge case of quarks) the octet of gluons, acquire masses spontaneously, then both fermion numbers F and $(B + xL)$ must be violated spontaneously, as the associated gauge particles acquire their masses. The important remark here is that even though B, L and F are conserved in the basic Lagrangian, they are inevitably and *unavoidably* violated spontaneously.***

Instead of the maximal symmetry G, one may of course choose to gauge a *subgroup* $\mathcal{G} \subset G$. But as long as the subgroup $\mathcal{G}$ assigns quarks and leptons into one irreducible multiplet, there are only two alternatives open. Either the subgroup $\mathcal{G}$ still possesses an effective fermion number F and/or

---

see G. Steigman, Erice Workshop Proceedings, March, 1980.) The alternative of ν_R acquiring a heavy Majorana mass is not relevant to proton decay.

*For the case of the lepton number being the 4-th color (Ref. 2), this generator is $\propto B_q - 3L = 3(B - L)$.

**From the limits on Eötvos type experiments, one knows that no massless gauge particle couples to B, L or F leading to an effective four fermion coupling $\gtrsim G_{\text{Newton}} \times 10^{-8}$

***Truly the argument above demands violations of linear combinations $F = (3B + L)$ and $(B + xL)$, i.e., at least B or L must be violated. Several authors have remarked that it is possible to introduce global quantum numbers within the quark-lepton unification hypothesis, which would preserve proton stability (see Ref. 11). But all simple models end up with an unstable proton (Refs. 1-3).

(B + xL) among its generators.* In this case these must be violated *spontaneously* for reasons stated above. Or the gauging of subgroup $\mathcal{G}$ leads to a "squeezing" of gauges of the maximal symmetry G such that one and the same gauge particle couples, for example, to the diquark $(\bar{q}^c q)$ as well as to the quark-lepton $(\bar{q}\ell^c)$ currents.** In this case, baryon, lepton, and fermion numbers are violated *intrinsically* through the gauge interaction itself. One way or another, some linear combinations of B and L must be violated, the basic reason in either case being the same--i.e., the appearance of quarks and leptons in the *same* symmetry multiplet.

Spontaneous and intrinsic violations of B, L, and F can, in general, lead to similar predictions for proton decay. But the two cases would differ characteristically from each other at superhigh temperatures, where intrinsic violations would acquire their maximal gauge strength with the superheavy gauge masses going to zero, while spontaneous violations would, in fact, vanish.***

2. MODELS OF GRAND UNIFICATION

It is useful to see the interrelationships between different types of unification models. The simplest realization of the idea of quark-lepton unification is provided by the hypothesis that *"lepton number is the fourth color."*[2] For a single family of (u,d) flavors, the corresponding multiplet is

$$(F_e)_{L,R} = \begin{bmatrix} u_r & u_y & u_b & u_\ell = \nu_e \\ \\ d_r & d_y & d_b & d_\ell = e^- \end{bmatrix}_{L,R} \tag{5}$$

with r, y, b and ℓ denoting red, yellow, blue, and lilac colors, respectively. The corresponding local symmetry is

$$\mathcal{G} = SU(2)_L \times SU(2)_R \times SU(4)'_{L+R} \quad , \tag{6}$$

---

*For example, $[SU(4)]^4$ and $[SU(6)]^4$ contain (B - L) as a local symmetry, but not F (though F is a global symmetry in the basic Lagrangian of these models). SU(16) operating on 16-folds of e, μ and τ family fermions (see Table I) is a subgroup of $[SU(16)]^3$ and SU(48). It contains B - L and F with $B = B_e + B_\mu + B_\tau$ and likewise for L and F.

**Examples of this type are SU(5) (Ref. 3) and SO(10) (Ref. 12). SU(5) does not contain B - L or F as local symmetries. SO(10) contains (B - L) but not F. Both SU(5) and SO(10) violate B, L and F intrinsically in the basic gauge Lagrangian.

***Alternatively, spontaneous violation may pass through a phase transition and increase in strength once the temperature exceeds a critical temperature T_c (Ref. 13).

where $SU(2)_{L,R}$ operates on the flavor indices $(u,d)_{L,R}$ and $SU(4)_{L+R}$ operates on the four color indices (r,y,b,ℓ). It is the SU(4) color symmetry which intimately links quarks and leptons. The symmetry $\mathcal{G}$ has three features: i) First, it is one of the simplest subunification models containing the low energy symmetry $SU(2)_L \times U(1) \times SU(3)'_{L+R}$ on the one hand and realizing quark-lepton unification on the other. It gauges through the SU(4) color symmetry (linking quarks and leptons) the combination $(B_q - 3L) = 3(B-L)$ as a local symmetry. ii) Second, it is nonabelian and thus provides a simple raison for the quantization of charges. iii) Its gauge structure is left-right symmetric.[14] (Indeed, the idea of lepton number as the fourth color requires that neutrinos must be introduced with left and right helicities* and thereby *the basic matter multiplet must be left-right symmetric*, given that quarks enter into the basic Lagrangian with both helicities.)

The symmetry $\mathcal{G}$, because of its simplicity, might be the right stepping stone toward grand unification. It should, of course, be viewed as a subunification symmetry, as it contains at least two gauge coupling constants--one[14] for $SU(2)_{L,R}$ and the other for $SU(4)'_{L+R}$. In other words, it should be regarded as part of a bigger unifying symmetry G possessing a single gauge coupling constant. There are a number of candidates for G which do contain the subunification symmetry $\mathcal{G}$. Table I provides a list of some of these symmetries.

$SU(2)_L \times SU(2)_R \times SU(4)'_{L+R} \rightarrow$

(1) $[SU(4)]^4 \cdots\rightarrow$ (4 flavors × 4 colors)$_{e,\mu}$ + (4 flavors × 4 colors)$_{\tau,\tau'}$ (Ref. 2)

(2) $[SU(6)]^4 \cdots\rightarrow$ (6 flavors × 6 colors) (Ref. 15)

- -

(3) Maximal symmetry for a family

$SU(16) \cdots\rightarrow \underset{\sim}{16}_e + \underset{\sim}{16}_\mu + \underset{\sim}{16}_\tau$

$16_e = [u,\nu_e : d,e^- \mid d^c,e^c : u^c,\nu^c]_L$ (Refs. 9,16)

- -

(4) $SO(10)^{**} \cdots\rightarrow \underset{\sim}{16}_e + \underset{\sim}{16}_\mu + \underset{\sim}{16}_\tau$ (Ref. 12)

(5) $E_6 \cdots\rightarrow \underset{\sim}{27}_e + \underset{\sim}{27}_\mu + \underset{\sim}{27}_\tau$ (Ref. 17)

Table I.***

---

*With ν_R being distinct from $\bar{\nu}_R$.

All the unifying symmetries listed in Table I are left-right symmetric. By contrast, one may consider the left-right asymmetric model[3] SU(5), which is the smallest grand unifying symmetry of all, with the multiplet structure

$$(\bar{5}+10)_e + (\bar{5}+10)_\mu + (\bar{5}+10)_\tau$$

in which the right-handed neutrinos $(\nu_{e,\mu,\tau})_R$ are missing. Note that in contrast to the models listed in Table I, the multiplet structure for SU(5) is reducible: $(\bar{5}+10)$ within one family.

Even if Nature is intrinsically left-right asymmetric, from the point of view of maximal gauging, the underlying family symmetry would be SU(15), for which the $(\bar{5}+10)$ form a single 15-fold. It is worth remarking that SU(15), SO(10) and SU(5) may all be viewed as subgroups of SU(16). The symmetry SU(16) gauges both fermion number F as well as B - L as local symmetries and thereby conserves both in the gauge Lagrangian before their spontaneous violation. SO(10), however, gauges B - L (since it contains $SU(2)_L \times SU(2)_R \times SU(4)'_{L+R} \approx SO(4) \times SO(6)$), but not the fermion number F. In fact, F is violated so far as the SO(10) gauge Lagrangian is concerned due to squeezing of SU(16) gauges in the sense mentioned earlier. By contrast, SU(5) gauges neither F nor B - L as local symmetries. For SU(5), like SO(10), F is explicitly violated in the gauge Lagrangian. Depending upon the choice of the Higgs structure and the fermion Higgs-Yukawa interactions, B - L may or may not be a good global quantum number for SU(5). (For example, the simplest SU(5) model with a $\underline{24}$ and $\underline{5}$ Higgs conserves (B - L); but this is not a general property of SU(5).)

In general, since SU(16) ⊃ SO(10) and SU(5), it may descend spontaneously to SU(2) × U(1) × SU(3) via SO(10) or SU(5) as intermediate steps. Alternatively, it may descend via $SU(8)_A \times SU(8)_B \times U(1)_F$, where $U(1)_F$ gauges fermion number and $SU(8)_A$ and $SU(8)_B$ operate on the octets of F_L and F_L^C of a given family. By introducing Higgs corresponding to *both* types of descent, one may descend directly to $SU_L(2) \times SU_R(2) \times SU(4)$ or even to $SU_L(2) \times SU_R(2) \times SU_{L+R}(3) \times U(1)$. Low energy phenomena including complexions for proton decay would depend upon which of these alternative routes is chosen. Some of these possibilities are exhibited in Fig. 1.

Maximal Symmetry and Spontaneous Violations of B, L and F

A maximal symmetry SU(2n), which puts n left-handed quarks and leptons as well as their charge conjugate fields within the same multiplet

---

** SO(10) is the simplest extension of $SU_L(2) \times SU_R(2) \times SU_{L+R}(4)'$ because the latter is isomorphic with SO(4) × SO(6).

*** The symmetries $[SU(n)]^4$ and SU(16) require the presence of mirror fermions for cancellation of anomalies (see discussion in Section 1).

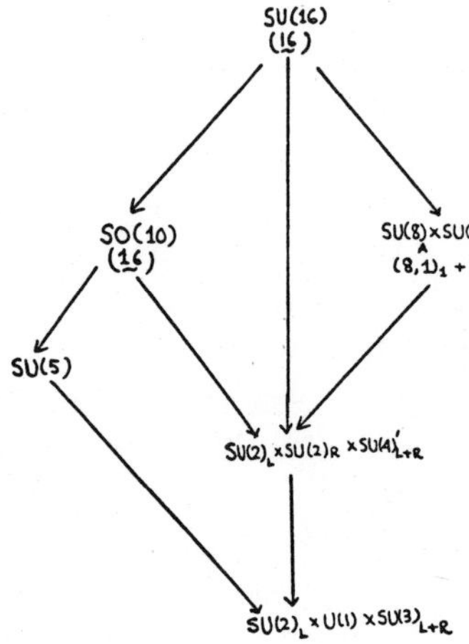

Figure 1. Some alternative routes for spontaneous descent of SU(16) to low energy symmetry. The subscripts ± 1 for $(\underset{\sim}{8},1)$ and $(1,\underset{\sim}{\bar{8}})$ denote the respective fermion numbers.

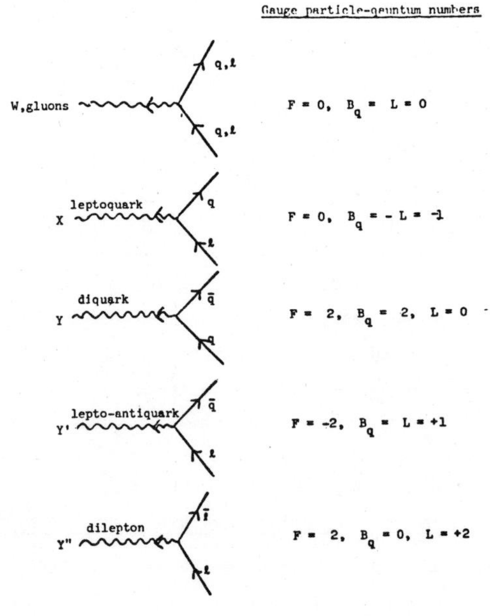

Figure 2. Gauge particles within a maximal symmetry. Here B_q, L, and $F \equiv B_q + L$ denote quark, lepton, and fermion numbers, respectively, as defined in Section 1.

$$F = [q,\ell \mid q^c,\ell^q]_L$$

generates the sets of gauge fields shown in Fig. 2.

We note that within maximal gauging, each of the gauge particles of the Lagrangian before spontaneous breaking carries definite baryon, lepton, and fermion numbers and thus these quantum numbers are conserved. The violations of these quantum numbers arise, however, as the gauge particles (barring the photon and the octet of gluons for the fractional charge quark case) acquire masses through spontaneous breakdown of the local symmetry G. The violations come about in two distinct ways.

a) *Gauge mixing*: Spontaneous symmetry breaking induces mixings[9] of gauge particles carrying different sets of values of B, L, and F, and this leads to violations of these quantum numbers. In particular, the $\Delta(B-L) = 0$, $\Delta F = -4$, and $\Delta(B+L) = 0$, $\Delta F = -2$ proton decays arise through the gauge mixings noted below:

Mixing	Symmetry Violation			Decay Mode
$Y \leftrightarrow \bar{Y}'$	$\Delta B_q = -3,$	$\Delta L = -1,$	$\Delta F = -4$	$p \to \bar{\ell}$ + mesons
$Y \leftrightarrow \bar{X}$	$\Delta B_q = -3,$	$\Delta L = +1,$	$\Delta F = -2$	$p \to \ell$ + mesons

These two kinds of spontaneously induced mixings are exhibited in Figs. 3a and 3b, respectively. For SU(16) the $Y \leftrightarrow \bar{Y}'$ mixing leads to $\Delta F = -4$ decays. Such mixings can be induced, for example, by Higgs of the type*

$$\Omega^{\{CD\}}_{\{AB\}}$$

where A, B, C, D range over 1-16. These violate B, L, F, but respect (B - L). If in addition, we also introduce Higgs of the adjoint ($\underset{\sim}{255}$) type ξ^B_A, SU(16) breaks down to $SU_L(8) \times SU_R(8) \times U_F(1)$. The descent to $SU_L(2) \times SU_R(2) \times SU(4)$ through both these types of Higgs, and the fact that X gauge bosons belong to SU(4), must ensure that

$$m_X^2 \lesssim \Delta^2_{YY'} \leq m_Y^2,\ m^2_{Y'} \quad ,$$

where $\Delta_{YY'}$ is the mixing parameter in the $Y - \bar{Y}'$ mass matrix.

The amplitude for a $\Delta F = -4$ transition (generated via Fig. 3a) is:

$$A(3q \to \bar{\ell})_{\Delta F = -4} \approx (g^2 \Delta^2_{YY'})/m_Y^2 m^2_{Y'} \quad .$$

---

*The detailed patterns of symmetry breaking as correlated with the helicities of particles involved in the decay will be presented in a paper with J. Strathdee.

This is bounded below by $g^2 m_X^2/m_Y^4$ and above by g^2/m_Y^2. The lower and upper bounds correspond to the minimal and maximal values of $\Delta^2_{YY'}$ mentioned above. (Note that the maximal value $\Delta^2_{YY'} = m_Y^2 - m_{Y'}^2$ is the case, for example, for SO(10) which contains $Y_s = (Y + \bar{Y}')/\sqrt{2}$ type gauge particles only.) This is recoverable from SU(16) when $(Y - \bar{Y}')/\sqrt{2}$ gauge mesons are of infinite mass.

The $Y \leftrightarrow \bar{X}$ mixing leads to $\Delta F = -2$ transitions (Fig. 3b). This could arise, for example, from VEV's of a Higgs of the type $\Psi^{[CD]}_{[AB]}$. Writing $\langle\Psi\rangle$ as $\langle\Psi_1\rangle + \langle\Psi_2\rangle$, where $\langle\Psi_2\rangle$ conserves $SU_L(2) \times SU_R(2) \times SU(4)$ and $\langle\Psi_1\rangle$ violates $SU_L(2)$ as well as $SU_R(2)$ (with the corresponding components transforming as $I_L = 1/2$ and $I_R = 3/2$), it follows that

$$\langle\Psi_1\rangle \leq m_{W_L}/g \quad ,$$

while $\langle\Psi_2\rangle$ may be as large as* m_X/g. Thus,

$$A(3q \to \ell)_{\Delta F = -2} \approx g^2 \frac{\Delta_{XY}}{m_X^2 m_Y^2}$$

$$\leq g^2 \frac{m_{W_L} m_X}{m_X^2 m_Y^2} \qquad \left(\text{or} \quad g^2 \frac{m^2_{W_L}}{m_X^2 m_Y^2} \right) .$$

It is worth noting from the above that if $\Delta^2_{YY'}$ has its maximal value $\sim m_Y^2$, then a $\Delta F = -4$ amplitude would always dominate over a $\Delta F = -2$ amplitude by a factor $\geq (m_X/m_{W_L}) \gg 1$. The two amplitudes would be comparable, however, if $\Delta^2_{YY'}$ is much smaller than m_Y^2 (see next section).

b) The violations of B, L and F may also arise through spontaneously induced three point Yukawa transitions of the type $q \to \ell + \phi$ (see Figs. 4a and 4b)

$$q^i_{\alpha L} \to \nu_R + C^i_\alpha + \langle C^{*1}_4 \rangle + \langle \bar{A}^0 \rangle$$

$$q^i_{\alpha R} \to \nu_L + B^i_\alpha + \langle B^{*1}_4 \rangle + \langle A^0 \rangle \tag{7}$$

---

*We face here, as elsewhere, the problem of gauge hierarchies; i.e., the question, are all VEV's of Higgs fields always of the same order of magnitude? If this is the case, $\langle\Psi_2\rangle$ is also of order

$$\left(m_{W_L} / g \right) .$$

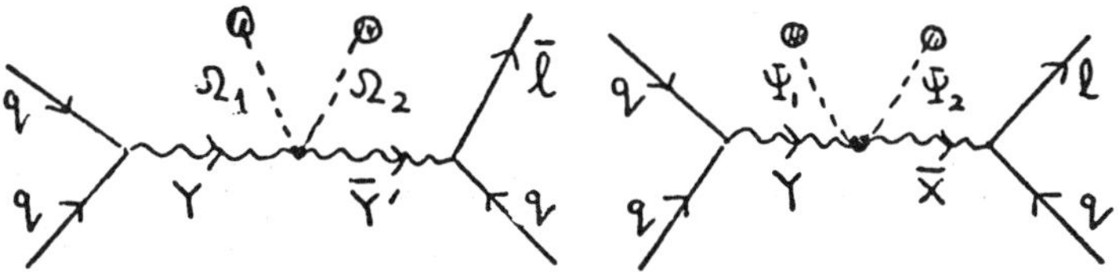

(a) (b)

($\Delta F = -4$) $p \to \bar{\ell} + \text{mesons}$ ($\Delta F = -2$) $p \to \ell + \text{mesons}$

Figure 3. *Spontaneous* violations of B, L and F in a maximal symmetry leading to gauge mixings. These induce, for example, $\Delta F = -4$ and -2 proton decays. $\Omega_{1,2}$ and $\Psi_{1,2}$ are Higgs fields (see text).

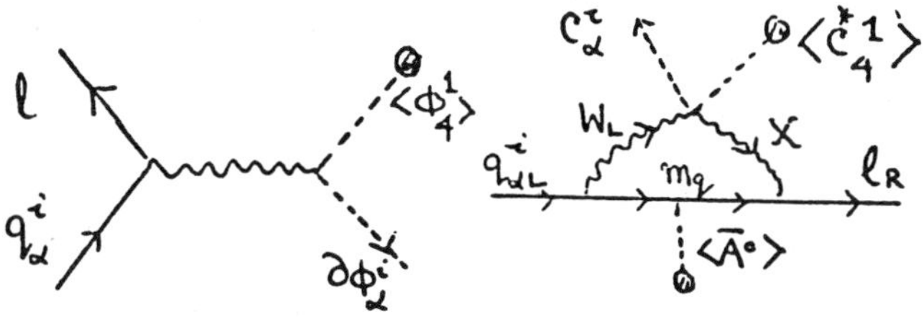

(a) (b)

Figure 4. *Spontaneous* violations of B, L (and, in general, F) leading to effective Yukawa transitions: $q_\alpha^i \to \ell + \phi_\alpha^i$. These transitions in third order induce $\Delta F = 0$ proton decays: $p \to 3\ell + \text{mesons}$ (and analogously $\Delta F = -6$ decays: $p \to 3\bar{\ell} + \text{mesons}$). See text.

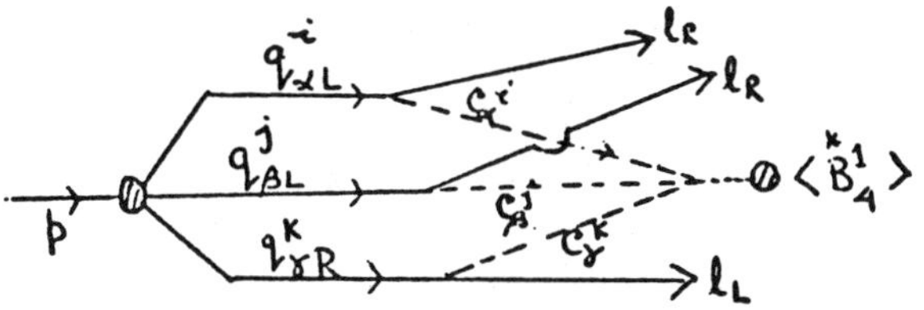

Figure 5. $\Delta F = 0$ proton decays through *spontaneous* violations of B and L. These utilize the effective Yukawa transitions of Fig. 4 thrice. Note that the mechanisms of Figs. 3, 4, and 5 apply to integer as well as fractionally charged quarks.

$$d_{\alpha L} \to e^-_R + C^1_\alpha + \langle \overset{*}{C}{}^1_4 \rangle + \langle A^0 \rangle \qquad (7)$$

$$d_{\alpha R} \to e^-_L + B^1_\alpha + \langle \overset{*}{B}{}^1_4 \rangle + \langle \bar{A}^0 \rangle \quad .$$

Here i and α denote flavor and SU(3) color indices, respectively. The fields A, B, and C, which are identical to those introduced in Ref. 2, transform as (2,2,1), $(1,2,\bar{4})$, and $(2,1,\bar{4})$, respectively, under $SU(2)_L \times SU(2)_R \times SU(4)_{L+R}$. Under SU(16), A belongs to a $[16 \times 16]_{\text{symmetric}}$ representation, while C and $B^\dagger$ together make a 16-fold. The fields C^1_4 and $\overset{*}{B}{}^1_4$ have the same quantum numbers within a 16-fold as ν_{eL} and ν^c_{eL}, respectively, while A^0 possesses

$$I_{3L} = - I_{3R} = - \frac{1}{2} \quad .$$

The VEV $\langle \overset{*}{C}{}^1_4 \rangle$ and $\langle A^0 \rangle$ are of order (m_{W_L}/g), while $\langle \overset{*}{B}{}^1_4 \rangle$ must be of order (m_{W_R}/g) or (m_X/g), whichever is lower. The effective Yukawa transitions (7), used thrice, induce ΔF = 0 proton decays (see Fig. 5) of the type[16,18]

$$p \to \nu_L \nu_R \ell_{L,R} + \pi\text{'s} \quad , \qquad (8)$$

where $\ell_{L,R}$ denotes either the charged or the neutral lepton. These transitions are made possible through quartic scalar interactions which permit, for example, $(C^1_1 + C^2_2 + B^2_3)$ to make a transition into $\overset{*}{B}{}^1_4$ and thereby disappear into vacuum through $\langle B^*_{14} \rangle \neq 0$ (see discussion later). Fermi statistics together with the color singlet nature of the proton inhibits all three leptons in the final state from having the same helicity[16] (see (8)).

An analogous mechanism induces the Yukawa transitions

$$q^i_\alpha \to \bar{\ell} + \psi^i_\alpha$$

which, in third order, induce proton decays satisfying ΔF = -6:

$$p = 3q \to (3\bar{\ell} + \pi\text{'s}) + 3\chi \to (3\bar{\ell} + \pi\text{'s}) + \langle \overset{*}{\chi}{}^1_4 \rangle \quad . \qquad (9)$$

In accord with the observation in Ref. 9, the above mechanisms show that all four modes for proton decay satisfying ΔF = 0, - 2, - 4 and - 6 can arise within a maximal symmetry G. Their relative rates would depend upon the associated gauge masses and the mixing parameters, which in turn depend upon the parent symmetry G as well as upon its breaking pattern.

Two common features of these mechanisms are worth noting:

i) They utilize only *spontaneous* rather than intrinsic violations of B, L, and F. None of these would be operative if the vacuum expectation values of all the relevant Higgs fields were set to zero.

ii) None of these mechanisms is tied to the nature of quark charges. *They hold for quark charges being either integral or fractional, with* SU(3) *color local symmetry either being broken spontaneously and softly or remaining exact.*

(We would like to make a small digression here. Our suggestion of quark-lepton unification of 1972 has been misunderstood in this regard, as though it is tied to integer charges for quarks. A bit of history is perhaps relevant. During the years 1972-74 almost everyone accepted fractional charges and absolute confinement. Our contention, however, was that *both* possibilities--fractional as well as integral charges for quarks--arise within the same unification hypothesis. For example, the hypothesis "lepton number is the fourth color" permits both charge patterns depending only upon the nature of spontaneous symmetry breaking.[2] Since it was a logical possibility, we built the theory of integer charges for quarks and possible "quark liberation" so that it can meaningfully be confronted with experiments. As far as we know, there does not exist any theoretical or experimental argument as yet providing unambiguous evidence for one quark charge pattern versus the other.* We therefore still keep our options open regarding the nature of quark charges and await experiment to settle this question. We stress, however, that the twin suggestions[1,2] of quark-lepton unification and consequent baryon number violation are not tied in any way to the nature of quark charges. They are more general.)

---

*Recent arguments of Okun, Voloshin, and Zakharov (Moscow preprint ITEP-79) favoring fractional charges for quarks do not take into account the facts that a) variation of electric charges for integer charge versus fractional charges as functions of momentum are governed by different renormalization group equations due to the presence of the color component in the former, which is absent in the latter, and that b) for a partially confining theory there exist singularities in the variable mass parameters in time-like regions even near the origin without requiring the existence of physical particles at such points. This will be elaborated in a forthcoming preprint. A second argument based on $\eta' \to 2\gamma$ (M. Chanowitz, *Phys. Rev. Lett.* ,1979) favoring fractional charges is subject to the uncertain PCAC extrapolation from 1 GeV^2 to zero for the case of ICQ. There is a third argument based on an empirical analysis of deep inelastic Compton scattering $\gamma p \to \gamma + X$ data (H.K. Lee and J.K. Kim, *Phys. Rev. Lett.* 40, 885 (1978), and J.K. Kim and H.K. Lee, preprint, 1979), which, by contrast to the previous two, favors integer over fractional charges. This argument is uncertain to the extent that the P_T involved in present experiments is not high enough to permit a legitimate use of the parton model. We must wait for unambiguous experiments--like the two photon experiments in $e^+e^- \to e^+e^- + \text{hadrons}$--to provide a definitive test. We understand that these will soon be completed.

Spontaneous Versus Intrinsic Violations of B, L and F

It is now instructive to compare violations of B, L and F that are spontaneous in origin (as outlined above) to those which are intrinsic. The latter arise in general *if* one chooses to gauge subgroups of the maximal symmetry defined by the fermion content. As mentioned earlier, examples of such subgroups are SU(5) and SO(10). For these cases, instead of the diquark current $(\bar{q}^c q)$ and $\overline{\text{(lepto-antiquark)}}$ current $(\bar{q}\ell^c)$ coupling to distinct gauge particles Y and $\bar{Y}'$, respectively, the two currents couple to one and the same gauge particle Y_s in the basic Lagrangian. This is equivalent to "squeezing" the two gauges associated with the two distinct currents mentioned above so that $Y_s \sim (Y + \bar{Y}')/\sqrt{2}$ coupling to the sum of the currents is present in the basic Lagrangian, but $Y_a \sim (Y - \bar{Y}')/\sqrt{2}$ coupling to the orthogonal combination is absent. (Equivalently, Y_a is assigned an *infinite* mass.) The exchange of the Y_s particle thus leads to a violation of B, L and F in the second order of the basic gauge interactions (see Fig. 6) and induces $\Delta F = -4$ decays

$$p \rightarrow \bar{\ell} + \text{mesons} \quad . \tag{10}$$

We see that intrinsic violations of B, L and F arising through squeezed gauging can, in general, lead to similar consequences for proton decay as the case of spontaneous violation arising for a maximal symmetry (compare Fig. 6 with Fig. 3a). To state it differently, if Y_s and Y_a are given masses in SU(16) spontaneously in the infinite limit for the mass of Y_a, we recover the predictions of SO(10). In this sense such predictions are contained in those obtained from SU(16).

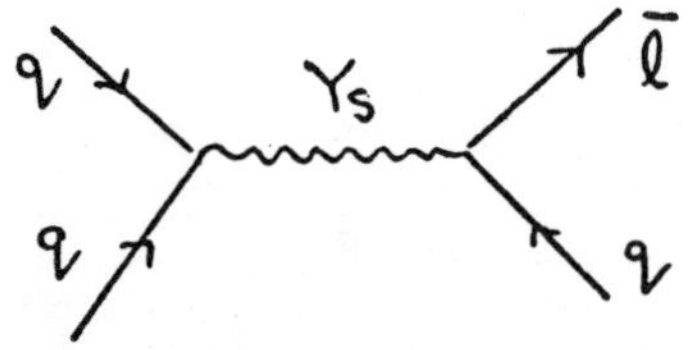

Figure 6. *Intrinsic* violations of B, L and F through gauge squeezing. Here, for example, $Y_S \equiv (Y + \bar{Y}')/\sqrt{2}$ is a gauge particle of the *basic* Lagrangian, but the orthogonal combination $(Y - \bar{Y}')/\sqrt{2}$ is absent, or effectively has infinite mass. Such gauge squeezings occur in SU(5) and SO(10).

The two cases--spontaneous versus intrinsic violations of B, L and F -- appear to possess an absolute distinction from each other at high temperatures within the range of temperatures between m_s and m_a and beyond m_a, where m_s and m_a are the masses of the combinations of the fields Y_s and Y_a.

3. CONDITIONS FOR RELEVANCE OF ALTERNATIVE PROTON DECAY MODES

We now proceed to obtain the "necessary" and sufficient conditions for alternative proton decay modes satisfying $\Delta F = 0,\ -2,\ -4$ and -6 to be relevant for either the forthcoming or the second generation proton decay searches. For this purpose, we shall consider only those mechanisms for proton decay which arise within a maximal symmetry, outlined in the previous section. Depending upon the decay modes* the experimental searches are expected to be sensitive to proton lifetimes varying between 10^{30} to 10^{33} years.

Now let us first observe the restrictions which arise from the effective low energy symmetry being $SU(2)_L \times U(1) \times SU(3)_{color}$. Weinberg and Wilczek and Zee[19] have shown that the effective proton decay interactions based on operators of lowest dimension, which is six,** automatically conserve $B-L$, if they are constrained to satisfy the low energy symmetry $SU(2)_L \times U(1) \times SU(3)_{color}$. Based on this observation, they have concluded that proton decay should be dominated by the $\Delta F = -4$ modes (e.g., $p \to e^+\pi^0$, $\bar{\nu}\pi^+$, etc.) which conserve $B-L$. In drawing this conclusion, they were motivated by the assumption that the theory possesses essentially *only two* mass scales $m_{W_L} \sim 100$ GeV and*** $M_Y \sim 10^{14} - 10^{15}$ GeV, in which case, the alternative decay modes $\Delta F = 0,\ -2,\ -6$ --requiring higher dimensional operators and/or violation of $SU(2)_L \times U(1)$ --would be damped at least by a factor $\sim (m_{W_L}/m_Y)$ compared to the "allowed" $(B-L)$ conserving decay modes in the amplitudes.

There are, however, good reasons why one may consider departures from this assumption.

The most important is that there do exist grand unification models such as $[SU(4)]^4$, $[SU(6)]^4$ and their extended maximal versions involving fermion number gauging (such as $SU(32)$, $SU(48)$ or the smaller tribal group $[SU(16)]^3$), which do permit intermediate mass scales filling the gap between 10^2 and 10^{15} GeV with the lightest leptoquark gauge particle X as light as $10^4 - 10^5$ GeV.[2,10] It is precisely because of the existence of these

---

*For the two-body modes such as $p \to e^+\pi^0$ and $n \to e^+\pi^-$ satisfying $\Delta F = -4$ as well as $n \to e^-\pi^+$ satisfying $\Delta F = -2$, the forthcoming experiments may be sensitive to proton lifetimes $\lesssim 10^{33}$ years. For the multiparticle modes such as $p \to e^- + 2\nu + \pi^+\pi^+$ and $n \to e^- + 2\nu + \pi^+$ satisfying $\Delta F = 0$, the sensitivity might be two or three orders of magnitude lower, while for $\Delta F = -6$ modes such as $p \to e^+ + \bar{\nu}_1 + \bar{\nu}_2$, the sensitivity may lie in between (see Ref. 7 for details).

**Thus these can include, in general, only operators of the form $qqq\ell$ and $qqq\bar{\ell}$ which, respectively, induce only $\Delta F = -4$ (e.g., $p \to e^+ + \pi^0$) and $\Delta F = -2$ (e.g., $p \to e^- + \pi^+\pi^+$) decays. The $\Delta F = 0$ and $\Delta F = -6$ decays would involve, in any case, higher dimensional operators with a minimum of six fermion fields (dimension $\gtrsim 9$).

***Here Y is used in the generic sense to denote a superheavy gauge particle coupling to different sorts of $F = \pm 2$ currents.

intermediate mass scales that models of Refs. 2 and 9 have permitted all along alternative proton decay modes.* This we elaborate below.

The second reason why such intermediate mass scales are worthy of consideration is purely experimental. They provide the scope for discovery of new physics through tangible evidence for quark-lepton unification in the conceivable future, especially if there exist leptoquark (X) gauge particles in the 10-100 TeV region.

The third reason is that if these intermediate mass scales do exist, they would permit $\Delta F = 0, -2,$ and -6 modes, whose rates may, in general, even exceed the rate of the $\Delta F = -4$ mode $(p \to \pi^0 + e^+)$. Experiments must therefore be designed to look for such modes.

One other reason for the existence of intermediate mass scales (with successive steps perhaps differing by powers of α or α^2) is that it may make it easier to understand the problem of the gauge hierarchy.** And finally, existence of intermediate mass scales may also account for the departure*** of present experimental $\sin^2\theta_W \approx 0.23 \pm 0.01$ from the "canonical" theoretical value of ≈ 0.20.

With these to serve as motivations for the existence of intermediate mass scales, let us first present a scenario for the hierarchy of gauge masses. This is depicted in Fig. 7. We discuss later how such a scenario can be realized within maximal symmetries in accord with renormalization group equations for the gauge coupling constants. The characteristic feature of this scenario is that the leptoquark gauge particles (X) are rather light, characterizing the fact that they belong to the lower subunification symmetry $SU(2)_L \times SU(2)_R \times SU(4)'_{L+R}$. The fermion number ± 2 gauge particles Y, Y', and Y", defined already, range in mass between 10^{10} and 10^{15} GeV with all possible mutual orderings including the possibility that they may all be nearly degenerate.

We now wish to argue that the $\Delta F = 0$ and -2 modes involving the decays $p \to 3\ell + \text{pions}$ and $p \to (e^- \text{ or } \nu) + \text{pions}$ would be relevant to forthcoming proton decay searches for the following set of values of the X and Y gauge particles:

---

*Several authors (Ref. 20) have recently considered the possibility of intermediate mass scales, permitting Higgs rather than gauge particles to acquire such masses and introducing Yukawa interactions to induce new complexions for proton decay. In view of the relative arbitrariness of Yukawa couplings, we pursue the consequences which follow from the gauge interactions and the Higgs self-coupling only, subject to spontaneous symmetry breaking. Recently Weinberg has extended his analysis (Ref. 21) permitting intermediate mass scales. We understand that H.A. Holden and A. Zee (Ref. 21) have made a similar analysis, though we have not seen their preprint.

**This is only a conjecture at present and needs to be further investigated.

***The weight of this remark is dependent upon further refinements in the measurements of $\sin^2\theta_W$.

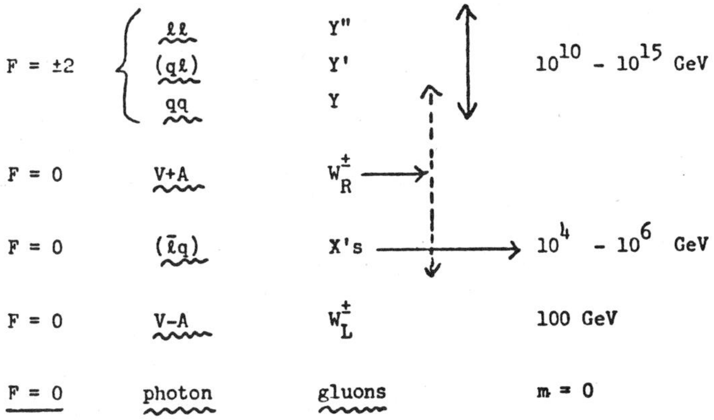

Figure 7. A scenario for gauge masses arising within a maximal symmetry. The masses of Y, Y', and Y" range between $10^{10}-10^{15}$ GeV with all possible mutual orderings including the possibility that they are degenerate. $W_R^{\pm}$ can be heavier or lighter than X's.

$$\Delta F = 0 \ldots \rightarrow \quad \text{Need} \quad m_X \approx 10^4 - 10^5 \text{ GeV}$$

$$\Delta F = -2 \ldots \rightarrow \quad \text{Need} \quad \begin{cases} m_Y \approx 10^{10} - 10^{12} \text{ GeV} \\ m_X \approx 10^4 - 10^5 \text{ GeV} \end{cases} \tag{11}$$

(Later we show that these requirements are met within a class of maximal symmetries.)

$\Delta F = 0$ Modes $(p \rightarrow 3\ell + \pi\text{'s})$

These decays occur for either integer or fractionally charged quarks as follows. Each quark makes a virtual transition to a lepton + a Higgs field ϕ_α^i; the three Higgs fields generated thereby combine to annihilate into vacuum through a VEV $\langle B_4^{*1} \rangle \neq 0$ (see Figs. 4 and 5 and discussion in the previous section). The preferred configuration corresponds to one quark being right handed, which emits a B_α^i field, and the other two being left handed, which emit appropriate components of C_α^i fields.* Furthermore, it turns out, owing

*Recall B and C transform as $(1,2,\bar{4})$ and $(2,1,\bar{4})$, respectively, under $SU(2)_L \times SU(2)_R \times SU(4)$. Together C and $B^\dagger$ make a 16-fold of SU(16). C and B fields have the same quantum numbers as the fermion fields F_L and F_R. Thus, only C_4^1 and B_4^1 possess non-zero VEV, which give masses to $W_L^{\pm}$ and $W_R^{\pm}$, respectively (see Section 2). With integer charges for quarks, additional components of C and/or B multiplets can acquire non-zero VEV (see Ref. 2); but these do not materially alter the complexions of proton decay.

to selection rules, that one of the quarks must proceed via a tree (Fig. 4a) and the other two via loops (Fig. 4b).* The corresponding amplitudes (suppressing spinors) are given by

$$M^{L,R}_{tree}\ (\text{Fig. 4a}) = \left(\frac{f_X^2}{m_X^2}\right)(m_q)\left(\langle C_4^{*1}\rangle, \langle B_4^{*1}\rangle\right)$$

$$M^{L,R}_{loop}\ (\text{Fig. 4b}) = \left(\frac{f_X^2}{m_X^2}\right)\left(\frac{g^2 m_q}{8\pi^2}\right)\left(\langle C_4^{*1}\rangle, \langle B_4^{*1}\rangle\right)\ \ln\left(\frac{m_X^2}{m^2_{W_{L,R}}}\right) \tag{12}$$

where the superscripts L and R and the sets of parameters $(\langle C_4^{*1}\rangle, m_{W_L})$ and $(\langle B_4^{*1}\rangle, m_{W_R})$ go with the transitions of left and right-handed quarks, respectively.

Using these amplitudes for $q \to \ell + (\text{Higgs})$ transitions, taking $m_X \approx (3 \text{ to } 10) \times 10^4$ GeV, and allowing a reasonable range for the Higgs parameters (which are constrained by the gauge masses, see Ref. 18 for details), the proton decay rate (ignoring all but $\Delta F = 0$ modes) is found to be[18]

$$\tau_p \approx 10^{28} - 10^{34} \text{ years} \quad . \tag{13}$$

The main point worth noting is that *the* $\Delta F = 0$ *modes become important if the leptoquark gauge boson* X *has a mass* $\approx (10^4 - 10^5$ GeV). A similar mass range is also obtained in Ref. 21.

$\Delta F = -2$ Modes $(p \to e^-\pi^+\pi^+$, etc.)

These decays arise through $\bar{X} \leftrightarrow Y$ mass mixing (see Fig. 3b and also Section 2). Since such a mixing violates $SU(2)_L \times U(1)$, the corresponding mixing (mass)^2 denoted by Δ^2_{XY} must be proportional to a VEV $\lesssim m_{W_L}$. Taking $\Delta^2_{XY} = m_{W_L} m_X$ (or conservatively $m^2_{W_L}$), and $m_X \approx 10^5$ GeV (as before), we see that the $\Delta F = -2$ proton decay interaction viewed as an effective four fermion interaction would have a strength $(\Delta^2_{XY})/(m_X^2 m_Y^2)$. This would exceed the canonical value $\approx 10^{-29}$ GeV^{-2} for $m_Y \approx 10^{13}$ GeV (or 3×10^{11} GeV), leading to a proton lifetime in the range of $10^{31} - 10^{32}$ years.

We thus see that for

$$m_X \approx 10^4 - 10^5 \text{ GeV} \quad \text{and} \quad m_Y \approx 3\times 10^{11} - 10^{13} \text{ GeV}$$

---

*This is under the assumption that leptoquark gauge particles (X') coupling to cross currents of the form $(\bar{e}u)$ and $(\bar{\nu}d)$ are much heavier than those coupling to $(\bar{\nu}u)$ and $(\bar{e}d)$, which are denoted by X. This situation emerges automatically if the unifying symmetry G descends to low energy symmetries via $SU(2)_L \times SU(2)_R \times SU(4)'_{L+R}$. See Ref. 18 for a more general discussion.

(the precise number depending upon Δ^2_{XY}), $\Delta F = 0$ *as well as* $\Delta F = -2$ *could coexist* with comparable rates and be relevant to forthcoming proton decay searches.

$\Delta F = -4$ Modes ($p \to e^+\pi^0$, etc.)

These arise through $Y \leftrightarrow \bar{Y}'$ mass mixing (see Fig. 3a). Denoting the mixing (mass)^2 by $\Delta^2_{YY'}$, the corresponding amplitude is $\Delta^2_{YY'}/(m^2_Y \sim m^2_{Y'})$. As explained in Section 2, $\Delta^2_{YY'}$ can be as large as $m^2_Y \sim m^2_{Y'}$. This is the case, for example, if SU(16) descends to low energy symmetries without passing through $SU(8) \times SU(8) \times U(1)$. If $\Delta^2_{YY'}$ has its maximum value $m^2_Y \approx m^2_{Y'}$ (as is the case, for example, for SO(10)), then the $\Delta F = -4$ amplitude would have a strength $\approx 1/m^2_Y$ and a proton lifetime $\approx 10^{30}$ years would require $m_Y \approx 10^{14}$ GeV. However, with the perfectly feasible possibility of $\Delta^2_{YY'}$ smaller than m^2_Y, there is the possibility that a proton lifetime of 10^{30} years can be compatible with Y and Y' being lighter. For example, if $\Delta^2_{YY'} = (10^{10}\text{ GeV})^2$, then a proton lifetime of 10^{30} years would be compatible with $m_Y \approx m'_Y \sim 10^{12}$ GeV. We are interested in this possibility because for such values of the Y mass (and with $m_X \approx 10^4 - 10^5$ GeV), the $\Delta F = -2$ and the $\Delta F = 0$ modes become relevant as well. Thus we see that a gauge mass pattern

$$m_X \sim 10^4 - 10^5 \text{ GeV}, \qquad m_Y \approx m_{Y'} \approx 10^{12} \text{ GeV} \tag{14}$$

$$(\Delta^2_{YY'})^{\frac{1}{2}} \approx 10^{10} \text{ GeV}, \qquad (\Delta^2_{XY})^{\frac{1}{2}} \approx m_{W_L} m_X \tag{15}$$

would permit the possibility that $\Delta F = 0,\ -2,$ *and* -4 modes can coexist and be relevant to present searches. The possible coexistence of the $\Delta F = -6$ mode depends upon some further considerations, which we shall not pursue here.

Our task now is to show that such mass patterns as outlined in (15) can be realized within unifying symmetries in accord with renormalization group equations as well as observed values of $\sin^2\theta_W$ and α_S.

4. SOLUTIONS TO HIERARCHY EQUATIONS FOR A CLASS OF UNIFYING SYMMETRIES

Perturbative renormalization group equations for the running coupling constants of a spontaneously broken unifying symmetry permit, in general, solutions for the gauge masses[6] that exhibit a hierarchy. We refer to these equations as hierarchy equations and ask: Do there exist solutions to these equations within some class of unifying symmetries which permit

(a) $M_X \sim 10^4 - 10^5$ GeV

(b) $M_Y \sim 10^{11} - 10^{15}$ GeV (16)

(c) $\sin^2\theta_W \approx 0.23$ (16)

(d) $\alpha_S(m_W) \approx 0.14$.

We know that the answer is negative for SU(5) and SO(10).

Now to see how a "light" leptoquark gauge particle X with a mass $\approx 10^4 - 10^5$ GeV can be realized in the first place, it is instructive to recall the case of the two-family $[SU(4)]^4$ model, which possesses a single gauge coupling constant because of discrete symmetry between the four SU(4) factors. This symmetry, depending upon the nature of SSB, can break via two alternative chains

$$[SU(4)]^4 \xrightarrow{M_1} SU(2)_L^{I+II} \times U(1) \times SU(3)'_{L+R}$$

$$[SU(4)]^4 \xrightarrow{M_2} SU(2)_L^{I+II} \times U(1) \times SU(3)'_L \times SU(3)'_R \qquad (17)$$

Here $SU(4)^{flavor}_{L,R}$ acts on $(u,d,c,s)_{L,R}$ flavors. $SU(2)^I_L$ and $SU(2)^{II}_L$ act on $(u,d)_L$ and $(c,s)_L$ doublets, respectively; $SU(2)^{I+II}_L$ is their diagonal sum. The gauge particles of $SU(2)^{I+II}_L$ are related to those of $SU(2)^{I,II}_L$ by

$$W_L^{I+II} = (W_L^I + W_L^{II})/\sqrt{2} \quad .$$

Thus, if g is the symmetric gauge coupling constant of each SU(4) factor, the coupling constant g_2 of the low energy symmetry $SU(2)^{I+II}_L$ would approach $g/\sqrt{2}$ in the symmetry limit. Likewise, the coupling constant g_3^V for vector color $SU(3)_{L+R}$ would also approach $g/\sqrt{2}$, since it is obtained by diagonal summing of $SU(3)'_L$ and $SU(3)'_R$. By contrast, the coupling constant g_3^C for chiral color* $SU(3)_L \times SU(3)_R$ (relevant to the lower chain) would approach g in the symmetric limit. Thus,

$$\text{vector color} \quad g_2 = g_3^V = \frac{g}{\sqrt{2}} \quad \text{(symmetric limit)}$$

(18)

$$\text{chiral color} \quad g_2 = \frac{g}{\sqrt{2}} \; ; \quad g_3^C = g \quad \text{(symmetric limit)}$$

*The chiral color symmetry must break to vectorial color eventually. But if this breaking takes place by a mass scale $< m_{W_L}$, one can ignore the effect of such a breaking for studies of renormalization group equations at momenta $\geq m_{W_L}$.

It can be shown that this difference of a factor of $1/\sqrt{2}$ between flavor versus color coupling constants, translates into a factor ≈ 2.5 that multiplies the logarithm of M_1/μ. It alters drastically the determination of the unification mass M_1 and one obtains[10]

$$\text{For vector color} \begin{cases} M_1 \approx 10^{15} \text{ GeV} \\ \sin^2\theta_W \approx 0.20 \end{cases}$$

$$\text{For chiral color} \begin{cases} M_1 \approx 10^{6} \text{ GeV} \\ \sin^2\theta_W = \frac{2}{7} + \frac{10}{21\,\alpha_S} \approx 0.30 \end{cases} \tag{19}$$

The mass of X is about a factor of 10 lower than M_1. Thus, for $M_1 \approx 10^6$ GeV, m_X is $\approx 10^5$ GeV, as desired. However, the case of the two family $[SU(4)]^4$ descending via chiral color is now excluded experimentally, since it yields too high a value for $\sin^2\theta_W$ (≈ 0.30) compared to the experimental value of ≈ 0.23.

Nevertheless, the above example provides the clue for low mass unification. The idea is to create through spontaneous descent a dichotomy between low energy flavor versus color coupling constants such that the former is lower than the latter in the symmetric limit.* This is best illustrated by the three family symmetry[22] $[SU(6)]^4$, which operates on six flavors (u,d,c,s,t,b) and six colors. There are three leptonic colors rhyming with three quark colors (r, y and b). There are the six observed leptons plus twelve unobserved heavy leptons in the model.** Here the low energy flavor $SU(2)_L$ is obtained by diagonal summing of *three* SU(2)'s, which respectively act on (u,s), (c,s), and (t,b) -doublets. Thus $g_2 = g/\sqrt{3}$ in the symmetric limit. In this case, even if the low energy color symmetry is vectorial $SU(3)'_{L+R}$,

$$g_2 = \frac{g}{\sqrt{3}} < g_3^V = \frac{g}{\sqrt{2}}$$

in the symmetric limit. This, together with the fact that the bare value of

---

*This ingredient hastens the "meeting" of the color and the SU(2) flavor coupling constants. There is a second ingredient which can speed up the "meeting" of SU(2) and U(1) coupling constants. This is realized through a lowering of the bare value of the weak angle $\sin^2\theta_0$ from the canonical value 3/8. This is the case for $[SU(6)]^4$ where $\sin^2\theta_0 = 9/28$ on account of the presence of the extra leptons. However, for SU(16) or $[SU(16)]^3$, $\sin^2\theta_0$ has the "canonical" value 3/8.

**This is without counting the mirror fermions.

the weak angle[23] $\sin^2\theta_0 = 9/28$ (rather than 3/8), leads again to a low unification mass for the descent

$$[SU(6)]^4 \xrightarrow{M_1} SU(2)_L^{I+II+III} \times U(1) \times SU(3)'_{L+R} \quad ,$$

i.e.,

$$M_1 \approx 10^6 \text{ GeV} , \qquad \text{i.e.,} \qquad m_X \approx 10^5 \text{ GeV} \quad . \tag{20}$$

In this case, one furthermore obtains a desirable value for the weak angle[23]

$$\begin{aligned} \sin^2\theta_W &= \frac{5}{24} + \frac{19}{36}\frac{\alpha}{\alpha_s} \\ &\approx 0.235 \quad . \end{aligned} \tag{21}$$

We thus see that quark-lepton unification could take place through lepto-quark gauge interactions at an energy scale 10^5 GeV. For the chiral descent

$$[SU(6)]^4 \rightarrow SU(2)_L^{I+II+III} \times U(1) \times SU'_L(3) \times SU'_R(3)$$

the unification mass could be lower still ($\approx 10^4$ GeV).

What about the masses of the fermion number $F = \pm 2$ gauge particles Y, Y', and Y" arising within a maximal symmetry? To obtain a scenario in which the masses of these gauge particles lie in the range of $10^{10} - 10^{15}$ GeV, while the X's are as light as $\approx 10^4 - 10^5$ GeV, we proceed as follows.* Assume (following the illustrations for $[SU(4)]^4$ and $[SU(6)]^4$) that each individual family defines a distinct SU(2) within the parent symmetry G; these distinct SU(2)'s combine (or, following a terminology used before, they are *"squeezed"*) through spontaneous symmetry breaking by a relatively heavy mass scale** M_L to yield the diagonally summed SU(2), which is the SU(2) of low energy electroweak symmetry. Thus allowing for q left-handed families, we envisage the descent

$$[SU(2)_L]^q \xrightarrow[SSB]{M_1} SU(2)_L \quad . \tag{22}$$

Recall that for $[SU(4)]^4$, q = 2, while for $[SU(6)]^4$, q = 3. If the theory is left-right symmetric, there would be the corresponding "squeezing" of

---

*The discussions to follow are based on a forthcoming paper by B. Deo, J.C. Pati, S. Rajpoot, and Abdus Salam (Ref. 24).

**To realize the known universality of different families in electroweak interactions and to preserve the GIM mechanism up to its known accuracy, M_L should exceed about 10^5 GeV.

$[SU(2)_R]^q$ into a single $SU(2)_R$ or even $U(1)_R$ through a heavy mass scale M_R

$$[SU(2)_R]^q \xrightarrow{M_R} SU(2)_R \quad \text{or} \quad U(1)_R \quad . \tag{23}$$

In general, the parent symmetry G may contain distinct SU(4) color symmetries* as well, which are distinguished from each other either through the helicity of fermions on which they operate, or through the family attribute, or both. For generality, assume that there are p SU(4) color symmetries within G. To be specific, we shall furthermore assume that these are vectorial L + R symmetries. (The generalization to chiral SU(4) color is straightforward.) These p $SU(4)_{L+R}$ symmetries are "squeezed" through SSB to a single** $SU(4)_{L+R}$ by a heavy mass scale M_4. The single $SU(4)_{L+R}$ subsequently descends also spontaneously to $SU(3)^{color}_{L+R} \times U(1)_{L+R}$ via a heavy mass scale M_3. The leptoquark gauge particles X' receive their mass through M_3 with $M_X \approx M_3/10$. Thus, the color sector may break as follows:

$$[SU(4)_{L+R}]^p \xrightarrow{M_4} SU(4)_{L+R} \xrightarrow{M_3} SU(3)_{L+R} \times U(1)_{L+R} \quad . \tag{24}$$

In short, the scenario which we are led to consider for the sake of obtaining intermediate mass scales, and thereby signals for grand unification at moderate energies, is that: The families define distinct SU(2)'s and possibly even distinct SU(3) or SU(4) color symmetries at the level of the parent symmetry. The distinction is lost, and thereby universality of families defined by discrete symmetries $e \leftrightarrow \mu \leftrightarrow \tau$ emerges at low energies due to spontaneous symmetry breaking.***

---

*The fourth color is lepton number.

**Alternatively, $[SU(4)]^p$ may descent first to $[SU(3)]^p \times [U(1)]^p$, which subsequently descends to $[SU(3)] \times U(1)$. This is considered in Ref. 24.

***For instance, taking only two families e and μ, there are two distinct W's (W_e and W_μ) in the basic Lagrangian. Due to hierarchical SSB $(W_e - W_\mu)/\sqrt{2}$ acquires a heavy mass $\geq 10^5$ GeV, but $(W_e + W_\mu)/\sqrt{2}$ acquires a mass only of order 100 GeV. Hence, the low energy $e \leftrightarrow \mu$ universality. Such a picture is logically feasible, since tests of $e \leftrightarrow \mu$ universality in weak interactions extend, at best, up to 10 to 30 GeV of center-of-mass energies. *Do there exist additional W's and Z's which couple to differences of e and μ currents?* Tests of such family universality should provide an important motivation for building high energy accelerators in the 1-100 TeV region.

Such family distinctions are not realized within smaller symmetries such as SU(5), SO(10), and SU(16). But they do exist within symmetries such as

$$[SU(4)]^4,\ [SU(6)]^4,\ [SU(5)]^3 = SU(5)_e \times SU(5)_\mu \times SU(5)_\tau \subset SU(15)$$

and likewise,

$$[SO(10)]^3 \quad \text{and} \quad [SU(16)]^3 \subset SU(48) \quad .$$

We are aware that the symmetries of the latter kind are gigantic. But then, Nature appears to be proliferated anyway beyond one's imagination at the quark-lepton level. Why are there families at all? If families proliferate, why not the gauge mesons? At the present stage of our ignorance, there is no basic reason why the family universality should be an exact principle for all energies. The gigantic symmetries are the price one is paying for believing quarks and leptons are fundamental entities. We return to this problem toward the end.

With these remarks to serve as motivations, we consider the possibility that the parent symmetry G breaks spontaneously to low energy components as follows:[24]

$$\begin{array}{ccccccc}
G & \xrightarrow{M} & [SU(2)_L]^q & \times & [SU(2)_R]^q & \times & [SU(4)_{L+R}]^p \times U(1) \\
 & & \downarrow M_L & & \downarrow M_R & & \downarrow M_4 \\
 & & SU(2)_L & & U(1)_R & & SU(4)_{L+R} \\
 & & & & & & \downarrow M_3 \\
 & & & & & & SU(3)_{L+R} \times U(1)_{L+R} \\
 & & & & \underbrace{\qquad\qquad\qquad\qquad\qquad}_{} & & \\
 & & & & \downarrow m_{W_L} & & \\
 & & & & SU(2)_L \times U(1) \times SU(3)_{L+R} & &
\end{array} \tag{25}$$

Such a hierarchy leads to the following two equations via the renormalization group equations for the coupling constants:

$$\frac{\sin^2\theta_0 - \sin^2\theta_W}{\alpha\cos^2\theta_0} = -\frac{11\,\ln_e A}{6\pi\cos^2\theta_0} \xrightarrow[M]{\text{single}} \frac{11}{3\pi}\ln\frac{M}{\mu} \quad ,$$

$$\frac{q}{\alpha_s} - \frac{p}{\alpha}\sin^2\theta_W = -\frac{11\,\ln B}{6\pi} \xrightarrow[M]{\text{single}} -\left(\frac{3q-2p}{2}\right)\frac{11}{3\pi}\ln\frac{M}{\mu}\sin^2\theta_0 \ ,$$

$$A \equiv D^{-1}\left[\left(\frac{M}{M_L}\right)^{2q}\left(\frac{M_L}{m_{W_L}}\right)^{2}\left(\frac{M}{M_R}\right)^{2q}\left(\frac{M}{M_4}\right)^{8p/3}\left(\frac{M_4}{M_3}\right)^{8/3}\right]$$

$$B \equiv D^{-1}\left[\left(\frac{M}{M_4}\right)^{4p}\left(\frac{M_4}{M_3}\right)^{4}\left(\frac{M_3}{m_{W_L}}\right)^{3}\right]$$

$$D = \left(\frac{M}{M_L}\right)^{2q}\left(\frac{M_L}{m_{W_L}}\right)^{2} \qquad (26)$$

The reductions shown on the right sides of the two top equations correspond to a single state descent

$$G \xrightarrow{M} SU(2)_L \times U(1) \times SU(3)_{L+R}$$

for which

$$M_L = M_R = M_4 = M_3 = M \gg m_{W_L} \quad .$$

We now ask, are there solutions to these equations for some p and q which correspond to the constraints on gauge masses as well as $\sin^2\theta_W$ and α_S as listed in the relation (16)? (Note that chiral color corresponds to $p = 2$.) We find[24] that there is no solution satisfying constraint (16) for $p = q = 1$; such values of p and q correspond to SU(5) and SO(10). There is also no solution for $p = 2$, $q = 1$ which corresponds to SU(16). But there do exist solutions for $p = q = 2$; for $p = 2$, $q = 3$, and for $p = q = 3$. Such values of p and q can be obtained, for example, within $[SO(10)]^3$, $[SU(16)]^3$. These solutions and the corresponding coexistence of alternative proton decay modes are listed below:

$p = q = 2$:

- $M \sim 10^{15}$ GeV, $\quad M_4 \sim M_L \sim M_R \sim 10^{12}$ GeV
- $M_3 \sim 10^{5.5}$ GeV $\rightarrow$ $M_X \sim 10^{4.5}$ GeV
- Thus $\Delta F = 0$ and $\Delta F = -4$ can coexist, but $\Delta F = -2$ is suppressed.

$$p = 2,\quad q = 3 \quad \begin{cases} M \approx M_4 \approx 10^{12}\ \text{GeV}, \quad M_R \approx 10^{10}\ \text{GeV} \\ M_L \approx 10^{7}\ \text{GeV}, \quad M_3 \approx 10^{5}\ \text{GeV} \Rightarrow M_X \sim 10^{4}\ \text{GeV} \\ \Delta F = 0,\ -2\ \textit{and}\ -4\ \text{can coexist} \end{cases}$$

$$p = q = 3 \quad \begin{cases} M \approx M_R \sim 10^{12}\ \text{GeV}, \quad M_4 \approx 10^{9}\ \text{GeV}, \quad M_L \approx 10^{6}\ \text{GeV} \\ M_3 \approx 10^{5}\ \text{GeV} \Rightarrow M_X \sim 10^{4}\ \text{GeV} \\ \Delta F = 0,\ -2\ \textit{and}\ -4\ \text{can coexist} \end{cases} \tag{27}$$

We thus see that within maximal symmetries permitting intrinsic family distinctions, the proton can decay through alternative decay modes as claimed in the introduction.* It is also worth noting that for three families with $p = q = 3$, the family universality of weak interactions can disappear at an energy scale of order 100 TeV, corresponding to $M_L \approx 10^6$ GeV.

5. A SUMMARY OF THE FIRST PART

We raised two questions:

i) Is it conceivable that the basic idea of quark-lepton unification may be tested tangibly through manifestation of exotic quark-lepton interactions in the conceivable future? This has been answered in the affirmative. A number of unification models permit at least the leptoquark X gauge particles to possess a mass in the 10 - 100 TeV region. We need Isabelle, ISR and the immediate successors thereof, as well as improved cosmic ray studies, to see the effects of the X particle. These may be seen, for example, through enhanced lepton pair production in pp and $\bar{p}p$ processes.

---

*Recently, Weinberg (Ref. 21) has noted that in addition to the $\Delta(B-L) = 0$, $\Delta F = -4$ mode $(p \to e^+\pi^0)$ there can be only *one* other proton decay mode satisfying either $\Delta F = 0$, $\Delta F = -2$, or $\Delta F = -6$. He was led to this observation by arguing that the $\Delta F = 0,\ -2,$ or -6 processes, which are mediated by intermediate mass scales $M_I \ll M \sim 10^{14}$ GeV would have rates $\sim\alpha T$ or $\alpha^2 T$ in the early Universe at temperatures in the range $M \gg T \gg M_I$. Such processes with rates exceeding the rate of expansion of the Universe would be in thermal equilibrium, and therefore wipe out any baryon excess generated in earlier epochs (due to $\Delta F = -4$ processes), unless a specific linear combination $(B + aL)$ is absolutely conserved. We observe that these arguments apply only if B, L, F violations are intrinsic rather than spontaneous. For the latter case, the violations disappear for temperatures $T > M_I$. Thus any baryon excess generated before this epoch is not wiped out. Thus there is no conflict between coexistence of $\Delta F = 0,\ -2$ *and* -4 modes for proton decay on the one hand and the observation of baryon excess on the other.

ii) Can the variety of complexions for proton decay outlined in Section 1 exist and coexist? This question has also been answered.

To summarize:

1) The idea of quark-lepton unification is not tied to the nature of quark charges.

2) Proton decay is central to the hypothesis of quark-lepton unification. It is a reasonable expectation within most models that the lifetime of the proton should lie within the range of 10^{28} - 10^{33} years.

3) Proton decay modes can provide an important clue to the underlying design of grand unification. For example, observation of a $\Delta F = 0, -2,$ or -6 mode at any level within the conceivable future will signal the existence of intermediate mass scales, which in turn will reflect upon the nature of the parent symmetry G. In particular, the observation of the $\Delta F = 0$ mode will strongly suggest the existence of new interactions in the 10 to 100 TeV region. Thus, a search for such decay modes, if need be through second and third generation experiments, would be extremely important in that such searches would have implications for building high energy accelerators.

4) Observation of proton decay will strongly support the idea that quark and leptonic matters are ultimately of the same kind, though this has no bearing on the question of whether quarks and leptons represent the ultimate constituents of matter.

This leads to the second part of our considerations where we indicate the directions in which some of the changes might occur for the unification hypothesis, if quarks and leptons are viewed as composites of more elementary objects--the preons, and also how the preons may bind.

6. PREONS*

To resolve the dilemma of quark-lepton proliferation, it was suggested in 1974 that quarks and leptons may define only a stage in one's quest for elementarity.[26,27] The fundamental entities may more appropriately correspond to the truly fundamental *"attributes"* (charges) exhibited (or yet to be exhibited) by Nature. The fields carrying these fundamental attributes we called *"PREONS"*. Quarks and leptons** may be viewed within this picture as composites

---

*This section follows a recent paper by J.C. Pati (Ref. 25). See also remarks by Abdus Salam, Concluding talk, EPS Conference, Geneva, 1979.

**For simplicity let us proceed with the notion that lepton number is the fourth color (Ref. 2). In this case, the composite structure is as follows: $(q_u)_{r,y,b} = u + (r,y \text{ or } b) + \zeta$, while $\nu = u + \ell + \zeta$, etc. Within the preon idea leptons may, however, differ from quarks by more than one attribute. For example, we may have $\nu = u + \ell + \dot{\zeta}$, where $\dot{\zeta} \neq \zeta$. Such variants will be considered elsewhere.

of a set of preons consisting, for example, of m elementary *"flavons"* (f_i) plus n elementary *"chromons"* (C_α). The flavons carry only flavor, but no color, while the chromons carry only color, but no flavor. If both flavons and chromons carry spin ½, one needs to include a third kind of spin-½ attribute (or attributes) in the preon set, which for convenience we shall call *"spinons"* (ζ); these serve to give spin ½ to quarks and leptons,* but may in general serve additional purposes. The quarks and leptons are in the simplest case composites of one flavon, one chromon, and one spinon, plus the "sea". If the μ and τ families are viewed to differ from the e family only in respect of an "excitation quantum number" or degeneracy quantum number, which is lifted by some "fine or hyperfine" interaction, then only seven preons consisting of (u,d,r,y,b,ℓ and ζ) suffice to describe the 24 quarks and leptons of three families (and possibly others yet to be discovered).

For this reason, the preon idea appears to be attractive. But can it be sustained dynamically? The single most important problem which confronts the preon hypothesis is this: What is the nature and what is the origin of the force which binds the preons to make quarks and leptons?

Our first observation, following the work of one of us,[35] is that ordinary "electric" type forces** --abelian or non-abelian--arising within the grand unification hypothesis are inadequate to bind preons to make quarks and leptons unless we proliferate preons much *beyond* the level depicted above.

The argument goes as follows: Since quarks and leptons are point-like--their sizes are smaller than 10^{-16} cm, as evidenced (especially for leptons) by the $(g-2)$ experiments--it follows that the preon binding force F_b must be strong or superstrong at short distances $r \lesssim 10^{-17} - 10^{-18}$ cm corresponding to running momenta $Q \gtrsim 1$ to 10 TeV. (Recall for comparison that the chromodynamic force generated by the $SU(3)_{color}$ symmetry is strong ($\alpha_c \gtrsim 1$) only at distances of order 1 Fermi, which correspond to the sizes of the known hadrons.) This says that the symmetry generating the preon binding force must lie outside of the familiar $SU(2) \times U(1) \times SU(3)_{col}$ symmetry.

Now, consistent with our desire to adhere to the grand unification hypothesis, we shall assume that the preon binding force F_b derives its origin either intrinsically or through the spontaneous breakdown of a grand unifying symmetry G. Thus, either the basic symmetry G is of the form $G_k \times G_b$ with G_k generating the known electroweak-strong forces and G_b generating the preon-binding forces (in this case G_k and G_b are related to each other by a discrete symmetry, so as to permit a single gauge coupling constant); or the

*With the spinon present the flavons and chromons can carry integer spin 0 or 1.

**By "electric" type forces we mean forces whose effective coupling strength is of order $\alpha \approx 1/137$ at the unification point M.

unifying symmetry G breaks spontaneously as follows:

$$G \xrightarrow{\text{SSB}} G_k \times G_b \times [\text{possible } U(1) \text{ factors}] \quad . \tag{28}$$

In the second case, G_k need not be related to G_b by discrete symmetry. But in either case, G_k contains the familiar $SU(2)_L \times U(1)_{EW} \times SU(3)_{color}$ symmetry and therefore the number of attributes (N_k) on which G_k operates needs to be *at least* 5. This corresponds to having two flavons (u,d) plus three chromons (r,y,b). To incorporate a leptonic chromon ℓ and possibly also the spinon ζ, N_k may need to be at least 7; but for the present we shall take conservatively $N_k \geq 5$.

Now consider the size* of G_b. On the one hand, the effective coupling constant $\bar{g}_b$ of the binding symmetry G_b is equal to the effective coupling constant $\bar{g}_c$ of the familiar SU(3)-color symmetry (up to embedding factors[10] like $1/\sqrt{2}$ or $1/\sqrt{3}$, etc.) at the unification mass scale $M \gg 10^4$ GeV. On the other hand, $\bar{\alpha}_b \equiv \bar{g}_b^{-2}/4\pi$ needs to exceed unity at a momentum scale $\mu_b \geq 1$ to 10 TeV, where the chromodynamic coupling constant $\bar{\alpha}_c \ll 1$. It therefore follows (assuming that the embedding factor mentioned above is unity) that G_b is much larger than SU(3).** Using renormalization group equations for variations of the coupling constants $\bar{\alpha}_b$ and $\bar{\alpha}_c$, one may verify that G_b minimally is SU(5) and correspondingly the dimension N_b of the space on which G_b operates is minimally five.

Now the preons $\{P_i\}$ which bind to make quarks and leptons must be non-trivial with respect to *both* G_k and G_b. Since each of G_k and G_b requires for their operations a space which is *minimally* five dimensional, it follows that the number of preons N_p needed (under the hypothesis above) is minimally $N_k \times N_b \geq 25$:

$$N_p \geq N_k \times N_b \geq 5 \times 5 = 25 \quad . \tag{29}$$

We may consider relaxing the assumption that the embedding factor is unity. This would permit the ratio $[\bar{g}_b(\mu)/\bar{g}_c(\mu)]_{\mu=M}$ to be a number like $\sqrt{2}$ or $\sqrt{3}$, for example. In turn, this can result in a reduction in the size of G_b. But simultaneously such a step necessitates an increase in the size of G_k or effectively of the number N_k with the result that the minimal number of preons needed, $N_p \geq N_k \times N_b$, is not reduced below twenty-one.

---

*In these considerations we assume all along that the conventional perturbative renormalization group approach applies to the variations of all running coupling constants down to such momenta, where they are small $(g_i^2/4\pi \lesssim 0.3)$. (See Ref. 6.)

**This, incidentally, excludes the possibility that G_b is abelian.

This number 25 (or 21) representing the minimal number of preons needed already exceeds or is close to the number of quarks and leptons which we need at present, which is 24. And if we include, more desirably, the leptonic chromon ℓ and the spinon ζ in the preonic degrees of freedom, the number of preons needed would increase to 35 (or 27).

Such a proliferation of preons defeats from the start the very purpose for which they were introduced--economy. In turn, this poses a serious dilemma. On the one hand, giving up the preon idea altogether and living with the quark-lepton system as elementary runs counter to one's notion of elementarity and is thus unpalatable. On the other hand, giving up the grand-unification hypothesis is not aesthetically appealing.

Because of this impasse, it has recently been suggested[25] that the preons carry not only electric but also magnetic charges and that their binding force is magnetic in nature. The two types of charges are related to each other by the familiar Dirac-like quantization conditions[28,29] for charge-monopole or dyon systems, which imply that the magnetic coupling strength $\alpha_m \equiv g_m^2/4\pi$ is $O(1/\alpha_e) \approx O(137)$ and thus is superstrong. In other words, the magnetic force can arise through an abelian U(1)-component within the unification hypothesis (as remarked further at the end), and yet it can be superstrong. This is what gives it the power to bind preons into systems of small size without requiring a proliferation. Quarks and leptons do not exhibit this superstrong force because they are magnetically neutral (see remarks below).

We first discuss the consistency of this idea with presently known phenomena from a qualitative point of view and later indicate the possible origin of this magnetic force.

1) Since the electric fine structure constant $\alpha_e = e^2/4\pi$ varying with running momentum remains small $\approx 10^{-2}$ almost everywhere (at least up to momenta $\sim 10^{14}$ GeV and therefore up to distances $\sim 10^{-28}$ cm), the magnetic "fine structure" constant $\alpha_m \equiv g^2/4\pi$ related to α_e by the reciprocity relations is superstrong even at distances as short as 10^{-28} cm (if not at $r \to 0$). It is this strong short-distance component of the magnetic force which makes quarks and leptons so point-like with sizes $r_0 \ll 10^{-16}$ cm. Their precise size would depend upon the dynamics of the superstrong force which we are not yet equipped to handle. For our purposes, we shall take r_0 to be as short as perhaps $1/M_{planck} \sim 10^{-33}$ cm but as large as perhaps 10^{-18} cm (i.e., $r_0 < 10^{-18}$ cm).

2) Quarks and leptons do not exhibit even a trace of the superstrong interactions of their constituents because they are magnetically neutral composites of preons and their sizes are small compared to the distances $R \geq 10^{-16}$ cm which are probed by present high energy experiments.

3) We mention in passing that had we assumed, following Schwinger,[29] that quarks (rather than preons) carry magnetic charges, we would not

understand why they interact so weakly at short distances as revealed by deep inelastic ep-scattering.

4) Due to their extraordinarily small sizes, it can also be argued[25] that low energy parameters such as $(g-2)$ of leptons would not show any noticeable departures from the normal expectations. Similar remarks apply to the P and T violations for quarks and leptons, which would be severely damped in spite of large P and T violations for preons carrying electric and magnetic charges.

What can be the possible origin of magnetic charges of preons? The origin could, perhaps, be topological.[30,31] Spontaneous breaking of the non-abelian preonic local symmetry G_p to lower symmetries may generate monopoles or dyons. Such a picture would be attractive if, in particular, it could generate spin-½ monopoles (in addition to spin 0 and spin 1) and assign electric and magnetic charges to the originally introduced spin-½ fields and their topological counterparts.

There is a second alternative, which is the simplest of all in respect of its gauge structure. Assume that the basic Lagrangian of the preons is generated simply by the abelian symmetry $U(1)_E \times U(1)_M$. The $U(1)_E$ generates the "electric" and $U(1)_M$ the "magnetic" interactions of preons. Subject to subsidiary conditions, the theory generates only one photon coupled to electric as well as magnetic charges.[32] The charges are constrained by the Dirac quantization condition. *In this model, the basic fields are only the spin-½ preons and the spin-1 photon.* The strong magnetic force binds preons to make spin-½ quarks and leptons as discussed earlier. Simultaneously, it makes spin-1 and spin-0 composites of even numbers of preons (including anti-preons), which also have very small sizes like the quarks and leptons. The spin-0 and spin-1 fields carry charges and interact with quarks and leptons as well as among themselves. The use of a recently suggested "theorem"[33] would then suggest that their effective interactions must be generated from a local non-abelian symmetry of the Yang-Mills type, which is broken spontaneously, in order that they may be renormalizable. The spin-0 composites will now play the role of Higgs fields.[34] It is amusing that if this picture can be sustained, the proliferated non-abelian quark-lepton gauge structure $G_{(q,\ell)}$, with the associated spin-½, spin-1, as well as spin-0 quanta, may have its origin in the simplest interaction of all: Electromagnetism defined by the abelian symmetry $G_P = U(1)_E \times U(1)_M$.

The idea of the magnetic binding of preons and its origin needs to be further developed. What is argued here is that within the unification context a magnetic binding of preons appears to be called for if we are not to proliferate preons unduly.[35]

We wish to thank Dibhuti Deo, Victor Elias, Subhas Rajpoot, and especially John Strathdee for several stimulating discussions. The research of J.C.P. was supported in part by the U.S. National Science Foundation and in part by the John Simon Guggenheim Memorial Foundation Fellowship.

REFERENCES

1. J.C. Pati and Abdus Salam, "Lepton hadron unification" (unpublished), reported by J.D. Bjorken in the *Proceedings of the 15th High Energy Physics Conference* held at Batavia, Vol. 2, p. 304, September, 1972; J.C. Pati and Abdus Salam, *Phys. Rev.* D8, 1240 (1973).

2. J.C. Pati and Abdus Salam, *Phys. Rev. Lett.* 31, 661 (1973); *Phys. Rev.* D10, 275 (1974); *Phys. Lett.* 58B, 333 (1975).

3. H. Georgi and S.L. Glashow, *Phys. Rev. Lett.* 32, 438 (1974).

4. J.C. Pati, *Proc. Seoul Symp.* 1978.

5. T. Goldman and D. Ross, Caltech Preprint (1979).

6. H. Georgi, H. Quinn, and S. Weinberg, *Phys. Rev. Lett.* 33, 451 (1974).

7. See L. Sulak, *Proceedings of the Erice Workshop* (March 1980) and M. Goldhaber, *Proceedings of the New Hampshire Workshop* (April 1980). The Harvard-Purdue-Wisconsin experiment is the other parallel experiment of comparable magnitude in progress (private communications, D. Cline and C. Rubbia).

8. F. Reines and M.F. Crouch, *Phys. Rev. Lett.* 32, 493 (1974).

9. J.C. Pati, Abdus Salam, and J. Strathdee, *Il Nuovo Cimento* 26A, 77 (1975); J.C. Pati, *Proc. Second Orbis Scientiae*, Coral Gables, Fla., p. 253, Jan. 1975; J.C. Pati, S. Sakakibara and Abdus Salam, ICTP, Trieste, Preprint IC/75/93, unpublished.

10. V. Elias, J.C. Pati, and Abdus Salam, *Phys. Rev. Lett.* 40, 920 (1978).

11. See M. Gell-Mann, P. Ramond, and R. Slansky, *Rev. Mod. Phys.* 50, 721 (1978); P. Langacker, G. Segre, and H.A. Weldon, *Phys. Lett.* 73B, 87 (1978).

12. H. Fritzsch and P. Minkowski, *Ann. Phys.* (NY) 93, 193 (1975); H. Georgi, *Proc. Williamsburg Conf.*, 1974.

13. See, for example, R.N. Mohapatra and G. Senjanovic, *Phys. Rev.* D20, 3390 (1979).

14. J.C. Pati and Abdus Salam, *Phys. Rev.* D10, 275 (1974); R.N. Mohapatra and J.C. Pati, *Phys. Rev.* D11, 566 (1975); G. Senjanovic and R.N. Mohapatra, *Phys. Rev.* D12, 1502 (1975).

15. J.C. Pati and Abdus Salam, unpublished work (1975); J.C. Pati, *Proc. Scottish Univ. Summer School*, 1976.

16. J.C. Pati and Abdus Salam, forthcoming preprint IC/80/72, ICTP, Trieste.

17. F. Gursey, P. Ramond, and P. Sikivie, *Phys. Lett.* 60B, 177 (1976).

18. B. Deo, J.C. Pati, S. Rajpoot, and Abdus Salam, Preprint in preparation; Abdus Salam, *Proc. EPS Conf.*, Geneva, 1979.

19. S. Weinberg, *Phys. Rev. Lett.* 43, 1566 (1979); F. Wilczek and A. Zee, *Phys. Rev. Lett.* 43, 1571 (1979).

20. See F. Wilczek and A. Zee, Preprint UPR-0135 T; R.N. Mohapatra and R. Marshak, Preprint VPI-HEP-80/1,2; S. Glashow (unpublished).

21. S. Weinberg, Preprint HUTP-80/A023; H.A. Weldon and A. Zee, Preprint, 1980.

22. See Ref. 15.

23. V. Elias and S. Rajpoot, ICTP, Trieste, Preprint IC/78/159.

24. B. Deo, J.C. Pati, S. Rajpoot, and Abdus Salam, manuscript in preparation. For some initial ingredients of this work, see Abdus Salam, *Proc. EPS Conf.*, Geneva, 1979.

25. J.C. Pati, University of Maryland Preprint TR 80-095, April, 1980; "Magnetism as the origin of preon binding", to be published.

26. J.C. Pati and Abdus Salam, *Phys. Rev.* D10, 275 (1974); *Proc. EPS Int. Conf. on High Energy Physics*, Palermo, June, 1975, p. 171 (ed. A. Zichichi); J.C. Pati, Abdus Salam, and J. Strathdee, *Phys. Lett.* 59B, 265 (1975).

27. Several authors have worked on composite models of quarks and leptons with an emphasis on classification rather than gauge unification of forces. K. Matumoto, *Progr. Theor. Phys.* 52, 1973 (1974); O.W. Greenberg, *Phys. Rev. Lett.* 35, 1120 (1975); H.J. Lipkin, *Proc. EPS Conf.*, Palermo, June 1975, p. 609 (ed. A. Zichichi); J.D. Bjorken (unpublished); C.H. Woo and W. Krolikowski (unpublished); E. Nowak, J. Sucher, and C.H. Woo, *Phys. Rev.* D16, 2874 (1977); H. Terezawa, Preprint, Univ. of Tokyo, September 1979, INS-Rep-351 (a list of other references may be found here). H. Harari, *Phys. Lett.* 86B, 83 (1979) and M.A. Shupe, *Phys. Lett.* 86B, 78 (1979) have recently proposed the most economical model of all, but with a number of dynamical assumptions, whose bases are not clear.

28. P.A.M. Dirac, *Proc. Roy. Soc.* (London) A133, 60 (1931); *Phys. Rev.* 74, 817 (1948).

29. J. Schwinger, *Phys. Rev.* 144, 1087 (1966); D12, 3105 (1975); *Science* 165, 757 (1969); D. Zwanziger, *Phys. Rev.* 176, 1489 (1968); R.A. Brandt, F. Neri, and D. Zwanziger, *Phys. Rev. Lett.* 40, 147 (1978).

30. G. 't Hooft, *Nucl. Phys.* B79, 276 (1974); A.M. Palyakov, *JETP Lett.* 20, 194 (1974).

31. C. Montonen and D. Olive, *Phys. Lett.* B72, 117 (1977); P. Goddard, J. Buyts, and D. Olive, *Nucl. Phys.* B125, 1 (1977).

32. The formalism may follow that of D. Zwanziger, *Phys. Rev.* D3, 880 (1971).

33. M. Veltman (unpublished); See J. Ellis, M.K. Gaillard, L. Maiani, and B. Zumino, Preprint LAPP-TH-15, CERN Th 2841.

34, In such a picture there would be a natural reason why electric charge may be absolutely conserved and correspondingly the photon may remain truly massless, despite spontaneous symmetry breaking, since the photon is distinguished by the fact that it is responsible for the very existence of the composite Higgs particles which trigger spontaneous symmetry breaking.

35. The arguments presented here are independent of the detailed structure of the preon model. If the suggestions of H. Harari (*Phys. Lett.* 86B, 83 (1979)) and M. Shupe (*Phys. Lett.* 86B, 87 (1979)) for generating a "degenerate" color degree of freedom can be sustained dynamically, it would be an attractive economical picture. The arguments presented here would suggest that such an economical set of preons (or pre-preons) must carry magnetic charges and that their binding must owe its origin to such charges.

EVIDENCE FOR NEUTRINO INSTABILITY*

F. Reines

Department of Physics
University of California
Irvine, CA 92717

Abstract

We report evidence for neutrino instability obtained from the analysis of data taken on the charged and neutral current branches of the reaction

$$\bar{\nu}_e + d \begin{pmatrix} \nearrow n + n + e^+ & \text{c.c.} \\ \searrow n + p + \bar{\nu}_e & \text{n.c.} \end{pmatrix}$$

at 11.2 meters from a 2000 MW Savannah River reactor. Tests of the results show them to be consistent with data on $\bar{\nu}_e + p \rightarrow n + e^+$ also taken at 11.2 meters and 6.0 meters.

If we assume that, for the low energies involved (< 10 MeV), $\bar{\nu}_e$ reflects contributions from only two neutrino states of masses m_1 and m_2, then the deuteron data taken by themselves imply limits on $\Delta = (m_1^2 - m_2^2)$ and on $\sin^2 2\theta$, where θ is the "mixing angle" of Pontecorvo.

As part of the University of California, Irvine program in reactor neutrino physics, we have made a measurement of the cross-sections for the charged and neutral current branches of the interactions of fission reactor neutrinos with deuterons:

$$\bar{\nu}_e + d \begin{matrix} \nearrow n + n + e^+ & \text{c.c.} \\ \searrow n + p + \bar{\nu}_e & \text{n.c.} \end{matrix} \qquad (1)$$

In our experiment, which is described in greater detail elsewhere,[1] 268 kg of D_2O is exposed to the neutrino flux from a 2000 MW reactor located 11.2 meters distant. Immersed in the D_2O are ^3He filled neutron proportional counters. The D_2O target is enclosed in a lead and cadmium shield and surrounded by liquid scintillator anticoincidence counters. Table I is an updated summary of data taken since the work reported by Pasierb et al.[1]

*Joint work with H.W. Sobel and E. Pasierb. Work supported in part by the United States Department of Energy

Observed Number of Neutrons	Reactor "on" (events/day)	Reactor "off" (events/day)	"on" - "off" (events/day) (reactor associated)
1	425 ± 2	351 ± 3	74 ± 4
2	54.00 ± 0.85 (74.4 days live)	50.16 ± 1.02 (47.8 days live)	3.84 ± 1.33

Table 1. Data summary for $\nu_{reactor}$ + d experiment.

From these data, the measured detection efficiences averaged over the detector: $\overline{\eta^2} = 0.112 \pm 0.009$ for double neutron events,[2] $\overline{\eta} = 0.32 \pm 0.02$ for single neutron events, and an estimate of the reactor-associated neutron background from the $\bar{\nu}_e + p \to n + e^+$ reaction on protons in the D_2O and in the anticoincidence detectors, we deduce the ratio of charged to neutral cross sections $(\bar{\sigma}_{ccd}/\bar{\sigma}_{ncd})_{expt.}$ for $\bar{\nu}_e$ on deutrons.

The measured total singles counting rate $R_{1n} = 74 \pm 4/d$ is related to the neutral current single neutron rate R_{1n}^{ncd} by the expression

$$R_{1n} = R_{1n}^{ccd} + R_{1n}^{ncd} + R_{1n}^{ccp}$$

where

R_{1n}^{ccd} = the single neutron count rate detected from the charged current reaction on the deuteron,

R_{1n}^{ncd} = single neutron count rate due to the neutral current reaction on the deuteron,

R_{1n}^{ccp} = $9.5 \pm 3/d$ is the calculated single neutron count rate due to the reaction of $\bar{\nu}_e$ on protons, and other smaller background contributions.

R_{2n}^{ccd} = $3.84 \pm 1.33/d$ is the double neutron count rate detected from the charged current reaction on the deuteron.

Let R^{ccd} and R^{ncd} be the true charged and neutral current rates for reactor neutrinos and deuterons. Then,

$$R_{2n}^{ccd} = \overline{\eta^2}\, R^{ccd} \,, \qquad R_{1n}^{ccd} = 2(\overline{\eta'} - \overline{\eta'^2})\, R^{ccd} \,,$$

$$R_{1n}^{ncd} = \overline{\eta'}\, R^{ncd} \,.$$

In addition, $\overline{\eta'} = 0.89\ \overline{\eta}$, $\overline{\eta'^2} = 0.89^2\ \overline{\eta^2}$, where the factor 0.89 arises from a choice of ^{3}He counter discriminator settings to reduce the reactor independent single neutron background. Since the ratio of the charged to neutral current cross-sections per fission $\bar{\nu}_e$ on deutrons is

$$\left(\frac{\bar{\sigma}_{ccd}}{\bar{\sigma}_{ncd}}\right)_{expt.} = \frac{R^{ccd}}{R^{ncd}}$$

where

$$\bar{\sigma} = \frac{\int_{E_1}^{\infty} \sigma(E)n(E)\,dE}{\int_{E_1}^{\infty} n(E)\,dE}$$

and $n(E)$ is the neutrino spectrum and E_1 is the energy above which we choose to measure the cross-section, $\sigma(E)$, then

$$\left(\frac{\bar{\sigma}_{ccd}}{\bar{\sigma}_{ncd}}\right)_{expt.} = \frac{R_{2n}^{ccd}}{\frac{\overline{\eta^2}}{\overline{\eta'}}\,(R_{1n} - R_{1n}^{ccp}) - 2\left(1 - \frac{\overline{\eta'^2}}{\overline{\eta'}}\right) R_{2n}^{ccd}}$$

Inserting numerical values, we find

$$\left(\frac{\bar{\sigma}_{ccd}}{\bar{\sigma}_{ncd}}\right)_{expt.} = 0.191 \pm 0.073 \quad .$$

We note that this ratio is determined from measurements taken concurrently and that various geometrical and instrumental stability factors cancel, including, incidentally, the ν flux.

We now compare this experimentally determined ratio with that which is to be expected on theoretical grounds, i.e., with

$$\left(\frac{\bar{\sigma}_{ccd}}{\bar{\sigma}_{ncd}}\right)_{th.}$$

so forming a ratio of ratios

$$\mathscr{R} = \frac{\left(\frac{\bar{\sigma}_{ccd}}{\bar{\sigma}_{ncd}}\right)_{expt.}}{\left(\frac{\bar{\sigma}_{ccd}}{\bar{\sigma}_{ncd}}\right)_{th.}} \qquad (2)$$

The denominator ratio has some very interesting properties:

a) It is independent of the reactor neutrino absolute normalization.

b) It is insensitive to the precise shape of the reactor neutrino spectrum.

c) $\bar{\sigma}_{ncd}$ is independent of neutrino type and for the low energy reactor neutrinos (< 10 MeV) is purely axial vector and hence does not involve the Weinberg angle.

d) The coupling constants are well known,[3] i.e., to better than 10%. Their ratio is known to $\lesssim 5\%$.

e) $\bar{\sigma}_{ccd}$ is dependent only on the $\bar{\nu}_e$ portion of the total neutrino flux incident on the detector.

As a result, a value of $\mathcal{R}$ below unity would signal the instability of $\bar{\nu}_e$ as it traversed the distance (centered in this deuteron experiment at 11.2 meters) from its origin to the detector.

Evaluating $(\bar{\sigma}_{ccd}/\bar{\sigma}_{ncd})_{th.}$ by integration over the $\bar{\nu}_e$ spectrum from fission calculated by either Avignone and Greenwood[4] or Davis et al.[4] we find

$$\text{Avignone spectrum:} \qquad \left(\frac{\bar{\sigma}_{ccd}}{\bar{\sigma}_{ncd}}\right)_{th.} = 0.44$$

$$\text{Davis spectrum:} \qquad \left(\frac{\bar{\sigma}_{ccd}}{\bar{\sigma}_{ncd}}\right)_{th.} = 0.42$$

Accordingly, we find

$$\mathcal{R} = 0.43 \pm .17 \quad \text{or} \quad 0.45 \pm .17 \quad ,$$

a 3.2 to 3.4 standard deviation departure from unity.

Other Reactor Data

Having found a value for $\mathcal{R}$, we now turn to an examination of other reactor data to data to test for consistency. Thus far, several tests have been made.

1) We have performed an analysis of the reaction $\bar{\nu}_e + p \rightarrow n + e^+$ measured with a different detector at the same 11.2 meter distance from the reactor.[5] This analysis has yielded a preliminary $\bar{\nu}_e$ spectrum. Although use of this spectrum, which is different from those of Davis or Avignone, introduces relative errors between the two experiments, the value for $\mathcal{R}$ is independent of the normalization and becomes $0.53 \pm .20$, a 2.4 standard deviation effect.

2) All ratios of experimentally determined rates to predicted rates, other than that for $\mathcal{R}$, are markedly dependent on the reactor neutrino

spectrum. For this reason, we do not consider the precise values of these ratios to be in themselves significant. They can, however, be used to test consistency with $\mathscr{R}$, and to this end, data from several reactor experiments: the 11.2 meter proton[5] and deuteron[1] data mentioned above, and the 6.0 meter proton data from an earlier experiment[6] were employed.

Comparisons of the observed data with predictions using the various reactor neutrino spectra are given in Table II.

Distance from core (meters)	Reaction	Neutrino Detection Threshold (MeV)	Ratio		
			Avignone Spectrum	Davis Spectrum	Measured $\bar{\nu}_e$ Spectrum (preliminary)
11.2	ccd/ncd	4.0 (ccd) 2.2 (ncd)	.43 ± .17	.45 ± .17	.53 ± .20
11.2	ncd	2.2	.84 ± .13	1.09 ± .16	1.20 ± .20
11.2	ccd	4.0	.36 ± .13	.49 ± .18	.64 ± .24
11.2	ccp	4.0	.76 ± .08	.99 ± .10	-
11.2	ccp	6.0	.46 ± .05	.66 ± .07	-
6.0	ccp	1.8	.70 ± .10	.91 ± .13	-
6.0	ccp	4.0	.83 ± .12	1.05 ± .16	-

Table II. Summary of results for the ratio $\bar{\sigma}_{expt.}/\bar{\sigma}_{th.}$

Interpretation of the Data

The idea of neutrino oscillations was proposed in analogy with the K° system by Pontecorvo.[7] If we interpret these results in terms of neutrino oscillations,[8] and only two base states are assumed to be involved for these low energy neutrinos,[9] then we find from the value of $\mathscr{R}$ a relationship between the allowed values of $\sin^2 2\theta$ (where θ is the mixing angle) and $\Delta = m_1^2 - m_2^2$ where m_1 and m_2 are the masses to be associated with these two states. The allowed regions are shown in Figure 1.

In the same way, allowed regions can be drawn for each of the experiments listed in Table II.[10] In these cases, however, any conclusions are once again heavily dependent upon the chosen reactor neutrino spectrum. If we use the Avignone spectrum, there is an overlapping region consistent with all of the experiments. This yields

$$0.5 \leq \sin^2 2\theta \leq 0.8 \qquad (32° < \theta < 22°)$$

and

$$0.8 \leq \Delta(\text{ev}^2) \leq 1.0 \quad .$$

If we use the Davis spectrum, there is no overlapping region at the level of one standard deviation.[11] It is evident that further experimentation with $\bar{\nu}_e$ at reactors, e.g., measurements of the $\bar{\nu}_e$ spectrum as a function of distance[12] via the reaction $\bar{\nu}_e + p \rightarrow n + e^+$ and further study of the reaction $\bar{\nu}_e + d$,[13] would be helpful in further elucidating the effect here observed[14] as would experiments at accelerators and with cosmic rays[15] where the higher energies would stretch out the distance scale and elicit more details of the phenomenon.

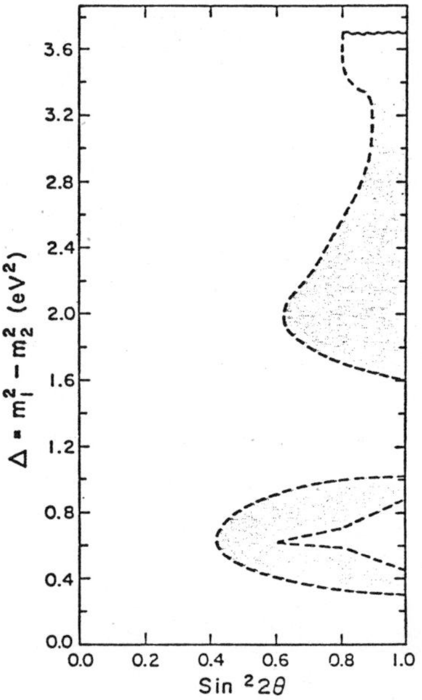

Figure 1. Allowed regions of Δ and $\sin^2 2\theta$ for $R = 0.43 \pm 0.17$.

$$\equiv \frac{\left(\frac{CC}{NC}\right)_{expt.}}{\left(\frac{CC}{NC}\right)_{th.}} \qquad \begin{cases} CC: & \bar{\nu} + d \rightarrow e^+ + n + n \\ NC: & \bar{\nu} + d \rightarrow \bar{\nu} + n + p \end{cases}$$

↑ with no mixing

We wish to acknowledge helpful discussions with members of the University of California, Irvine Neutrino Group, in particular, W.R. Kropp and H.H. Chen, and with L.W. Alvarez, R.M. Ahrens, F. Avignone, F. Boehm, W.A. Fowler, S. Glashow, A. Hahn, J. Learned, R. Nix, A. Salam, J.J. Sakurai, J. Schultz, D. Silverman, A. Soni, S. Weinberg, J. Weneser, J. Vuilleumier, and P. Vogel.

T. Godbee was most helpful in operating the equipment and film scanning. Special thanks are due to the late August A. Hruschka for his mechanical design and construction of the detector.

We are grateful to our hosts at the Savannah River Plant of the E.I. Dupont de Nemours Company where these measurements were made.

REFERENCES

1. E. Pasierb, H.S. Gurr, J. Lathrop, F. Reines, and H.W. Sobel, *Phys. Rev. Lett.* 43, 96 (1979). As will emerge from the present discussion, it is more nearly correct to label the neutral current branch as

$$\nu_{reactor} + d \rightarrow n + p + \nu_{reactor} \quad .$$

2. We wish to thank A. Hahn for pointing out an effect which changed this quantity by about 10%.

3. R.M. Ahrens and L. Gallaher, *Phys. Rev.* D20, 2714 (1979), and private communication (1980).

4. F.T. Avignone, III and Z.D. Greenwood (Univ. of South Carolina Preprint, Feb. 1979); B.R. Davis, P. Vogel, F.M. Mann, and R.E. Schenter, *Phys. Rev.* D19, 2259 (1979). Recent data on fission fragments (neutron rich rubidium and cesium isotopes) far off the line of stability, directly measured by M. Epherre et al., *Phys. Rev.* C19, 1504 (1979) near the peaks in the fission fragment mass distribution, give masses somewhat higher than those calculated by Janecke (J. Janecke, *At. Nucl. Data Tables* 17, 455 (1976) and used by Avignone and Davis. Accordingly, the $\bar{\nu}_e$ energies are slightly larger than those given by both groups. We are indebted to R. Nix for a discussion of this point.

5. The positron spectrum from the reaction $\bar{\nu}_e + p \rightarrow n + e^+$ was measured concurrently with the operation of the elastic scattering detector

$$(\nu_{reactor} + e^- \rightarrow \nu_{reactor} + e^-) \ :$$

F. Reines, H.S. Gurr, and H.W. Sobel, *Phys. Rev. Lett.* 37, 315 (1976). Analysis of this portion of the data is proceeding and will be published elsewhere.

Because the elastic scattering of reactor neutrinos is suspected to be sensitive to neutrino instability, we are subjecting our published and subsequent data to careful scrutiny.

6. F.A. Nezrick and F. Reines, *Phys. Rev.* 142, 852 (1965).

7. B. Pontecorvo, *Soviet Physics JETP* 26, 984 (1968); V. Gribov and B. Pontecorvo, *Phys. Lett.* 28B, 493 (1969).

8. For a discussion of various theoretical aspects of low energy neutrino interactions, see the prescient article by H. Fritzsch in *Fundamental Physics with Reactor Neutrons and Neutrinos*, T. Von Egidy (ed.), Inst. of Physics, Bristol and London (1978), p. 117.

9. We note that the negative results of R. Davis, *Phys. Rev.* 97, 766 (1955), and private communication (1980) who looked at a nuclear reactor for $\nu + {}^{39}Cl \rightarrow {}^{37}Ar + e^-$ and saw $< 1/50$ of that which would be produced if the reactor ν were ν_e, can in the light of the present results be interpreted as ruling against lepton → antilepton oscillation, i.e., $\bar{\nu}_e \not\rightarrow \nu_e$.

10. An extended discussion of this interesting topic of possible ranges of Δ allowed by the deuteron and other data is in preparation.

11. Data on $\bar{\nu}_e + p \rightarrow n + e^+$ obtained by the California Institute of Technology-Grenoble group at a distance of 8.7 meters from the 57 MW research reactor of Institut Laue-Langevin seems to be consistent with these conclusions. (Private communication, F. Boehm, P. Vogel, April, 1980).

12. A moveable detector designed for this purpose by S.Y. Nakamura, W.R. Kropp, H.W. Sobel, and F. Reines is nearing completion. See also S.Y. Nakamura et al., Contribution to ν-76, Aachen.

13. An experiment to study reaction (1) in more detail is under development by a Georgia Tech-University of South Carolina group. In addition to detecting the neutrons they propose to measure as well the energy spectrum of the product protons, so yielding $\mathscr{R}$ as a function of energy. (T. Ahrens and T.P, Lang, *Phys. Rev.* C3, 979 (1971); T.P. Lang et al., in *Neutrinos 78*, E.C. Fowler (ed.), Purdue University, 1978, p. C68).

14. F. Reines, in *Unification of Elementary Forces and Gauge Theories*, D.B. Cline and F.E. Mills (eds.), Harwood, London, 1978, p. 103.

15. Some evidence for neutrino oscillations may have been obtained earlier, in the atmospheric neutrino experiments conducted in a deep South African gold mine by the Case Institute/University of Witwatersrand/ University of California, Irvine group, F. Reines, Atmospheric neutrinos as signal and background, *Proceedings of the XVI Cosmis Ray Conference*, Kyoto, August, 1979. In this paper attention was called to the paucity of neutrinos in the horizontal direction where the ratio of the observed to the expected flux of product muons was deteremined to be

$$0.62 \pm {}^{0.21}_{0.12} \quad .$$

Since the experimental value was determined to ± 10%, the major uncertainty in this ratio arises from the predicted $\nu_\mu(\bar{\nu}_\mu)$ fluxes. A reexamination of the assumptions suggests that the uncertainty assigned to the predicted result represents an overestimate and that it could, in fact, be less, yielding a ratio of

$$0.62 \pm {}^{0.17}_{0.10} \quad .$$

A more precise evaluation is underway. These observations are of particular interest since they bear on the role of higher energy neutrinos in the oscillation phenomenon. A detailed description of these experiments is given in M.F. Crouch, P.B. Landecker, J.F. Lathrop, F. Reines, W.G. Sandie, H.W. Sobel, H. Coxell, J.P.F. Sellschop, *Phys. Rev.* D18, 2239 (1978).

REVIEW OF NEUTRINO OSCILLATION EXPERIMENTS

John M. LoSecco

Physics Department
University of Michigan
Ann Arbor, MI 48109

and

High Energy Physics Laboratory
Harvard University
Cambridge, MA 02138

A number of sensitive experiments have recently been done, or are in progress, to search for evidence for neutrino oscillations. If observed, these would be clear indications for the violation of the individual lepton type quantum numbers and evidence for flavor changing currents of a kind found in grand unification models.

In general, the experiments start with a pure neutrino beam of a specific type $(\bar{\nu}_e, \nu_\mu,$ etc.). The beam evolves in time and a search is made for the presence of other neutrino types or an attenuation of the initial neutrino beam.

For example, in a two flavor theory, if one starts with a pure ν_μ beam, of momentum p, after traveling a distance ℓ the beam will have a ν_e content given by:

$$|\nu_e(\ell)|^2 = \frac{1}{2}\sin^2 2\xi \left[1 - \cos\left(\frac{m_1^2 - m_2^2}{2}\ \frac{\ell}{p}\right)\right]$$

ξ is related to the degree of lepton number violation and m_1 and m_2 are the neutrino mass eigenvalues. To get oscillations, one needs both neutrino masses and some lepton number violation. As the ν_e content grows, the ν_μ content decreases.

For small masses, or small oscillation lengths:

$$|\nu_e(\ell)|^2 \propto \left(\frac{\ell}{p}\right)^2$$

$$(m_1^2 - m_2^2)^{\frac{1}{2}} \cong 2\left[\frac{\hbar c}{\sin^2 2\xi}\ \left(\frac{p}{\ell}\right)\right]^{1/2} \left[\frac{N(\nu_e)}{N(\nu_\mu)}\right]^{1/4}$$

So different experiments can be compared by the ratio of drift distance to neutrino momentum.

Reactor experiments, which are discussed in greater detail elsewhere in the proceedings,[1] employ the lowest momentum neutrinos. The useable spectrum runs from 1.8 MeV to about 7 MeV, but only a few percent ($\sim 4\%$) are above 3 MeV. It is purely $\bar{\nu}_e$. Since the neutrino source is isotropic, reasonable drift distances are limited by the $1/\ell^2$ decrease in the flux. Since the neutrino energy is below threshold for all but neutral current, and electron charged current final states, the occurrence of oscillations must be inferred from a, possibly momentum dependent, difference between expected and observed event rate. More recently, the neutral current process[1] has been used to monitor the presence of other species of ν.

The drift distances in these experiments are from 6 to 11.2 meters.

Very sensitive experiments can be done that will give a positive signal and permit one to measure ξ. In general, these employ ν_μ beams and look for the presence of electrons.

A LAMPF[2] experiment has just been completed that is the first experiment to study the neutrinos in μ decay. The neutrino source was the decay

$$\mu^+ \rightarrow e^+ \nu_e \bar{\nu}_\mu$$

at rest. The spectra for both ν_e and $\bar{\nu}_\mu$ are well known, extending from about 25 MeV to 53 MeV. To distinguish ν_e from $\bar{\nu}_e$ two different detectors were used. The reaction

$$\bar{\nu}_e + H \rightarrow e^+ + n$$

was looked at in water and the reactions

$$\bar{\nu}_e + d \rightarrow e^+ + n + n$$

$$\nu_e + d \rightarrow e^- + p + p$$

were looked at in D_2O.

The results for the $\bar{\nu}_e$ search are shown in Fig. 1. The dashed curve represents the expected signal if all the $\bar{\nu}_\mu$ had oscillated to $\bar{\nu}_e$. With 90% confidence, less than 6.5% $\bar{\nu}_e$ was observed. This converts to a mass limit, for maximal mixing, of $< .8$ eV. Or, for complete oscillations, this implies $\xi < 11°$ (90% c.l.).

The momentum and distance scale typical of this experiment is about 35 MeV at 10 meters. $P/\ell \approx 3.5$ MeV/m.

In principle, the D_2O data (Fig. 2) can set a limit on electron neutrino oscillations. The ratio of observed to predicted events is

$$R_D = 1.09 \pm .37 \quad ,$$

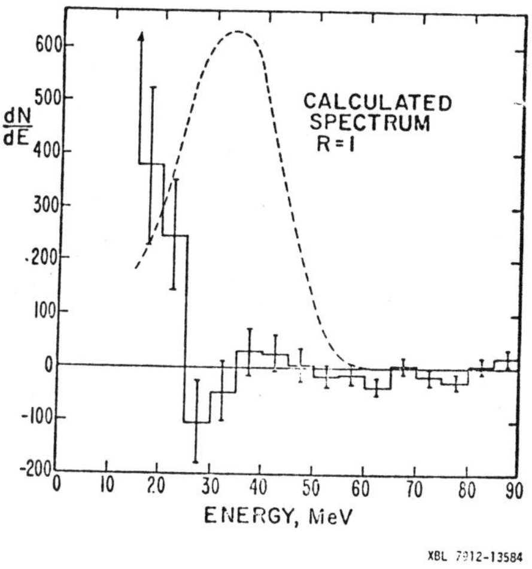

Figure 1. Search for the presence of $\bar{\nu}_e$ in the neutrinos from μ^+ decay. The dashed curve represents 100% $\bar{\nu}_\mu \to \bar{\mu}_e$ transitions.

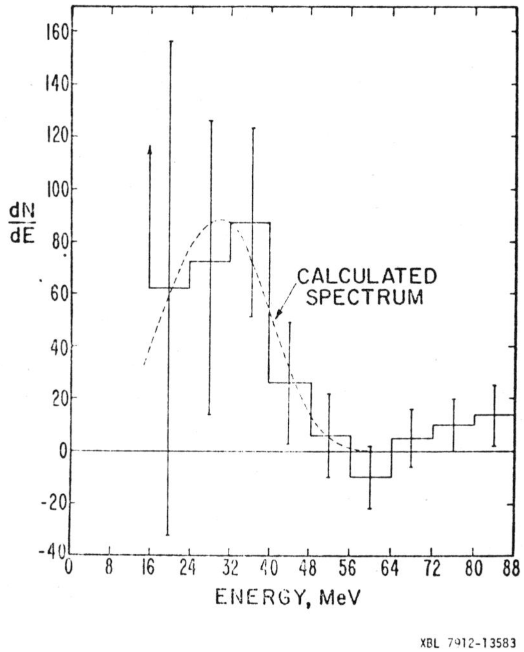

Figure 2. Comparison of predicted (dashed curve) versus observed spectrum for ν_e from μ^+ decay.

so $R_D > .61$ at 90% confidence limit. This converts to a mass limit for maximum mixing of the ν_e beam of

$$\Delta_e \equiv (m_1^2 - m_2^2)^{1/2} < 1.29 \text{ eV} \qquad (90\% \text{ c.l.}) \quad .$$

Or, for complete oscillations, this implies $\xi < 30°$.

The limitations to the sensitivity of this experiment are beam associated and cosmic ray induced background. The event rates are high and in principle background rejection could be improved.

The Gargamelle group[3] has searched for neutrino oscillations represented by ν_e and $\bar{\nu}_e$ events in their high energy data. Some ν_e and $\bar{\nu}_e$ are present from μ and K decay. These limits are about 1 eV, for maximal mixing. For complete oscillations, these data place limits of about $\xi < 10°$ with better than 90% confidence.

We are currently analyzing our experiment that should be a factor of 5 more sensitive to $\nu_\mu \to \nu_e$ oscillations.[4] This experiment makes use of a specially constructed low energy ν_μ beam at the Brookhaven A.G.S. The ν_μ's are produced from π decay from a 1.5 GeV/c primary proton beam. This yields a beam with a narrow momentum spread (Fig. 3) from 150 to 250 MeV. The distance to the detector is from 100 to 150 meters, so $P/\ell \approx 1.5$ MeV/m. This experiment searches for a positive ν_e signal from a pure ν_μ beam. Since the ν_μ beam is above muon production threshold, in principle the ν_μ flux attenuation can also be observed.

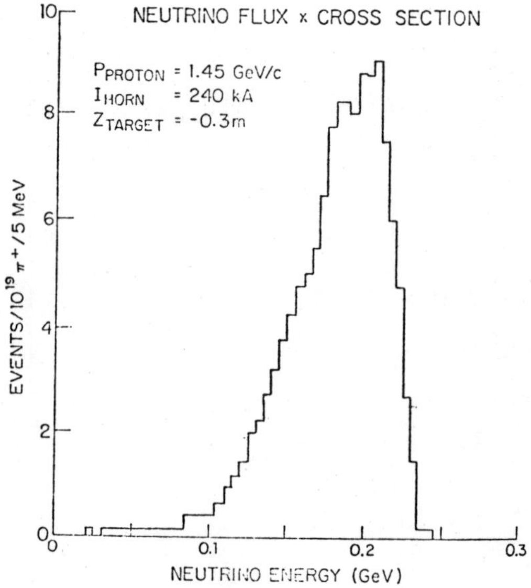

Figure 3. The pion decay neutrino spectrum used in the Brookhaven neutrino oscillations experiment.

The discrimination between μ and e events is accomplished in several ways. Most simply, because of the 106 MeV mass of the muon events involving a ν_μ will deposit only between 50 and 150 MeV in our 30 ton detector (Fig. 4). ν_e events, on the other hand, will deposit from 150 to 250 MeV of visible energy. We also have the capability of observing the muon decay which is a distinctive signature. The detector is segmented in such a way that electromagnetic showers caused by ν_e interactions can be identified and distinguished from the minimum ionizing track of a muon.

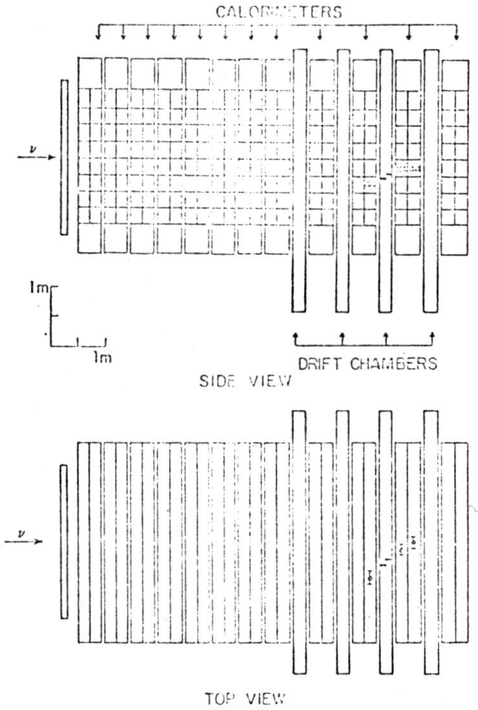

Figure 4. The Brookhaven neutrino detector

Because of the low energy of the beam, the primary background to this experiment will be cosmic rays. We make use of the timing structure of the beam to reject non-beam associated events. We can also do an out of time subtraction.

An exposure of 10^{19} protons should yield an expected rate of 40 neutrino events. At present, about half of the analysis is done.

This experiment should be able to set limits on the oscillations of $\nu_\mu \rightarrow \nu_e$ down to masses on the order of .3 eV. This is considerably better than any previous limits on $\nu_\mu \to \nu_e$ mixing. It may also be sensitive to the mixing parameter to $\xi > 13°$.

The experiments discussed here are predominantly concerned with $\nu_\mu - \nu_e$ mixing. They are sensitive to other classes of mixing, but cannot fully describe the process. Oscillation with ν_τ, or as yet unobserved neutrino types, can be bounded by the agreement between theory and experiment for experimental event rates.

REFERENCES

1. F. Reines et al., these proceedings.

2. S.E. Willis et al., *Phys. Rev. Lett.* 44, 522 (1980); Suzanne E. Willis, thesis, Yale University, 1979 (unpublished); P. Nemethy, private communication. We would like to thank Dr. Nemethy for copies of Figs. 1 and 2 and for permission to reproduce them here.

3. J. Bleitschau et al., *Nucl. Phys.* B133, 205 (1978).

4. E. Egelman et al., A study of the time evolution of a long lived ν_μ beam. Brookhaven proposal #704 (January 1977); A. Soukas et al., *Proc. AGS Fixed Target Workshop*, page 233, presented by L. Sulak, B.N.L. Report 50947 (1978); L.R. Sulak et al., *Neutrino '77*, page 280, Nauka, Moscow, 1978.

THE IRVINE-MICHIGAN-BROOKHAVEN* NUCLEON DECAY FACILITY:
STATUS REPORT ON A PROTON DECAY EXPERIMENT
SENSITIVE TO A LIFETIME OF 10^{33} YEARS
AND
A LONG BASELINE NEUTRINO OSCILLATION EXPERIMENT
SENSITIVE TO MASS DIFFERENCES OF HUNDREDTHS
OF AN ELECTRON VOLT

L. Sulak**

Randall Laboratory
University of Michigan
Ann Arbor, Michigan 48109

Abstract

We have studied the properties of, and the expected backgrounds in, a totally active 10,000 ton water Cerenkov detector located deep underground and sensitive to many of the conjectured decay modes of the nucleons in it. Identification of (π,μ) and (e,γ) secondaries, good energy resolution, and good angular resolution provide sufficient background rejection in the detector under construction to permit one to obtain significant information about several decay channels, should they be observed. If no events were recorded in the device in one year, a lower limit of $\sim 10^{33}$ years would be placed on the partial lifetime for the most distinct nucleon decay modes. Depending upon the decay channel, this is ~ 3 orders of magnitude longer than previous measurements, and is at or beyond the level suggested by many unifying theories. The sensitivity predicted for this instrument is within an order of magnitude of that achievable in an arbitrarily large detector of this general type, since known background from atmospheric neutrinos imposes an inherent limit.

We also detail the capabilities of a search for neutrino oscillations sensitive to low energy atmospheric ν_e's and ν_μ's using as a baseline the diameter of the earth. A flux independent asymmetry in the up/down ratio of the two neutrino species is the primary signal. The full 10,000 water Cerenkov detector is necessary; smaller detectors would have insufficient statistical power. For a two year exposure, the detector provides a several standard

*Supported in part by the U.S. Department of Energy. The members of the collaboration are the following: M. Goldhaber, Brookhaven National Laboratory; B. Cortez, G. Foster, L. Sulak, Harvard University and the University of Michigan; C. Bratton, W. Kropp, J. Learned, F. Reines, J. Schultz, D. Smith, H. Sobel, C. Wuest, Univsity of California at Irvine; J. LoSecco, E. Schumard, D. Sinclair, J. Stone, and J. Vander Velde, University of Michigan.

**Also Visiting Scholar, Harvard University, Cambridge, MA 02138.

deviations signal for maximal mixing of either species over the mass difference range of 10^{-3} to 10^{-1} eV. The upper end of this range coincides with the lower end of conceivable accelerator-based searches for neutrino oscillations.

1. INTRODUCTION

The Irvine-Michigan-Brookhaven Collaboration is currently building a detector which ultimately will either measure nucleon decays at a lifetime level of 10^{33} years, or establish a lower limit several times longer than that. If no events were observed in it, the limits set would be more than three orders of magnitude greater than existing ones. If the nucleon should decay at a rate which is fortunately close to the present limits (in particular for the modes favored by the unifying theories, $N \to \ell\pi$), one could record ~ 1000 event per year, study branching ratios, and, if muons were copiously produced, measure their polarization.

The detector consists of a 21 meter cube of water with phototubes placed in regular arrays on all six faces. Charged particle tracks (and π^0 mesons through their photon showers) are detected via the cones of Cerenkov light that they emit in water. Angles, energies, and particle types are determined with sufficient precision to identify nucleon decay and reject background at the required level. The detector is particularly sensitive to the decay modes expected in the unification models, e.g., $p \to e^+\pi^0$, $n \to e^+\pi^-$. The signature of these events, two "back-to-back" monoenergetic particles or showers of $\sim \frac{1}{2}$ GeV each, is distinctive relative to all backgrounds. Only atmospheric neutrino-induced $\Delta(3/2, 3/2)$ production, the limiting background, might rarely produce a similar pattern. The detector is also capable of observing baryon decays in a variety of other possible decay modes. We have checked the design with two independent Monte Carlo simulations of events and backgrounds. Detector performance and reconstruction of events are discussed.

Secondarily, the detector permits the study of other interesting phenomena taking place in its deep underground environment, chosen to shield out muon-induced background. The search for neutrino oscillations is discussed in a companion article in these proceedings, for example.

Mechanical and electronic construction is well underway. Phototubes and their associated hardware are being produced at the rate of 200 PM's per month. The mine site is in preparation. The full sized detector is expected to start operation with half of the PM's around the turn of the year.

This paper is an update of the experiment which has already been described elsewhere.[1]

2. CERENKOV PATTERNS

The initial configuration that we are constructing employs a 16×16 array of 5" hemispherical phototubes on each face of a 21 m cube of water. Charged particles with $\beta \geq 0.75$ in the detector emit Cerenkov light in cones of half-angle of $\leq 42°$ with respect to their direction. The light travels to the surface where it produces an elliptically shaped annular pattern of triggers on an array of PM tubes as ideally shown in Fig. 1a. For particles exiting the fiducial volume, the ring becomes a filled in ellipse of light. The energy (range) of a stopping particle is determined by the total number of photoelectrons collected. The distance from the decay vertex to the detector plane is measured by the outer radius of the ring, the range by the thickness of the ring, and the direction by the relative time that each PM fired. Due to the $\sim 90°$ opening angle of the cone and the long level arm across a ring, time differences (tens of ns) yield a good measurement of angles relative to the plane of PM's, as is ideally illustrated in Fig. 1b.

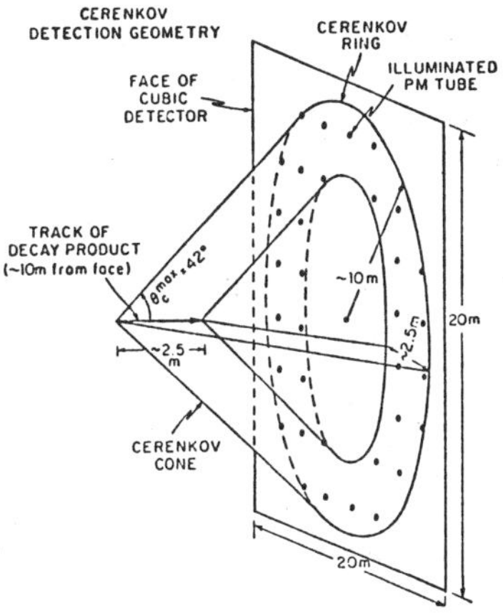

Figure 1a

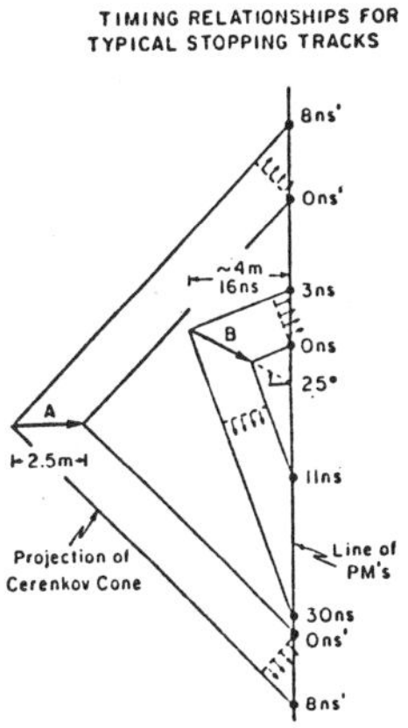

Figure 1b

In practice, the cones of light from decaying nucleons are modified by physical processes which smear the pattern. We have treated these processes in Monte Carlo simulations of various nucleon decay modes as well as for background processes. The calculations include Fermi motions, dE/dx, multiple scattering, electromagnetic shower development, Cerenkov light generation, spectral absorption in the water, photocathode spectral response, photoelectron capture efficiency, Poisson stratistics, PM time resolution, and the hemispherical geometry of the phototubes.

From studies of simulated nucleon decays, we find that for $N \to \pi\ell$ decays, charged pions and muons have typical radiating ranges of 75 to 175 cm, whereas $e^{\pm}$ and π^0 showers have relevant lengths of 100 to 300 cm. The ratio of Cerenkov light emitted, from a 500 MeV $\pi^{\pm}$, $\mu^{\pm}$, and $e^{\pm}$ (or π^0) is about

1 : 1.3 : 3. The $e^{\pm}$ or π^{0}'s generally produce a large number of short electron tracks (which are highly efficient Cerenkov radiators) and form a pattern of PM hits that seldom is a perfect ring. Figure 2 shows the phototube amplitude response pattern from a typical nucleon $p \to e^{+}\pi^{0}$ decay in the fiducial volume. Even though the three ring patterns are somewhat filled in, their circular nature is still apparent. The π^{0} decay shown is unusually asymmetric. Although we have not exhibited it in this figure, the timing information for each hit PM event supplies a powerful additional constraint on event reconstruction. Our computer codes reconstruct separate tracks from patterns such as these; the reconstructed errors for the event in Fig. 2 are indicated there.

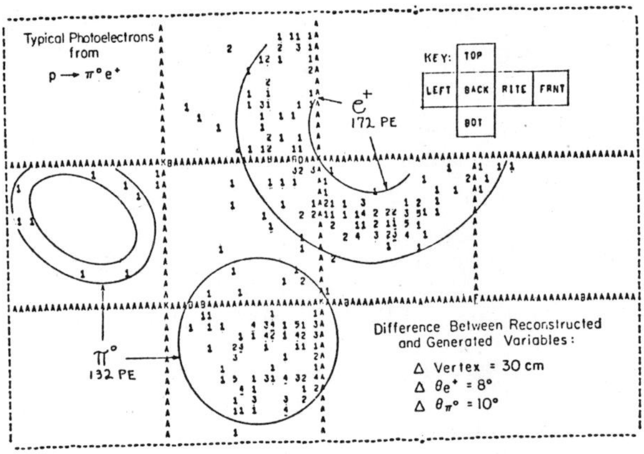

Figure 2

3. EVENT RECONSTRUCTION

The detector is optimally sensitive to nucleon decay signatures of the form $N \to \pi\ell^{\pm}$, where both the meson or its decay products and the charged lepton are capable of generating a Cerenkov signal. Various instrumental and physical factors limit our ability to recognize these modes and reject backgrounds. The observables to be measured are:

(1) The location of the decay vertex
(2) The total energy of the event
(3) The decay opening (collinearity) angle
(4) The energies of each of the decay products

The most significant physical effect is the Fermi motion of bound nucleons in oxygen. (Only 20% of the protons in water are free.) The Fermi motion distorts parameters (3) and (4), introducing deviations from collinearity as large as 30°, and momentum imbalances of $\leq$ 200 MeV/c. Averaged over the radiating portion of the path, multiple scattering introduces an angular error of 5° to 7° for muons and charged pions. A positron-induced shower will have an angular error of 10° to 12°, while those from π^0 decay gammas will have errors of 12° to 17°. Hence the resolution in collinearity angle ranges from about 5° for $\mu^+\pi^-$ modes to about 15° for $e^+\pi^0$. While these effects are smaller than those of Fermi motion for bound nucleons, they are significant in the free proton portion of the sample. The ultimate energy resolution per track is also limited by Fermi motion to $\sim$20%. The PM grid size (number of PM's used) has been chosen so that the instrumental smear is negligible compared to the physical smearing processes. The actual angular resolutions achieved by our first generation reconstruction programs are, in fact, already close to the physical limits.

To arrive at the ultimate number of phototubes (2400) necessary in a 10 kT detector, we have performed a comparison of different configurations of PM tube arrays, using idealized non-showering tracks. A comparison of the initially generated variables with those obtained after reconstructing the events shows that the angular error (2°) is much less than that expected from multiple scattering (7°), showering (12°), or Fermi motion (20°). The error in the inferred starting and ending positions of a track and in the total range (energy) is a strong function of the tube spacing. Using Cerenkov light yields consistent with long baseline tests discussed later, and the measured response of the phototubes (3 ns timing resolution), a detailed optimization shows that a 1.2 m tube spacing achieves an energy resolution for idealized tracks of $\sim$10% with 140 pe per e^+ or π^0 track. A larger tube spacing results in significant deterioration of the resolution because the grid size becomes comparable to the range of the tracks. When energy smearing due to shower fluctuations is included, the energy resolution becomes 15% per track. Figure 3 shows a typical energy distribution after reconstruction of a decay e^+ from a stationary proton. If the outputs of all PM's are summed to yield a measure of the decay energy, the Fermi motion smearing cancels and the anticipated monochromatic line at $\sim m_p$ has an error of $\sim$10% after reconstruction of events from the $\pi^0 e^+$ mode.

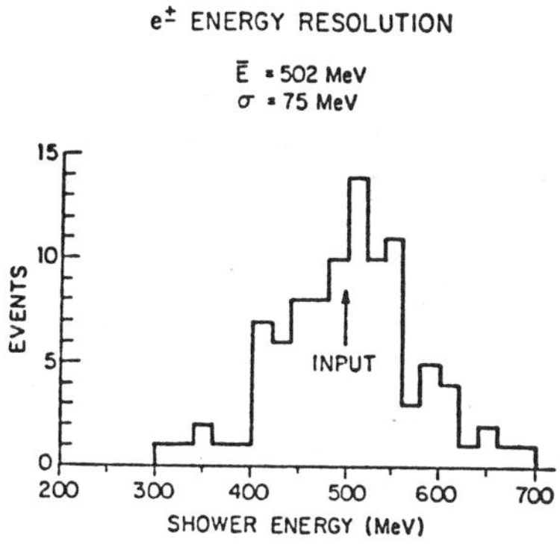

Figure 3

The energy (20%) and angular (20%) resolution achieved per track is the minimum required for background rejection to the 10^{33} year level. Since the physical smearing processes (and not the detector configuration) limit the resolution, collecting more photo-electrons by more PM's, addition of wave-shifter, use of mirrors, etc., provides no gain in resolution.

For non-showering tracks ($\mu^{\pm}, \pi^{\pm}$) an average number of ~ 50 photoelectrons per track appears necessary for reliable reconstruction. Fewer photoelectrons result in a significant loss of background rejection. We find that we can achieve 50 pe for a π^- track (from the $n \to e^+\pi^-$ mode) also using a 1.2 m spacing. Thus, an additional line of approach leads to the same tube spacing and the same number of tubes, ~ 2400 5" PM's, or an equivalent mix of 5" and 8" tubes.

The results of event reconstruction with full smearing show that the error on the vertex position (see Fig. 4) is $\sim 20(30)$ cm for the $\mu^+\pi^-$ ($e^+\pi^0$) mode, essentially independent of position in the detector. For sufficient rejection of entering cosmic ray muons and the neutral products of their interactions in the rock near the detector, we estimate that a veto region of thickness ~ 2 m is necessary to reject background at the required level. This leaves a 17 m cube fiducial volume containing 4000 tons of water with 2.5×10^{33} nucleons.

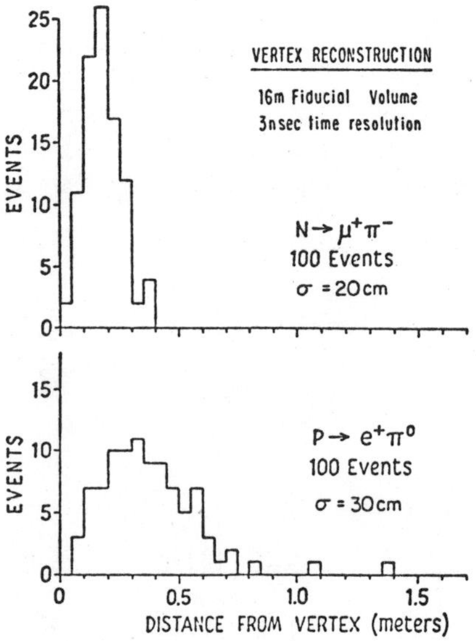

Figure 4

The separation of particles into the categories of "showering" $(\pi^0, e^{\pm})$ and "nonshowering" $(\pi^{\pm}, \mu^{\pm})$ is relatively simple. In Fig. 5 we show the distribution of total photoelectron yields from various tracks from nucleon decays. When the decays are from nucleons at rest (hydrogen nuclei, Fig. 5a), the resulting monoenergetic tracks give clearly separable peaks whose widths reflect the combined geometrical and Poisson fluctuations of the events. Although decays from oxygen nuclei (Fermi momentum = 235 MeV/c, Fig. 5b) have broadened peaks, they are still separable at the 95% level. The pattern of hits is also quite different for showering and nonshowering tracks so that an essentially complete separation of the two categories should be possible (except for a correction due to conversion of $\pi^- \rightarrow \pi^0$ via nuclear charge exchange). However, it appears that the majority of π^0 induced showers will be difficult to distinguish from $e^{\pm}$ showers of the same primary energy.

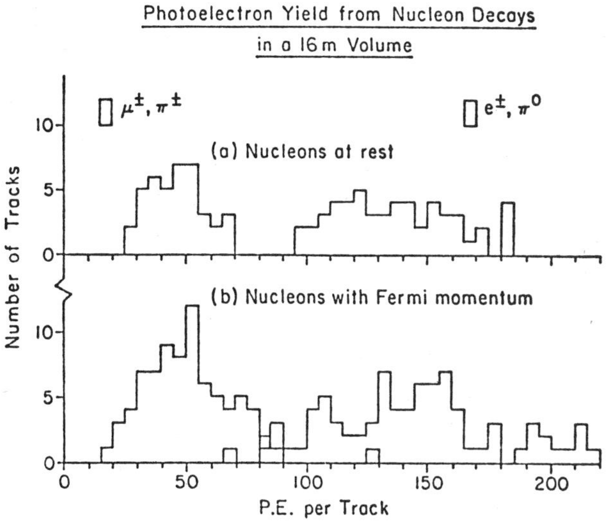

Figure 5

Positive identification of most π^+'s and μ^+'s, and $\sim$70% of μ^-'s will be achieved by detection of the light produced by the electrons from delayed μ decay. Typically, $\sim$10 PM's fire in coincidence over a 50 nsec period. Figure 6 shows the photoelectron yield from 40 MeV $e^{\pm}$ typical of a muon decay. By virtue of their temporal and spatial distributions, delayed hits are identifiable with the information produced by the stopping track from the event that triggered the detector. Two or more muon decays can be recognized as well, thus labelling multibody decay modes. It also appears that with $\sim$300 muon events, one could measure the muon polarization to $\sim$30% by the angular decay correlation. This would be important to ascertain the scalar vs. vector nature of the superheavy boson mediating the decay.

4. BACKGROUND REJECTION

Background which could simulate nucleon decay in the detector could be induced by three possible sources: Entering cosmic ray muons, either straight-through (10^8), stopping (10^6), interacting (10^5), or corner clipping (10^7); Entering hadrons produced in the neighboring rock by cosmic muons which miss

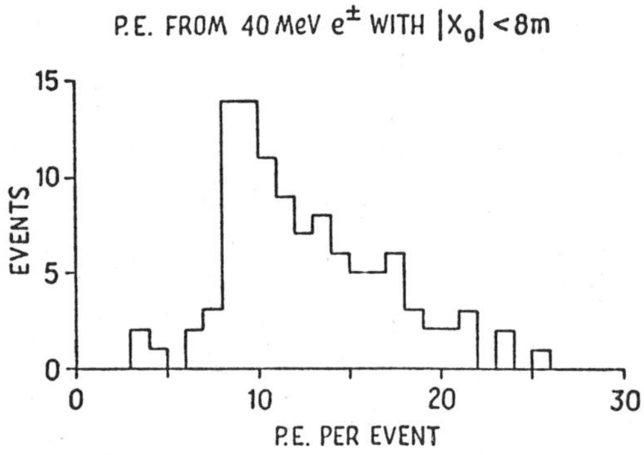

Figure 6

the detector (10^3); and neutrino-induced events inside the detector (10^3). The numbers in parentheses indicate the total rate per year, before any trigger requirement, energy cut, 2m fiducial cut, or back-to-back, equal-energy, 1 GeV requirement is made. Reduction to $\lesssim$ 1/year of each of these backgrounds is discussed below.

The muon flux at the depth of the experiment (1.5 km we) has been documented by measurements over several years at the proposed site. Muons that traverse the entire detector pose no problem. The signal is large (1100 pe as opposed to $\sim$300 for proton decay) and distributed over many more phototubes than a proton decay event. The spread of arrival times for far and near portions of the track is large, and a local "hot spot" of light is deposited in PM's near the entrance and exit point due to the 1/r divergence of the light. Hardware selected straight-through tracks will be recorded with the data for on-line monitoring and for calibration purposes, as well as other physics studies. Similarly, tagged entering muons which stop in the detector provide an important check of the ability of the apparatus to respond to low light level decays within it.

Those muons that interact in the detector produce other tracks, but in general they are at small angles to the initial track and the total pulse height will be even greater. An upper energy threshold can remove them from the trigger. However, the characteristics of these events give valuable information on the nature of similar interactions in the neighboring rock where the muon goes undetected, but the resulting hadrons penetrate the detector.

Muons that traverse only a small portion of the detector ("corner-clippers") and thus fall within our pulse-height cuts have a large portion of their path external to the fiducial volume. Events of this type can be recognized early in the analysis, since their signal is confined to a small, contiguous group of phototubes. A single software cut (or a trigger requirement) based on the number of phototubes hit and their pulse heights can reduce these events by a factor of ~ 1000 to the level of one per several minutes. The simplest of these cuts is evident from Fig. 7. There, distinct regions in the correlated plot of tube multiplicity vs. collected photoelectrons are occupied by nucleon decay events generated in the fiducial volume for three different decay modes. Cosmic ray muons entering the detector are clearly differentiated from the decay events by having fewer tubes fired per photoelectron due to their path near PM's at the surface of the detector. Cutting the events in the region highly populated by muons removes a negligible number of the nucleon decays ($\lesssim 5\%$). A further reduction in muon background is achieved by reconstructing the track's path through the veto region. The remainder are rejected by requiring a back-to-back pattern.

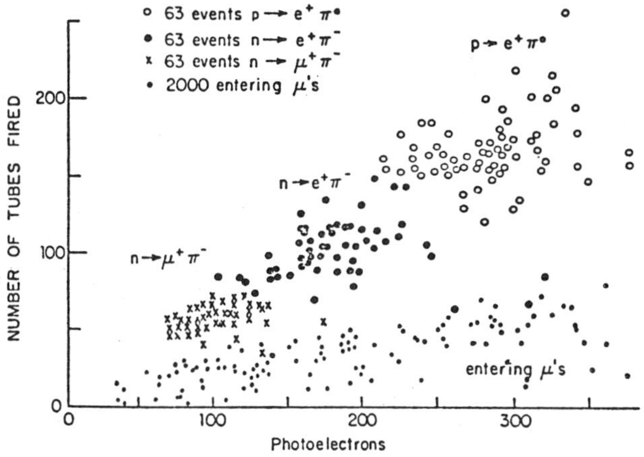

Figure 7

Neutral hadrons from muon interactions in the nearby rock (with a small additional contribution from neutrino interactions) can enter the detector. To mimic nucleon decay, they must either be produced at a large angle (rare) relative to the parent muon passing near a side, or at a small angle (and traverse a large amount of material in the veto region). The interactions of the hadrons that penetrate to the fiducial volume (~ 20/yr) are not expected to have the characteristic nucleon decay signature. Our level (i) vertex finding program is already capable of distinguishing incoming π^0 showers (Fig. 8a) from nucleon decays in the fiducial volume (Fig. 8a) by a simple cut on vertex position.

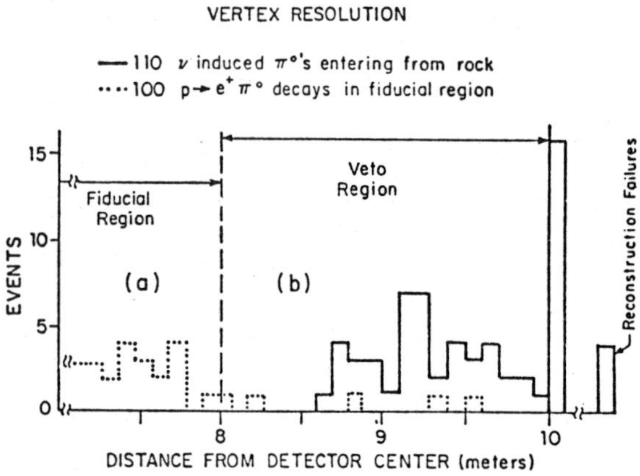

Figure 8

Events induced by neutrinos and anti-neutrinos with two radiating tracks are the limiting background to the experiment. Using estimates of the cosmic ray ν_μ and ν_e spectra[2] (with energies between 0.3 and 2 GeV) and known neutrino interaction cross-sections and q^2 distributions,[3] we have undertaken two independent Monte Carlo simulations of events of this type, including the effects of nuclear Fermi momentum and detector resolution.[4]

In general, recoil protons from such events are below Cerenkov threshold. (At higher energies, where relativistic protons are possible, the neutrino flux is too low to give a substantial background.) Thus elastic charged-current events do not contribute to the background. Double pion production in neutral current interactions is appreciable only at energies that give pulse heights and opening angles outside our range of acceptance.

We are left with single-pion production by charged current interactions, which occur at a rate of about 200 events per year, 140 from ν_μ and $\bar{\nu}_\mu$, and 60 from ν_e and $\bar{\nu}_e$. We simulate these Δ production events with a Breit-Wigner mass spectrum at 1232 MeV with a width of 110 MeV. We generated a sample corresponding to 25 years of data from our detector.

The ν_μ events that survive a total energy cut of 300 MeV window around the proton mass are plotted (Fig. 9) in terms of two configuration-dependent variables used to reject background, the angle between the two tracks, and the fraction of the total energy carried by one of them. The region inside the broken line is that defined by the Fermi motion from bound nucleon decays. Three events fall on the boundary of that region, with 7 or 8 more within one standard deviation in our resolution of these variables. On this basis, we would anticipate a background of somewhat less than one event per year in or near the allowed region for $\mu\pi$ channels and a factor of two fewer for $e\pi$ channels. Varying the shape of the Δ mass spectrum (Gaussian, Breit-Wigner, etc.) and the form of the q^2 distribution (dipole or e^{-aq^2}) has a significant effect on the number of background events which fall within one standard deviation of the Fermi motion region for nucleon decay. However, even in the worst case, the total background is estimated to be less than one event/year for all $N \to \pi\ell$ channels.

Kinematic Characteristics of Neutrino Background

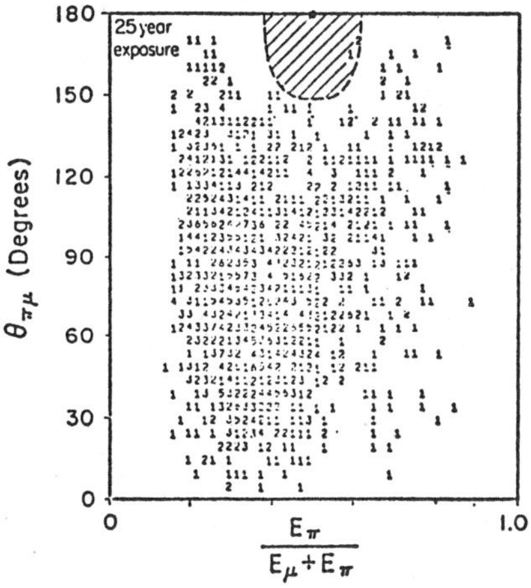

Figure 9

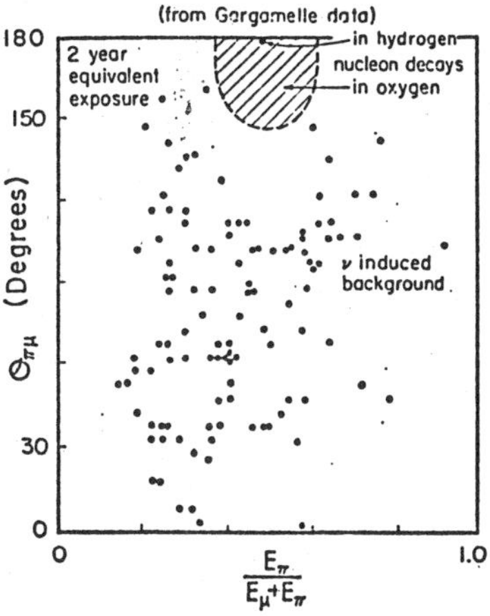

Figure 10

To provide a check of this Monte Carlo, the Gargamelle collaboration[5] applied our criteria to a sample of ν_μ events from their data tapes equivalent to two years of data collection with our detector (in terms of the number of interactions in the allowed energy range). All candidates topologically compatible with single-pion production were placed in the sample, whether they successfully fit this hypothesis or not. This retains any unanticipated reaction channels, as well as events distorted by nuclear reinteraction of pions exiting from a decaying nucleus. The sample is plotted in Fig. 10. Again, the results are compatible with an event rate of less than one per year, and the background only approaches the boundary of the nucleon decay region of the plot. The nuclear scattering effects in freon are worse than in water, so the background indicated is probably an overestimate.

The foregoing represents what we believe to be a realistic assessment of background expected in the detector. However, in the event that increased background rejection is required, we can increase the thickness of the veto region by suitable adjustment of the selection criteria. An increase in the effective veto thickness from the assumed 2 to as much as 4 meters provides for a further background reduction by a factor of 13 (with an attendant decrease in fiducial volume from 4000 to 1700 tons).

In the actual analysis, the precise kinematic cuts and the surviving background will be decay-mode dependent. In addition to shower-nonshower discrimination, ν_μ charged current background can be eliminated from searches for $e^\pm$ modes by the signal from the muon decay. This should be reliable enough to eliminate 70% to 80% of μ^+, and a somewhat smaller fraction of μ^-, since 22% of these are captured on oxygen. In addition, π^+ backgrounds in π^- final states can be eliminated by this technique. We are currently studying our sensitivity to decay modes involving other mesons (ρ, K, etc.) as well as multi-body modes. For example,

$$p \to e^+\omega^0, \quad \omega^0 \to \pi^+\pi^-\pi^0$$

has a signature comprised of one monoenergetic shower and two close showers consistent with a π^0, as well as one delayed muon decay pointing to the vertex. Although this is a fairly restrictive signature, it can be mimiced by

$$\nu_e N \to eN^*, \quad N^* \to \pi^0\pi^+X$$

or

$$nN \to \pi^0\pi^0\pi^+X \quad .$$

For the modes expected from various selection rules in the decay, Table I summarizes our best estimates of the sensitivity of the detector. The second column of the table shows that for each selection rule at least one characteristic proton decay exists that is unique to that selection rule. Thus, this detector has the sensitivity to unfold the physics of possible baryon decays, should they occur.

It is appropriate to mention two ancillary questions relative to the water Cerenkov scheme. First, are the nuclear rescattering effects of the pions born in oxygen nuclei important? Table II, which has been tabulated by R. Barloutaud, addresses this issue. Note that at least ~60% of the pions escape unscattered. Approximately 15% each are either elastically scattered or charge exchange scattered. Although these processes confuse the information relating to the branching ratio, they do not necessarily remove the events from the search for nucleon decay. In effect, pion charge exchange just

$\Delta(B-L)$	Proton Mode Signature	Decay Mode Proton	Decay Mode Neutron	Approximate Sensitivity
-4		$\nu\nu\nu\pi^+$	$\nu\nu\nu\pi^0$	$\gtrsim 10^{30}$ yr
	No line 1 shower 2 μ decays	$\mu\mu e^-\pi^+\pi^+$	$\nu\nu e^-\pi^+$	$\gtrsim 10^{30}$ yr
-2	line at 1 GeV 2 μ decays	$\ell^-\pi^+\pi^+$	$\ell^-\pi^+$	$\sim 10^{33}$ yr
		$\nu\pi^+$	$\nu\pi^0$	$\sim 3 \times 10^{31}$ yr
		$\nu_1\nu_2 e^+$	$\nu_1\nu_2 e^+\pi^-$	$\gtrsim 10^{30}$ yr
0	line at 1 GeV 2 body	$\ell^+\pi^0$	$\ell^+\pi^-$	$\sim 10^{33}$ yr
		$\bar{\nu}\pi^+$	$\bar{\nu}\pi^0$	$\sim 3 \times 10^{31}$ yr
		ℓ^+K^0	ℓ^+K^-	$\sim 10^{32}$ yr
+2	event deep in detector	$\overline{\nu\nu\nu}\pi^+$	$\overline{\nu\nu\nu}\pi^0$	$\gtrsim 10^{30}$ yr
		$\overline{\nu\nu}e^+$	$\overline{\nu\nu}e^+\pi^-$	$\gtrsim 10^{30}$ yr

<u>Table I</u>. Detector Sensitivity to Various Selection Rules

	No Internal Interaction	Elastic Scattering	Charge Exchange	Absorption	Ref.
^{12}C	0.64 ± 0.06	π^0 $0.14 \pm .04$	$0.11 \pm .03$	$0.11 \pm .03$	[6]
		π^- $0.19 \pm .04$	$0.06 \pm .02$		
^{16}O	0.58	—	—	—	[7]
^{27}Al	0.51	—	—	—	[7]
^{27}Al	0.68	π^0 0.11	0.14	0.06	[8]
		π^- 0.18	0.07		

<u>Table II</u>. Pion Interaction in Nucleus

"exchanges" decaying protons for decaying neutrons, and elastic scattering just spoils the back-to-back signature; the 1 GeV, equal-energy signature is still preserved. Typically, we only expect ~10% of the nucleon decays to be removed from the search by absorption.

A second question relates to the detection technique. Why a surface rather than a volume array? Although a volume array always has some tubes close to the track and these usually capture a large number of photoelectrons (however, with large fluctuations), we found in designing the detector that the *total number* of tubes fired by an event is far more important to reconstruction than the total number of photoelectrons. For each additional tube hit, three space coordinates, one time, and one pulse height are added to the number of constraints available to the reconstruction routine. Further, since most of the fiducial volume is near the faces of the cube (as well as most of interactions due to the entering, non-neutrino background), a surface detector has many more tubes lit up per trigger than a volume array outfitted with an equal number of PM's. Figure 11 shows the number of tubes hit by a 2 m long track going toward a wall as a function of its distance from the wall. Results for both surface and volume arrays of PM's are shown. The surface array typically has a factor of two more PM's hit than a volume array.

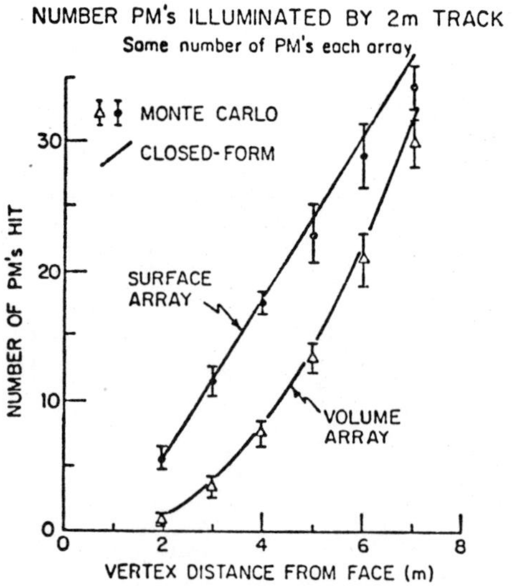

Figure 11

5. PROGRESS IN DETECTOR CONSTRUCTION

The underground site that we have chosen is located in the Morton Salt mine at Headlands Beach State Park, Fairport Harbor, Ohio. The depth is 1500 mwe ($\sim$550 m at density 2.7 g cm^{-3}). A tailor built cavity is currently being excavated to house our detector. Simultaneously, manufacture and assembly of detector components is proceeding apace. The complete electronic readout system--from vertical strings of 16 underwater PM tubes to mini-computer recording of data--is currently under systems development at the IBM 21 m vertical test tank at Ann Arbor. For each tube, the electronics digitizes one pulse height and records time on two scales. One time scale has sufficient resolution for the initial event reconstruction, while the other records delayed pulses from π, μ decays. Both pulse height and timing requirements are more modest than in typical accelerator experiments, and are particularly easy to meet in our constant 75°F thermal environment. A modest 3 ns timing resolution, 10% energy resolution, and a low factor of 15 dynamic range in energy per tube, and the low ($\sim$4 sec^{-1}) trigger rate permit an electronics design that is simple, economical, and highly reliable for operation in the unusual experimental site.

The housing and support structure for the photomultiplier tubes has been perfected in both the vertical and in a 10 m horizontal test tank as well as in other pools. Each tube can be positioned to a few centimeters in space and a few degrees in angle. Features include easy PM retrieval for above-water servicing, construction with corrosion-free materials, maintenance of high purity water, and a minimum of structure so as to permit easy redeployment of the detector.

The support structure is illustrated in Fig. 12. The tubes are positioned and accessed by a "conveyor belt" made of a parallel pair of nylon monofilaments. Any one of the 16 tubes on each string can be brought to the surface for maintenance by rotating the string. The support structure is stable because the tube modules are designed to be neutrally buoyant and to have no net torque when suspended in the water. Polyethylene sealed anchors maintain the positions of the strings.

After manufacture and assembly, the PM support structure, the PM housings, the cables, etc., are tested in the high purity water ($>$ 40 m attenuation length at 450 nm) of the vertical and horizontal test tanks for at least two days. In addition, a large number will have been operated for months in this simulated detector environment before installation at the site.

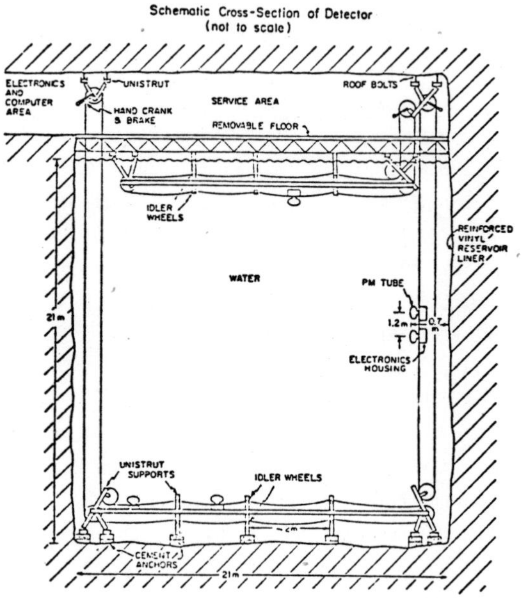

Figure 12

6. SUMMARY OF STATUS OF PROTON DECAY FACILITY

The two body nucleon decay signature is remarkably distinct in a straightforward, though massive, Cerenkov detector. The backgrounds are tractable up to a sensitivity of 10^{33} years. Civil construction of the underground laboratory is proceeding on schedule. Occupancy of the laboratory is expected in the fall of 1980 when 1200 five inch PM tubes, one-half the anticipated final detector, will be installed. Manufacture and long term underwater proofing of PM's, housings and electronics is progressing at a production rate of 200 PM detector modules per month.

7. CAPABILITIES OF A NEUTRINO OSCILLATION STUDY IN THE DETECTOR

Although dedicated to the search for nucleon decay, the apparatus comprises the largest mass cosmic neutrino detector yet proposed. Since the limiting backgrounds to the proton decay experiment are induced by atmospheric neutrino events, these events must be understood in some detail. In an attempt to use the information from these background events creatively, we have undertaken a feasibility study of a neutrino oscillation experiment using them.

The search for oscillations[9] of atmospherically produced neutrinos is straightforward in the detector. Unlike neutrino beam lines at accelerators, the muons produced in the upper atmosphere have flight paths long enough for them to decay. A typical production chain for 1 GeV neutrinos is

$$\pi^+ \to \mu^+ \nu_\mu$$
$$\quad \hookrightarrow e^+ \nu_e \bar{\nu}_\mu \quad .$$

Superficially, this suggests that the expected neutrino ratio should be $\nu_e : \nu_\mu = 1 : 2$. In fact, an integration of cross sections and calculated fluxes show that (in a detector that does not measure charge, as the one we consider) the ratio of electron to muon events is expected to be $\sim 1 : 2$. In principle, one would measure this ratio as a function of the neutrino flight path ℓ from the neutrino production point (directly as a function of zenith angle θ_Z) and as a function of neutrino momentum p. As depicted in Fig. 13 and plotted in Fig. 14, the dynamic range in ℓ is 3×10^3, from downward coming neutrinos produced ~ 5 km above the detector (which is located 0.5 km below the earth's surface) to upward coming neutrinos that have traversed the 13,000 km diameter of the earth. The dynamic range in p is insignificant relative to that in ℓ. On the lower end it is bounded by the onset of the cross-section for elastic neutrino scattering (see Fig. 15) and on the upper by the rapidly falling $(\propto E^{-2.5})$ energy spectra of neutrinos (see Figs. 16 and 17). Although one would like to measure $\nu_e : \nu_\mu$ as a function of ℓ/p to map out explicitly any oscillatory behavior, the statistics in even this massive detector will be a limiting factor. However, we will show that a comparison of the ratios for the upper and the lower hemispheres (equal solid angles) will be sufficient to search for neutrino oscillations out of either channel for large mixing (Pontecorvo) angles[10] ξ and eigenmass differences

$$\Delta m = \sqrt{m_1^2 - m_2^2}$$

between 10^{-3} and 10^{-1} eV.

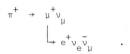

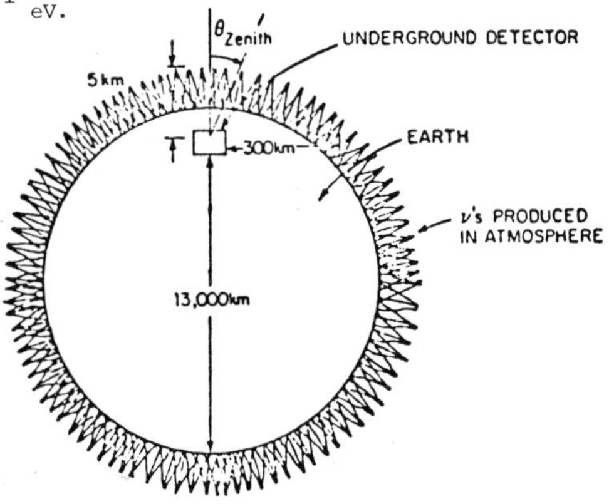

Figure 13

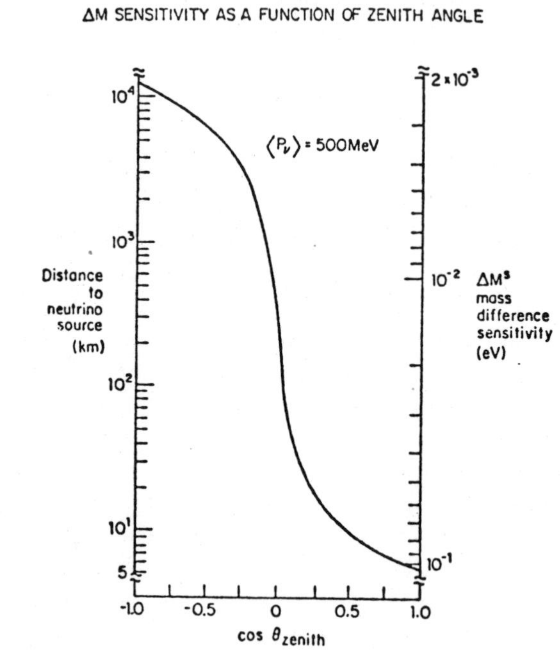

Figure 14

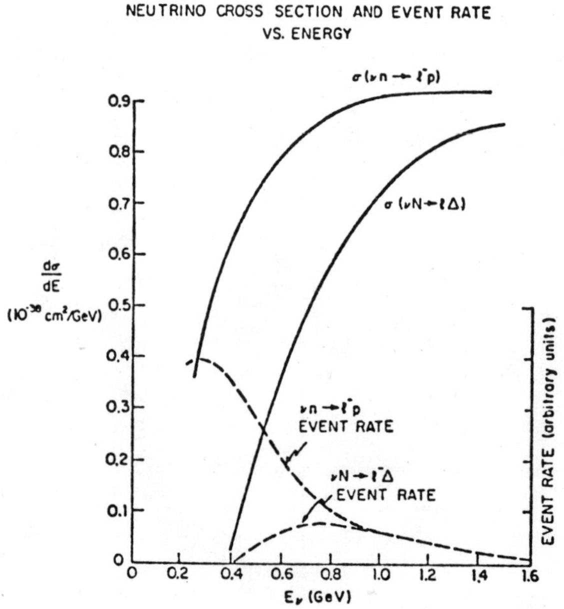

Figure 15

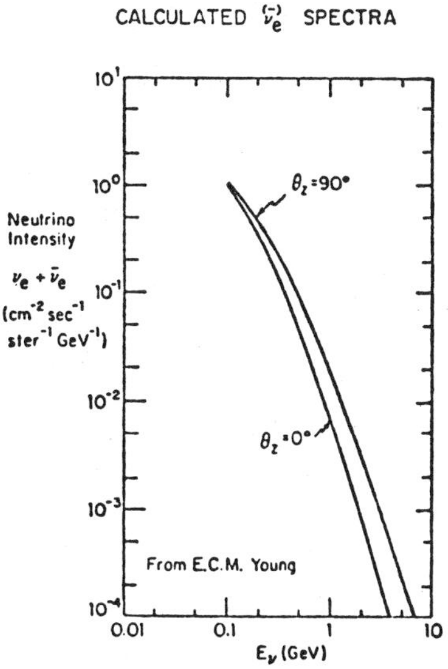

Figure 16

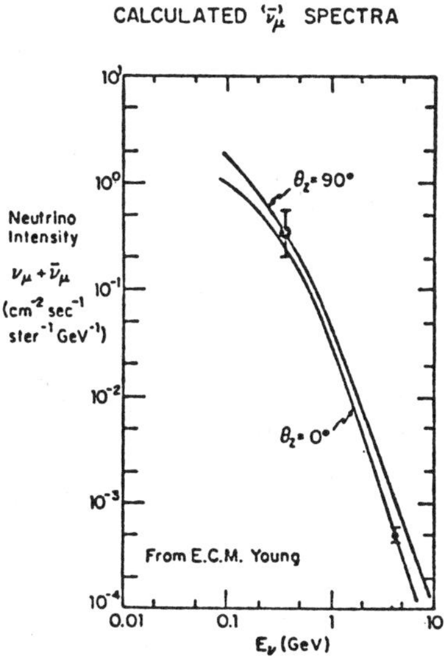

Figure 17

The evolution of an electron neutrino beam of intensity I_e that is mixing with another neutrino beam of intensity I_x is described by

$$I_e = I_e(0)\left[1 - \frac{\sin^2 2\xi}{2}\left(1 - \cos\frac{2\pi\ell}{L}\right)\right] + I_x(0)\,\frac{\sin^2 2\xi}{2}\left(1 - \cos\frac{2\pi\ell}{L}\right),$$

which is familiar from the two component evolution of the $K^\circ - \bar{K}^\circ$ system. The oscillation length L is

$$L(m) = 2.5\,\frac{p(\mathrm{MeV})}{\Delta m^2(\mathrm{eV}^2)},$$

and the characteristic measure of the proper evolution time in the rest frame of the neutrino system is proportional to ℓ/p. The sensitivity in Δm of any experiment is characterized both by ℓ/p and by statistics. The achievable value Δm^s is typically

$$\Delta m^s \simeq 2(N_e/N_x)^{1/4}\,(p/\ell \sin 2\xi)^{1/2}$$

for $N_e(N_x)$ observed events induced by ν_e and ν_x's. For maximal mixing ($\xi = 45°$), $p \sim 500$ MeV, and $\ell = 1.3 \times 10^7$ m, the ultimate sensitivity of an atmospheric neutrino experiment is 2×10^{-3} eV. Figure 14 shows the sensitivity as a function of both ℓ and of $\cos\theta_Z$. The $\cos\Theta_Z$ dependence is experimentally relevant since equal statistics come in equal intervals of solid angle. Note that the dynamic range in Δm^s spans the region from 10^{-3} eV to 10^{-1} eV. The upper limit of this range coincides with the lower limit of the best sensitivity of accelerator and reactor based experiments.[11,12]

For communication between the three neutrino states, the phenomenology is generalized to two eigenmass differences, three Pontecorvo angles, and one CP violation phase.

Recent reports of possible evidence for neutrino oscillations[13] and a non-zero neutrino mass[14] are consistent with the previous experimentally allowed regions of these variables, as analyzed in Ref. 12. In particular, a small mass difference ($\sim 10^{-2}$ eV) may exist between the ν_e and ν_μ, and a large one (~ 10 eV) between these two states and the ν_τ. The experiment we outline is sensitive both to 1) $\nu_e - \nu_\tau$ mixing for $\Delta m > 0.1$ eV by an unexpected loss of ν_e's relative to ν_μ's in the downward coming neutrino flux, and to 2) $\nu_e - \nu_\mu$ mixing of $\sim 10^{-2}$ eV by comparison of downward and upward ratios of e's to μ's.

8. SIMULATION OF NEUTRINO EVENTS IN THE DETECTOR

We have investigated the characteristics of neutrino-induced events in the I.M.B. detector to determine the event rates and the resolution in the direction of the incident neutrinos. To simulate events in the detector, several ingredients are necessary, including the neutrino flux and the charged-current cross sections. The atmospheric neutrino spectra for ν_e and ν_μ have been calculated some years ago by E.C.M. Young (Refs. 15 and 16, respectively) and are reproduced in Figs. 16 and 17, respectively. The curves are similar in shape; on average, the ν_μ's dominate the ν_e's by a factor of two. Also the relative composition of neutrinos vs. antineutrinos can be inferred from the μ^+/μ^- charge ratio (1.2) of cosmic ray muons at the earth's surface.[17] The zenith angle distribution[15,16] of the fluxes is also taken from Young. The flux decreases by 25% from $\theta_Z = 90°$ to $\theta_Z = 0°$. Although no observations of cosmic ray ν_e's have been reported, and only a handful of ν_μ events have been recorded, the errors in these flux calculations are only expected to be $\sim 30\%$ around 1 GeV since they are inferred from the observed muon flux at sea level. (With the vastly better knowledge of the π/K ratios, etc., today (relative to 1969 when the calculations were made), a reevaluation of the spectra is strongly encouraged.) The ν_μ charged current cross sections near 1 GeV are well known from Argonne and Gargamelle data.[3,4] The energy dependence for elastic scattering and $\Delta(3/2, 3/2)$ production is shown in Fig. 15. The products of flux and cross section, also shown in Fig. 15, demonstrate that only elastic charged-current scattering is statistically important. This is fortuitous, since otherwise charged pions from resonance production could confuse the μ/e event identification, which relies on stopping muon decay signatures.

Thus, the dominant signal in the detector for the oscillation experiment is

$$\nu_\mu n \to \mu^- p$$

and

$$\nu_e n \to e^- p \quad .$$

However, the recoil proton initiated by 1 GeV neutrinos is below Cerenkov threshold ($\beta = 0.75$) and therefore is invisible in a water Cerenkov detector. The correlation between the direction of the charged lepton and the incident neutrino will be used to measure the incoming neutrino direction. This correlation is shown as a function of the lepton total energy in Fig. 18. In this plot, the Fermi motion of the nucleons is assumed to be that of water, with $p_f = 220$ MeV/c. To insure that the angular correlation between outgoing charged lepton and the incoming neutrino is sufficient to sense an up/down asymmetry, a cut at ~ 0.3 GeV in total lepton energy is necessary. Events with this cut have the angular distribution of Fig. 19. The mean correlation angle is 45°. Although the energy cut eliminates 1/2 of the events from the data sample, the rejected events predominantly have low energies and therefore are more susceptible to background and to misidentification. After the charged-lepton energy cut, the mean neutrino energy of the accepted events is ~ 0.7 GeV. Neutrinos of this energy produce μ's and e's with a radiating range of typically a factor of three shorter than that encountered in two body, 1 GeV decays of the proton. Thus, the available containment volume in the detector for this experiment could be somewhat larger than that for the proton decay experiment. Rather than the 2m veto ragion there, we make a 1.2 m cut on all faces but the top, giving an anticipated containment volume of $\sim 4 \times 10^{33}$ nucleons. Conservatism suggests the added veto region on the top as insurance against the entrance of cosmic ray muons, which are inherently angular asymmetric toward the vertical.

The apparatus resolution in angle for the charged lepton ($\sim 20°$) is sharp enough to preserve any up/down asymmetry. Also, the energy cut is not significantly smeared by the energy resolution of the detector, which is $\sim 15\%/\sqrt{E(\mathrm{GeV})}$.

Muon events are distinguished from electron events by the delayed pulse produced by stopping decaying muons. In the ~ 10 μsec live period after each muon trigger, $\sim 90\%$ of μ^- decay events are expected to register in > 3 tubes at coincident times. These times must be consistent with the stopping point of the muon registered during the initial trigger. When compounded with the nuclear absorption probability of 20% in oxygen, the detection efficiency for stopping μ^-'s should be $\sim 70\%$.

The sensitivity of the experiment can be characterized by mean event rates expected under various oscillation hypotheses as is shown in Table III(p.186). The total number of fully contained events per year is expected to be 500 (1000)

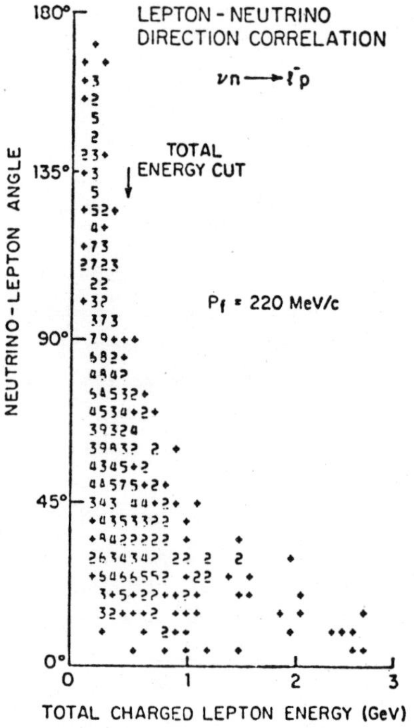

Figure 18

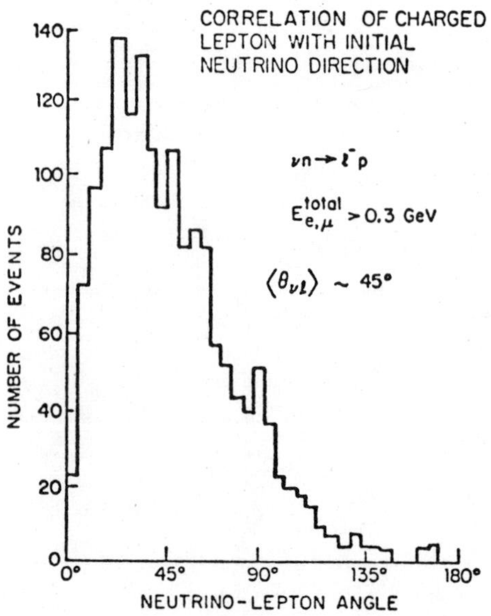

Figure 19

induced by ν_e (ν_μ). This rate is halved by the 0.3 GeV cut on charged lepton energy. Thus, for no oscillations, the e/μ event ratio should be 0.5 ± 0.07, whether initiated by down or up going neutrinos. Both down and up going e/μ ratios will be unity if ν_e, ν_μ oscillations exist with $L \lesssim 300$ km. The e/μ ratio will be unity only for upcoming neutrinos if $L \sim$ radius of the earth. And if muon neutrinos do not oscillate, but ν_e, ν_τ oscillations exist with a large mass difference, the e/μ event ratio will be 0.25 ± 0.04 for both up and down going events.

For e,μ oscillations, the sensitivity of the detector is characterized as a function of the mass difference in Fig. 20. The e/μ ratio is shown both for upward coming and downward going neutrinos. For two years of data the statistical power is ~four standard deviations for each ratio in the region between 0.006 eV and 0.15 eV. Thus, for e,μ mixing, the experiment is optimized in the low mass range, but the full 10 KT size of the detector is necessary to have sufficient statistical power.

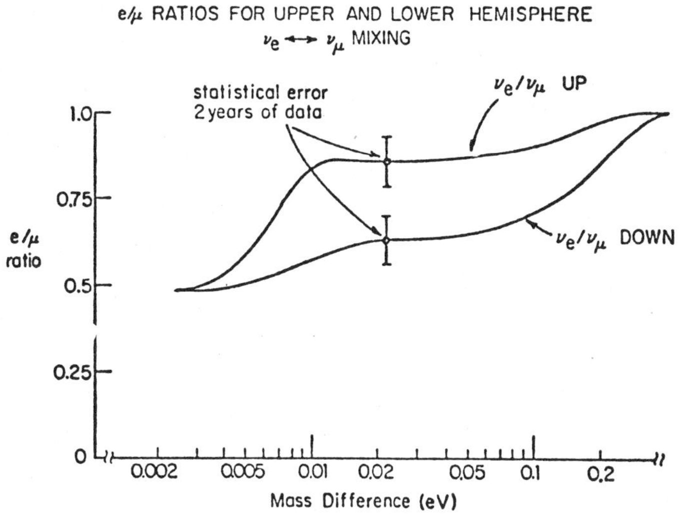

Figure 20

We should also consider the effect on vacuum neutrino oscillations induced by the different forward scattering amplitudes in the matter through which the upward coming neutrinos have passed. ν_e's can scatter from electrons by both neutral and charged currents, whereas ν_μ's and ν_τ's only have charge-current interactions. Wolfenstein has shown that both the vacuum Pontecorvo angles and the oscillation lengths are modified by the matter at oscillation lengths comparable to the earth's radius.[18] Since this results in exploring a somewhat different range of variables than in vacuum over the same distance, we ignore these effects in the current paper.

Neutrinos Mixing	Oscillation Length	Event Rates ν's down	Event Rates ν's up
ν_e, ν_μ	L > 13,000 km	125 ± 12 e 250 μ	125 e 250 μ
ν_e, ν_μ	13,000 > L > 300 km	125 e 250 μ	187 e 187 μ
ν_e, ν_μ	300 >> L	187 e 187 μ	187 e 187 μ
ν_e, ν_τ	13,000 > L > 300 km (ν_μ, L > 13,000 km)	125 e 250 μ	60 ± 8 e 250 μ

Table III. Typical Neutrino Event Rates Under Various Oscillation Hypotheses. Errors noted are purely statistical. Entries are events/year in the fiducial volume above a total energy cut of 0.3 GeV on the charged lepton.

In conclusion, the massive detector under construction for the baryon decay experiment may provide the most sensitive detector for oscillation searches between ν_μ and ν_e species, and may yield a signal for ν_e, ν_τ oscillations soon after turn on if the expected 0.5 ratio of ν_e/ν_μ is not observed.

Acknowledgments. During the last year our theoretical colleagues have continued to find more ways for the nucleon to decay. We take the opportunity to thank them here for many enlightening comments: J. Ellis, M.K. Gaillard, M. Gell-Mann, H. Georgi, S. Glashow, R.J. Goldman, C. Jarlskog, P. Langacker, M. Machacek, R. Mohapatra, D.V. Nanopoulos, J.C. Pati, H.R. Quinn, P. Ramond, D.A. Ross, A. Salam, R. Slansky, S.B. Treiman, S. Weinberg, F. Wilczek, L. Wolfenstein, F.J. Yndurain, and A. Zee. We apologize to them and their friends for neglecting to cite the voluminous literature they have generated on the subject of this report. In addition, conversations with C. Baltay, C. Bennett, J. Cronin, and C. Rubbia are gratefully appreciated.

We also acknowledge innumerable enlightening discussions with R. Barbieri, A. De Rujula, P. Frampton, A. Yildiz on the topic of neutrino oscillations. We also are deeply thankful to L. Wolfenstein for continued clarification of these matters. Conversations with R. Barloutaud revealed that he has independently done some of the calculations presented here. See Ref. 19. B. Cortez has been primarily responsible for the study of the feasibility of using the detector for neutrino oscillation searches.

REFERENCES

1. M. Goldhaber, et al., *Proceedings of the International Conference on Neutrino Physics*, Bergen, June 1979, C. Jarlskog (ed.), p. 121.

2. A.W. Wolfendale and E.C.M. Young, CERN 69-28, p. 95, 10 Nov. 1969; E.C.M. Young, in *Cosmic Rays at Ground Level*, A.W. Wolfendale (ed.), Institute of Physics Press, London and Bristol, 1973, p. 105; E.C.M. Young, private communication, 1977.

3. R.T. Ross, *Proceedings of the 1978 Conference on Neutrino Physics*, E.C. Fowler (ed.), p. 929; M. Derrick, *Proceedings of the 1977 Conference on Neutrino Physics*, H. Faissner, et al. (eds.), p. 374; and the references cited in the two works.

4. Experimental results with which the Monte Carlo has been compared include: E.C.M. Young, CERN Yellow Report 67-12, "Neutrino Charged Current Single Pion Production in Freon"; S.J. Barish, et al., *Phys. Rev. Lett.* 36, 179 (1976); J. Campbell, et al., *Phys. Rev. Lett.* 30, 335 (1973).

5. We would like to thank Drs. J. Morfin and M. Pohl for providing the data from the Gargamelle experiment.

6. Experimental results from Gargamelle for 250 MeV/c pions. B. Degrange, private communication via R. Barloutaud.

7. Monte Carlo results of T.W. Jones, private communication.

8. Monte Carlo study of C. Longuemare, private communication via R. Barloutaud.

9. For a review of the theory of neutrino oscillations, see S.M. Bilenky and B. Pontecorvo, *Phys. Reports* 41, 225 (1978).

10. We choose to refer to the neutrino mixing angle in this way since B. Pontecorvo has continually emphasized the open nature of this question.

11. The status of the most sensitive accelerator based oscillations search is described by the author in *Proc. BNL Workshop on the Fixed Target AGS Program*, BNL 50947, Nov. 8-9, 1978, p. 212.

12. A. De Rujula, et al., "A fresh look at neutrino oscillations", Ref. TH 2788-CERN; V. Barger, et al., "Mass and mixing scales of neutrino oscillations", coo-881-135 (Univ. Wisconsin preprint).

13. F. Reines, H. Sobel, and E. Pasierb, "Evidence for neutrino instability" (Univ. of Calif. at Irvine preprint), submitted to *PRL*.

14. Private communication; E.T. Tretyakov, et al., *Proceedings of the International Neutrino Conference*, p. 663, Aachen (1976), has a description of the experiment and a preliminary analysis of the data.

15. E.C.M. Young, in *Cosmic Rays at Ground Level*, Ref. 2, and private communication (1977).

16. A.W. Wolfendale and E.C.M. Young, Ref. 2.

17. M.G. Thompson, in *Cosmic Rays at Ground Level*, Institute of Physics Press, A.W. Wolfendale (ed.), London and Bristol (1973).

18. L. Wolfenstein, *Proceedings of the Clyde L. Cowan Memorial Symposium*, A. Saenz (ed.), Catholic University, April 25, 1978; *Phys. Rev. D* 17, 2369 (1978).

19. R. Barloutaud, "Test of neutrino oscillations in a large underground detector", 12 March 1980 (draft).

A SEARCH FOR BARYON DECAY:
PLANS FOR THE HARVARD-PURDUE-WISCONSIN
WATER CERENKOV DETECTOR

David R. Winn
Department of Physics
High Energy Physics Laboratory
Cambridge, Mass. 02138

ABSTRACT

We describe the plans for and expected sensitivity of the HPW proton decay detector. Particular emphasis is placed on reconstruction and signal/background studies. A brief discussion of future baryon number violation experiments concludes.

I. Introduction

We are motivated to attempt a search for baryon number violation by several considerations:

1. The successes of SU(5) and other unified models both in particle physics and as elegant theories[1] (however see Note 2);
2. Cosmological puzzles, for example $\bar{B}/B$, B/S and the electrical neutrality of galaxies;[3]
3. Baryon number conservation is not motivated by any fundamental principle, such as coupling to a long range force, or by a gauge principle;[4]
4. Proton decay does not require the approval of program committees, an operating accelerator or many other uncontrollable and unnatural acts of persons (the Deitron by definition has no failures);
5. There are many other interesting measurable phenomena which can be studied in underground apparatus suitable for proton decay experiments:
   a) high energy cosmic ray muon and neutrino interactions
   b) neutrino oscillations
   c) muon bundles and rare underground phenomena[5], such as the Kolar Gold Field events
   d) Galactic and cosmological events, such as super-nova neutrino bursts, or effects from exotic particles remaining from the early epoch.

II. Proton Decay, Detector Requirements and Detector Strategy

The proton lifetime τ_p is given in SU(5) by amplitudes similar to those represented by the diagrams shown in Figure 1:

$$\tau_p \approx \frac{k}{\alpha_g^2} \frac{m_x^4}{m_p^5} \approx 10^{30} - 10^{32} \text{ years}$$

(note dependence on θ_w, α_s, quark wave functions, number of generations, etc.). The branching ratio into charged antileptons and meson(s) is large. For example:

$$\text{B.R.} \left(\frac{p \to e^+ + (\pi^o \to \gamma\gamma)}{p \to \text{all}}\right) \approx 10 - 35\%$$

Briefly, these considerations imply one needs more than 10^{32} baryons to do a practical experiment; 1 kiloton contains 6 x 10^{32} baryons. Signal to noise requirements then force one to shield from cosmic rays, meaning going deep underground. Since neutrino interactions cannot be directly shielded and the neutrino flux is poorly known, the detector must also have discrimination on an event by event basis against neutrino interactions. With no discrimination they provide an interaction rate equal to a proton lifetime of less than 10^{31} years. This last requirement together with costs and the branching modes into electromagnetic energy have led us to design a totally active "target-detector" consisting of:

1. A large water Cerenkov volume detector/medium, in the present case arranged as a rectangular tank containing 1,000 tons of water, located 1,700 m.w.e. below the earth's surface; the Cerenkov shock in the water resulting from the moving decay products of water nucleons is the signal to be detected.
2. Large acceptance photomultipliers to detect Cerenkov light are embedded on the lattice sites of a cubic lattice throughout the tank volume; the lattice spacing is about 1 meter; mirrors line the tank to create an apparent infinite lattice; pulse height and relative time of arrival of light are recorded on each phototube; there are 800 5' p.m.t.'s.
3. A shield system surrounding the 6 sides of the tank tags entering ionizing radiation (a "veto"), "contains" events having exiting tracks, and absorbs particles entering the tank. In the present detector, the shield will consist of a layer of gas proportional counters and a water radiator/absorber layer instrumented with both phototubes and plastic waveshifter sheets in such a fashion as to discriminate exiting and entering particles by the direction of Č light.

Figures 2-4 show a schematic of the detector and the signal of proton decay. A more detailed technical description will follow in Sections III and IV.

III. Signal, Background & Reconstruction

Signal

The signal is indicated schematically on Figures 2 and 3. It consists of 2 roughly monoenergetic ($E \approx m_p/2$) back-to-back "cones" of light, one from the Cerenkov light from an e^+ electromagnetic shower and one broadened "cone" of light from, for example, the $\pi^o \rightarrow \gamma\gamma$ showers. Relevant properties of water are given in Table 1. Note that since the proton is at rest in the lab, the momenta of the e^+ and meson are equal and opposite being about 1/2 m_p and the total number of Cerenkov photons corresponds to $E = m_p$. The signal is then extracted by the total energy, by the "reconstruction" of the pattern of the cones both from pulse height and time of arrival of the light at each photodetector (i.e., these times increase monotonically with distance in opposite directions from the vertex, which is in the neighborhood of the earliest p.m. firing) and by the energy balance along the event axis.

Background

The backgrounds are shown schematically in Figure 5. Generally speaking, the backgrounds are moving in the lab, and thus have event topologies which are not back-to-back and are therefore distinguishable from proton decay by various strategies of pattern recognition. In addition, the bulk of the background events will have energies not equal to m_p. There are two broad classes, Types A and B.

Type A backgrounds consist of entering ionizing radiation, primarily muons, but also produced particles from deep inelastic reactions in the rock walls surrounding the detector. Various muon fluxes and properties underground are shown in the graphs of Figure 6. At 1700 m.w.e., the depth of this experiment, the muon rates are roughly 10^8/kton/year of which 5 x 10^6/kton/year stop and decay inside the detector. The average muon energy of traversing muons is about 200 GeV. Since the probability of a photonuclear reaction with $y > 0.1$ is about 10^{-8}/radiation length, the entering $n(k^o)$ rates can be computed to be about $10^{-4}(10^{-6})$ of the muon rate for $E_n, k^o > 1$ GeV. Thus although the topology of entering radiation is not back-to-back as in proton decay, rejection beyond pattern recognition and energy resolution is essential. A water layer shield/absorber together with ionization proportional counters provides additional rejection. Studies have indicated that only 10^{-2}-10^{-3} of the entering particles enter unaccompanied. Of those the majority are at shallow angles with the plane of the shield and will therefore be absorbed.

Type B backgrounds are neutrino events inside the detector. These are especially pernicious as they track the detector mass as does proton decay, and they cannot be directly shielded. Some neutrino reaction rates and properties are shown in Figures 7 and 8. As a rule of thumb:

1. $\nu_\mu/\nu_e \approx 2$; $\bar{\nu} \approx \nu$

2. $E_\nu \lesssim 2$ GeV

3. 140 events/kton/year with $E_\nu > 500$ MeV

4. For $m_p/2 < E_{vis} < 3\ m_p/2$ roughly

$\sim$ 1/3 are 1 track

$\sim$ 2/3 are multitrack, 1/2 being single pion production

$\sim$ 1% strange particle production.

When Fermi motion and pion rescattering and charge exchange are taken into account, single pion production is the most serious background. The technique to reject this background follows from the reconstruction where the neutrino events appear to be moving in laboratory, have $E \neq m_p$ or have a pattern different from proton decay events.

Monte Carlo and Reconstruction

To study the detector a proton decay Monte Carlo, background Monte Carlos and reconstruction programs were written. The reconstruction program reads the output of the Monte Carlo consisting of "digitized" phototube responses. Outlines of these programs are shown in Tables 2 and 3, which should clarify some of the above discussions of signal and background. Tables 4-7 show respectively the event descriptions, collected photoelectrons, asymmetry, sphericity (see Table 3) vertex errors, for various proton decay branching modes and for neutrino backgrounds. Figure 9 shows the comparison between distributing the photomultipliers on the surface of the water volume, and throughout the volume using similar photocathode area. Figure 10 shows the Cerenkov photon detection efficiency as a function of photodetector cross section, lattice spacing and water clarity. Our current detector has an average lattice spacing of 1 meter with 12 cm diameter p.m.t.'s. Figure 11 shows gnomonic projections (planar projections of the celestial sphere from the vertex) of the events in our phototube lattice. Figure 12 shows pulse height spectra for proton decay and for neutrino events. Figures 13 and 14 show the vertex and angular errors in the reconstruction.

As a recapitulation, the general technique to reject background and accept signal is given as follows:

1. pulse height - in the range m_p
2. shield - tag and absorb particles exiting or entering detector
3. pattern recognition - examples are A, S, vertex, tracks, opening angles, sum of momentum equal to zero.
4. delayed signals - π-μ-e
5. unique range of e, μ from proton decay
6. maximum likelihood
7. software event migration, angular and fiducial volume cuts
8. known background studies, such as stopping muons.

In the Monte Carlo and reconstruction as they presently stand, the "noise limits" achievable in terms of known backgrounds with simple uncorrelated cuts on energy, asymmetry and sphericity are:

Type A: less than 0.5 events/year

Type B: 0.5 - 1 event/kton/year

These background rates then imply an upper limit on proton decay of about $(0.5 - 1) \times 10^{33}$ years achievable by this type of detector independent of the total mass. However, this will improve by improving the reconstruction algorithms which are still in early stages of sophistication.

The "signal limits" are quoted in several ways. Making the simple cuts:

asymmetry $A < 0.4$

sphericity $S > 0.5$

Energy $E = mp \pm 0.2\ mp$

gives a detection efficiency for $p \to e^+\pi^o$ of 55% including nuclear effects, reconstruction efficiency and containment cuts. (These cuts are also the background rejection cuts). Including the p/n ratio in water implies S/N = 1/1 per kiloton per year at a $\tau_p(\text{B.R.}:p \to e^+\pi^o)^{-1} = 0.85 - 1.7 \times 10^{32}$ yr. If $(\text{B.R.}: p \to e^+\pi^o) = 30\%$, then it is possible to measure $\tau_p \lesssim 2.5 - 5 \times 10^{31}$ years with this detector in a one year exposure. However, this is quite conservative since neutron decays, other proton decays such as $p \to e^+(\eta,\omega,\rho)$ and muons from the floor are not included in this estimate. A "detector rate" in excess of the neutrino reactions would reinforce evidence for proton decay. Some of the cuts used for this analysis are shown in the scatter plots in Figures 15 and 16, of energy against opening angle and asymmetry, for proton decays and neutrino background. The neutrino events shown are from real data. Note that limits in this discussion are from modelling known parameters of the detector.

Other Physics

As indicated in the introduction, a number of other cosmic ray experiments can be extended with these detectors. Of particular interest are experiments to search for neutrino oscillations. Since the atmosphere acts as a shell source of neutrinos, the number of neutrinos going "down" into the detector is roughly the same as the number coming up from the earth; thus a difference in the rates up/down or in the types up/down may signal neutrino oscillations, or matter regeneration. For 2 component neutrinos:

$$\Delta m \sim \left(\frac{N_e}{N_\mu} \right)^{1/4} \left(\frac{p_\nu}{\ell \sin 2\theta} \right)^{1/2}$$

This mass difference is, of course, predictable in many unified models. Figures 17a, b indicate sensitivity of these detectors versus other possible oscillation experiments. Other rare phenomena involving muons and neutrinos can be detected; for example, muon bundles of very high multiplicity, indicating, perhaps, the production of new quantum numbers in H.E. air showers.

IV. Detector: Parameters and Status

Important parameters of the detector are listed below:

1. Location: U. of Utah cosmic ray station in Park City, Utah. 1700 m.w.e. depth; 14,000' horizontal tunnel. Country water 10,000 gal/min.
2. Detectors: 800 5" hemispherical bialkali 11 stage p.m.t. with no back reflector: EMI 9870B. δt FWHM = 7 nsec, gain to 10^7, lattice spacing $\approx$ 1 meter.
3. Mass: 950 tons, contained by steel wall; 6 x 10^{29} N/ton.
4. Water: λ_{max} = 20 meters; R/O-DI hypalon liner; mirror liner reflectance = 0.9.
5. Shield: 1600 wire gas prop. layer tag; 600 m^2 BBQ wave shifter with 1 meter H_2O radiator directional veto and absorber.

The detector is calibrated primarily by utilizing the muon trajectories tagged by the proportional counters; they measure muon tracks to ± 10 mrad.

When a muon trajectory is known, the relative time of arrival and number of photons of the resulting Cerenkov shock can be calculated for each photomultiplier independently from the p.m.t. response. In addition, the reconstruction can be tuned by comparing the fitted muon tracks with the independently tagged tracks. Secondary calibration systems using other light sources will also be employed.

The shield thus supplies:

1. Hadronic absorption
2. Veto
3. Tagging through-going and stopping entering muons
4. Tagging existing muons from neutrino reactions, muon inelastics and proton decay
5. Event containment
6. Calibration

There are several special technical features of the apparatus. One is the deadtimeless data acquisition system for each p.m.t. which uses dual bank switched read-simultaneous-write buffer memory to record a purely digital deadtimeless vernier TDC and a 6-bit flash-encoder type ADC. The hardware trigger uses a high speed analog multiplicity decision matrix and the directional veto layer. The pulse height level for electromagnetic showers corresponds to about 0.7 photoelectrons/MeV, giving about 20 p.e./muon decay and 110 p.e./meter of traversing muon track.

The detector is currently under construction with all major components designed and/or under construction. Data is expected to begin in early 1980. Figures 18 and 19 show test set-ups and mockups used to measure the parameters of the Monte Carlo. Figures 20-22 show some details of the detector and components as they will appear in the Park City site. For example, the apparatus shown in Figure 18 allowed us to confirm the rms angular dispersion of 5° of the Cerenkov cones.

V. Improving and/or Extending Proton Decay Experiments

If proton decay is not found by the currently scheduled detectors, a general list of options is the following:

1. Grow larger with $\rho = 1$ detectors or increase the density
2. Go deeper - for example, at 5,000', the muon flux is reduced by a factor of 300 compared to the Park City site
3. Improve the detection technique

Unfortunately 1-3 are somewhat mutually exclusive. Going deeper reduces the sizes of holes available since the rock strength is limited. The detection technique can be improved by building more finely divided detector elements to reduce the graininess of the "image". This is particularly important to reduce neutrino background which currently limits size in our detector, unless the reconstruction evolves. (In fact, there are a number of promising techniques in information theory and image recognition which can be applied to the signals from Cerenkov detectors).[6] Fine grain detectors are either small, or cost prohibitive, or both.

If proton decay is found, then a whole range of research opens up. It will be important to measure the partial rates, absolute rates, and polarizations of the decay products. Generally the previous discussion is still true; however, in this case, additional detection techniques can be contemplated:

1. Heavy shields, with true ionization calorimeters
2. Liquid argon image chambers
3. Calibration at accelerators
4. Order of magnitudes increase in photodetector density
5. High density media, or liquid hydrogen media (no Fermi motion)
6. Magnetic fields for muon polarization or muon momenta
7. Hybrid detector - Cerenkov prompt light and slow decay time scintillator for ionization calorimetry
8. Great depths

At the present time the HPW group is engaged in a search for a series of deep and large sites in the Western Hemisphere. Currently, water Cerenkov detectors require about one 5" pmt/ton, costing \$250/tube, plus electronics, or $\$1/cm^2$ of cathode area. This implies that a 100 kiloton detector (6×10^{34} nucleons) would cost more than \$25 million, requiring 100K channels of electronics! Thus our group has also embarked upon a plan to achieve the following device:

1. Increase in photocathode area by X10 - X100 per unit
2. Similar space-time resolution as a pmt
3. Price/cathode area lower by X10

To this end, W. Huffman at Harvard University has reliably constructed four standard Cs_3Sb semiconductor photocathodes similar to an S-11 photocathode. Two were constructed as cylinders with a coaxial 1 mil W wire. When filled with ultra pure 90% A, 10% CH_4 these functioned as p.m.t.'s with gas proportional gains up to 10^4. However these devices have a somewhat limited dynamic range and time resolution. We have thus devised a Large Area Light Amplifier. This device will consist of an outer cylindrical photocathode of a standard C_s-S_b type with a coaxial inner cylinder of a very fast phosphor; for example, ZnO(Ga) has a decay time of 0.4 nsec. Photo electrons are radially focussed and accelerated at high voltage onto the inner phosphor producing 15-20 photons/KeV in the phosphor. These photons are then reflected by a 20 nanometer Al film on the outer surface of the phosphor through the inner glass envelope onto an inner wavelength shifter rod coaxial with the inner (phosphor) and outer (photo cathode) glass cylinders. The resulting shifted photons are then detected by small (1" - 2" diameter) p.m.t. at each end. This device will scale "arbitrarily" along its axis, and provides good spatial resolution by pulse height/time division. We believe it will be possible to reach the goals outlined above. Figures 23-25 show the gas gain device and properties, and the large area light amplifier. We note that these devices may be useful in very large cosmic ray detectors of the DUMAND variety, and for calorimetric work on large scales at accelerators.

ACKNOWLEDGEMENTS

It is a pleasure to acknowledge my collaborators who have accomplished most of the work discussed in this note:

J. Blandino[1], U. Camerini[3], D. Cline[3], E.C. Fowler[2], W.F. Fry[3], J.A. Gaidos[2], J. Innvaer[3], W.A. Huffman[1], G. Kullerud[2], R.J. Loveless[3], A.M. Lutz[1], J. Matthews[3], R. McHenry[2], R. Morse[3], T.R. Palfrey[2], I. Orosz[3], D.D. Reeder[3], C. Rubbia[1], A. Sarracino[3], A.H. Szentgyorgyi[3], H. Wachsmuth[3], R.B. Willmann[2], C.L. Wilson[2]

In addition many ideas and problems have been pointed out to me by members of the Physics Department at Harvard University, particularly Paul Frampton, Howard Georgi, Sheldon Glashow, Roy Schwitters, Steven Weinberg, Richard Wilson and Asim Yildiz. For many pleasant discussions I am grateful.

[1]Harvard University, Cambridge, MA 02138

[2]Purdue University, Lafayette, IN 47907

[3]University of Wisconsin, Madison, WI 53706

REFERENCES

[1]Many references are found in J. Blandino et al., "A Decay Mode Independent Search for Baryon Decay Using a Volume Cerenkov Detector" and "A Multi Kilo Ton Detector to Search for Nucleon Decay", proposals to the Department of Energy, U. of Wisconsin preprints, 1979, 1980, and also the following references:

H. Georgi and S.L. Glashow, Phys. Rev. Lett. 32, 438 (1973).

"Baryon and Lepton Non Conserving Processes", S. Weinberg, Harvard Preprint, HUTP-79/A050; "The Future of Elementary Particle Physics", S.L. Glashow, HUTP-79/A059.

"On the Effective Lagrangian for Baryon Decay", J. Ellis, M.K. Gaillard and D.V. Nanopoulos, CERN Theory Preprint (1979) and private communication.

W.J. Marciano, Rockefeller University Preprint COO-2232B-173 (1979).

T.J. Goldman and D.A. Ross, Phys. Lett. 84B, 208 (1979).

H. Georgi, H. Quinn, S. Weinberg, Phys. Rev. Lett. 33, 451 (1974).

F. Wilczek, talk at the Fermilab Electron-Photon Conference, 1979.

[2]The success of "SU(5)" depends on the success of SU(3) and SU(2) x U(1); for example: m_{Higgs}? $m_{W,Z}$? Generations m_c, m_b', m_t' ? C-P violation? m_ν and ν oscillations? Desert? n electric dipole? confinement...? SU(5) monopoles? The successes in Reference 1 are, however, still of great encouragement.

[3]Cosmological implications are reviewed by G. Steigman in "The Proceedings of the Conference on Non-Conservation of Baryon Number", Dec. 1979, U. Wisconsin Publication and also in Reference 1.

[4]For a discussion and current proton decay limits, see, for example, F. Reines in "Unification of Elementary Forces and Gauge Fields", Ben Lee Memorial Conference, Fermilab Pub., October 1977.

[5]For example, $P(n)_2 \approx n^{-5}$ where n is the muon multiplicity; n has been observed as high as 13/m in 1 μsec. deep underground. The Kolar events have been ascribed to neutral heavy leptons, background or other phenomena.

[6]For example, Composite Classifier, Sequential Weight Incremental Factor, and Hierarchy Segmentation Techniques.

LIST OF TABLES

LIST OF FIGURES

TABLE 1

Properties of H_2O & Cerenkov Light

ρ = 1 ton/m^3	n = 1.33
L_{abs} = 78.8 cm	$\theta_c = 42.3^o$ (β = 1)
L_{rad} = 36.1 cm	photons/cm ∿ 200*
dE/dx = 2.03 MeV/cm	λ_{abs} ∿ 20 m*
% free p = 11%	Oxygen Fermi ∿ 220 MeV/c
% p = 56%	Oxygen π escape ∿ 65%*
E_{crit} = 90 MeV*	

* = rule of thumb

TABLE 2

Monte Carlo

Step	Description/Input
1a. Proton Decay Generation	Branching mode, Fermi motion Nuclear rescattering
1b. Neutrino Event Generation	ν data from ANL 12', Gargamelle, cosmic ray neutrino spectra
1c. Muon interactions and entering hadrons	Muon fluxes, particle production spectra in deep inelastics, accelerator shielding calculations, π, K scattering data
2. Track and Shower Generation	Multiple scattering, ionization Bremsstrahlung, pair production, Compton scattering, particle production spectra and cross sections; 1 cm segments to 5 MeV.
3. C Light	$1/\lambda^2$ spectrum, θ, $\delta\theta$, $\emptyset$; individual photons
4. Light Propagation	Spectral absorption in water, mirror reflectance, p.m. geometry
5. Detection	p.m. and electronic properties: δt, δ pulse height, noise, resolution

TABLE 3

Reconstruction

Step	Description
1. Total energy	Sum photoelectrons
2. Vertex	Neighborhood of earliest tube to fire. Minimize $\sum_i (\lvert \bar{d}_i \rvert - v(t_i - t_o))$ $\bar{d}_i = \bar{x}_i - \bar{x}_o$
3. Event Axis	Diagonalize the moment of inertia (tensor) $T^{\alpha\beta} = \sum_i P_i (d_i^2 \delta_{\alpha\beta} - d_i^\alpha d_i^\beta)$ $\alpha,\beta = 1,2,3,$ p_i = photoelectrons on i^{th} tube, the principal eigen-vector is event axis
4. Optimization	Do Step 3 for each ± hemisphere around origin. Weight tubes according to how closely they lie near the cones (conicity). Iterate Step 3 again changing vertex. Correct for phototube calibrations.
5. Define Asymmetry A and Sphericity S	$A = \frac{P_+ - P_-}{P_+ + P_-}$ $S = \frac{3}{2} \min \frac{\sum_i P_\perp^{i2}}{\sum \vec{p}^{\,i2}}$ where $P_\pm$ is total energy along event axis, and $\vec{P}_i = P_i \frac{(\vec{X}_i - \vec{X}_o)}{\lvert X_i - \bar{X}_o \rvert}$
6. Maximum likelihood	Hypothesis generation from Monte Carlo

TABLE 4

Decay Mode	Signature	Major Backgrounds
$p \to e + \pi^o$	E_e, $E_\pi \sim 460$ MeV $\theta_{e\pi} < 6^o$ $E_{vis} \sim 940$ MeV	$\nu_e N \to e\pi X$ + nuclear scatter
$p \to e + \eta$ ($\hookrightarrow \gamma\gamma/3\pi^o$)	$E_e \sim 300$ MeV $E_{vis} \sim 940$ MeV	$\nu_e N \to en(\pi^o)X$
$p \to e + \omega$ $(\mu \to e)$*	$E_e \sim 150$ MeV $E_{vis} \sim 450$ MeV	$\nu_e N \to en(\pi)X$ $n > 1$
$p \to e + \rho^{o*}$	$E_e \sim 150$ MeV $E_{vis} \sim 380$ MeV	
$p \to \mu + \pi^{o*}$	E_μ, $E_\pi \sim 460$ MeV $\theta_{\pi\mu} < 10^o$ $E_{vis} \sim 650$ MeV	$\nu_\mu N \to \mu\pi X$ + nuclear scatter
$p \to \mu + K_1^{o*}$	$E_\mu \sim 300$ MeV $E_{vis} \sim 325$ MeV	$\nu_\mu N \to \mu\pi\pi X$ $\mu K_1^o X$
$n \to e + \pi^-$	$E_e \sim 460$ MeV $\theta_{e\pi} < 6^o$ $E_{vis} \sim 800$ MeV	$\nu_e N \to e\pi X$
$n \to e + \rho^-$	$E_e \sim 150$ MeV $E_{vis} \sim 470$ MeV	$\nu_e N \to e\pi\pi X$
$n \to \mu + \pi^{-*}$	E_μ, $E_\pi \sim 460$ MeV $E_{vis} \sim 615$ MeV	$\nu_\mu N \to \mu\pi X$

*Events with $\mu^+ \to e^+$ decays yielding $\sim$ 50 photoelectrons within a few microseconds

TABLE 5

Decay Type	Lepton	Meson	
$p \to e^+\pi^o$	$\langle E_{e^+}\rangle \sim 450$ MeV	$\langle E_{\pi^o}\rangle \sim 480$ MeV $\langle E_{\gamma 1,2}\rangle \sim 240$ MeV $\langle \theta_{\gamma 1 \gamma 2}\rangle \sim 45^o$	Back to back event. Energy between 1 forward and 2 backward in showers 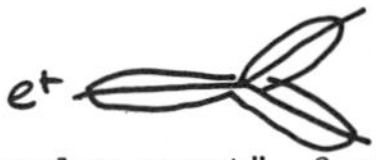
$p \to e^+\eta^o$	$\langle E^+\rangle \sim 310$ MeV	$\langle E_{\eta^o}\rangle \sim 630$ MeV $\langle E_{\gamma 1,2}\rangle \sim 320$MeV $\langle \theta_{\gamma 1 \gamma 2}\rangle \sim 125^o$	"Triangular event". 3 em showers of $\sim$ same energy 
$p \to e^+\rho^o$	$\langle E_{e^+}\rangle \sim 150$ MeV	$\langle E_{\rho^o}\rangle \sim 790$ MeV $\langle E_{\pi^\pm}\rangle \sim 400$ MeV $\langle \theta_{\pi^+\pi^-}\rangle \sim 150^o$	"Triangular event" 3 low energy. Few Cerenkov light tracks. e^+ from $\pi^+ \to \mu^+ \to e^+$ decay chain 
$p \to e^+\omega^o$	$\langle E_{e^+}\rangle \sim 145$ MeV	$\langle E_{\omega^o}\rangle \sim 800$ MeV $\langle E_{\pi^o_{\pm}}\rangle \sim 270$ MeV Isotropic 3 body decay	Isotropic event. 5 low energy. Few Cerenkov light tracks. e^+ from $\pi^+ \to \mu^+ \to e^+$ decay chain 
$p \to \mu^+\pi^o$	$\langle E_{\mu^+}\rangle \sim 465$ MeV	$\langle E_{\pi^o}\rangle \sim 475$ MeV $\langle E_{\gamma 1,2}\rangle \sim 235$ MeV $\langle \theta_{\gamma 1,2}\rangle \sim 45^o$	Back to back event. No light balance between low C light back and e.m. show. e^+ from μ^+ decay 

TABLE 6

Number of Collected Photoelectrons

(Non Wavelength Shifted Case)

Decay type	$p \to e^+\pi^o$	$p \to e^+\eta^o$	$p \to e^+\rho^o$	$p \to e^+\omega^o$	$p \to \mu^+\pi^o$
Lattice spacing					
1.0 m	662±106	655±99	269±60*	337±67*	453±100*
1.5 m	222±72	220±62	79±26**	111±52**	141±47**

* not including 18±10 ph. el. from e^+ (from $\pi^+ \to \mu^+ \to e^+$ decay chain)
(<Pe> = 32±13 MeV)

** not including 6±3 ph. el. from e^+ (from $\pi^+ \to \mu^+ \to e^+$ decay chain)

N.B.: If large phototube fluctuations are removed, the error in pulse height goes to a factor 0.6 of those listed, and the pulse height goes to a factor of 0.9 of those listed.

TABLE 7

Non Wave Length Shifted Case

Decay Mode & Neutrino Backgrounds	<ASYM>	<SPH>	<D>* m
$p \rightarrow e^+ \pi^o$	.16±.1	.52±.06	.11±.11
$p \rightarrow e^+ \eta^o$	.17	.75	.35
$p \rightarrow e^+ \rho^o$	.24	.67	.35
$p \rightarrow e^+ \omega$	.32	.78	.37
$\nu N \rightarrow X$	.7	.4	.5
$\nu N \rightarrow \mu \pi p$	.5±.2	.6±.1	.5

*RMS error on vertex location

N.B.: Simple "pattern recognition" cuts to accept/reject proton decay/neutrino events are: $A < 0.4$

$S > 0.5$

$E = m_p \pm 0.2 m_p$

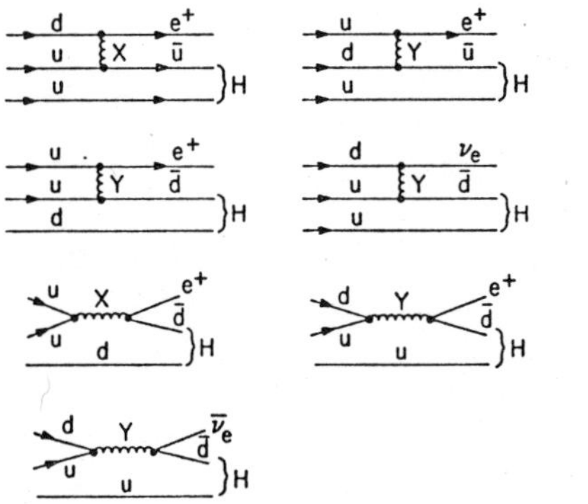

QUARK DIAGRAMS FOR PROTON DECAY

FIGURE 1.

SCHEMATIC OF PROTON DECAY DETECTOR

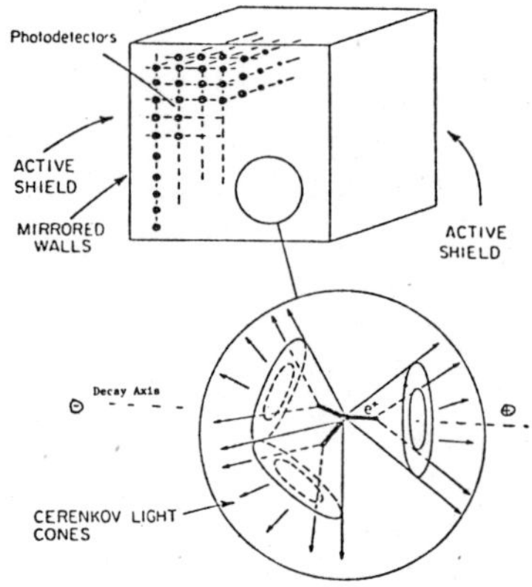

FIGURE 2.

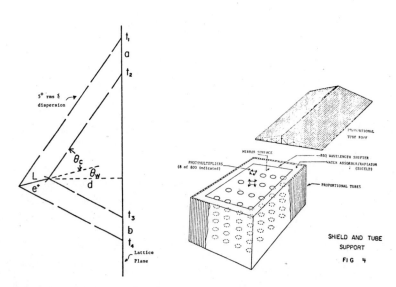

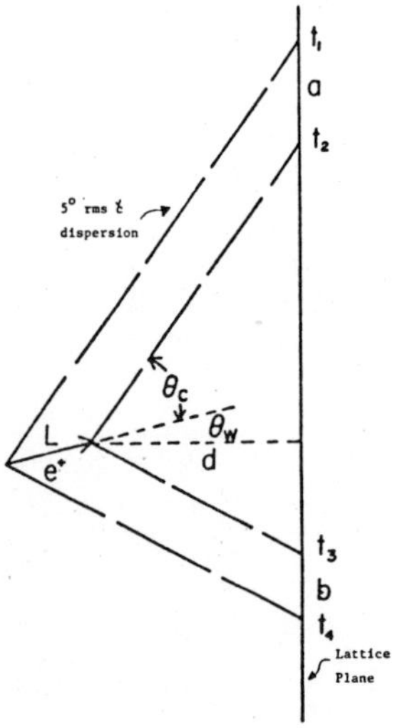

Cherenkov Cone from Track Segment

FIGURE 3.

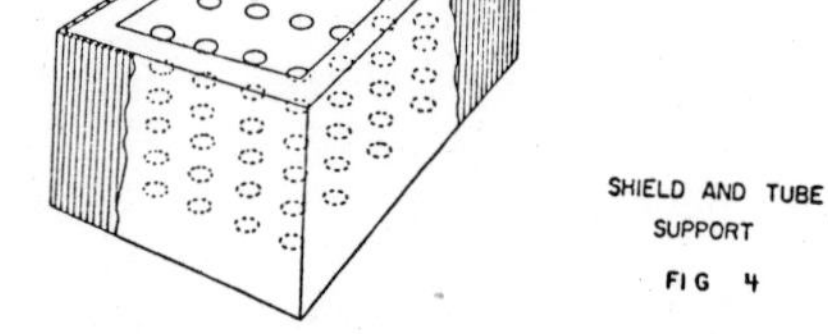

SHIELD AND TUBE
SUPPORT

FIG 4

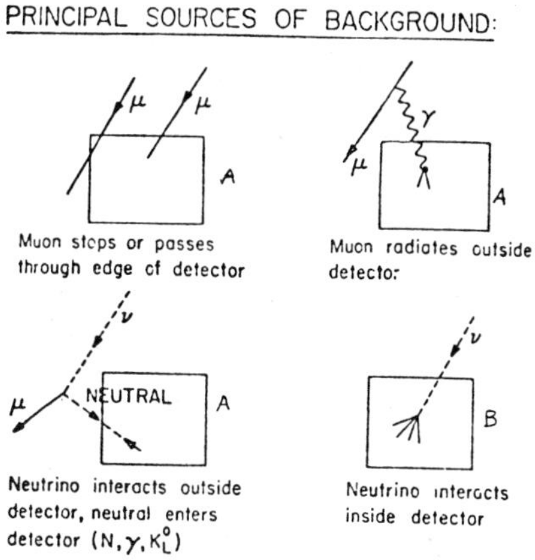

FIGURE 5

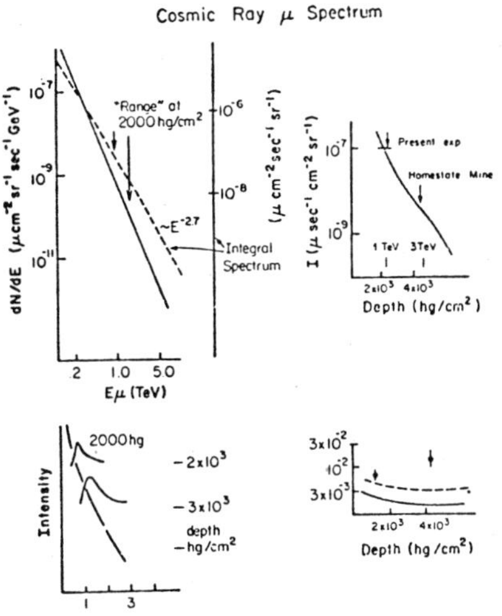

FIGURE 6.

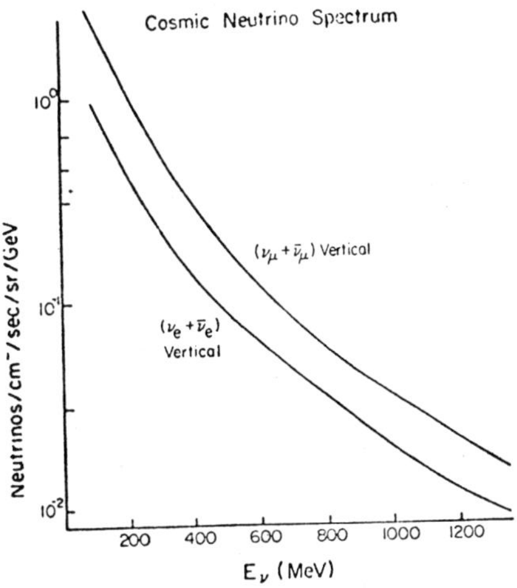

FIGURE 7.

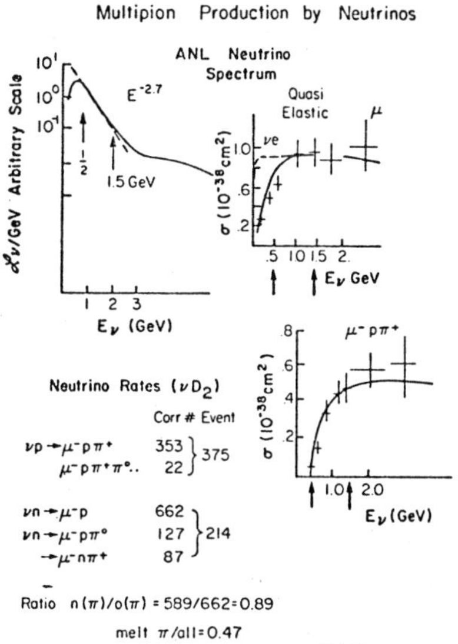

FIGURE 8.

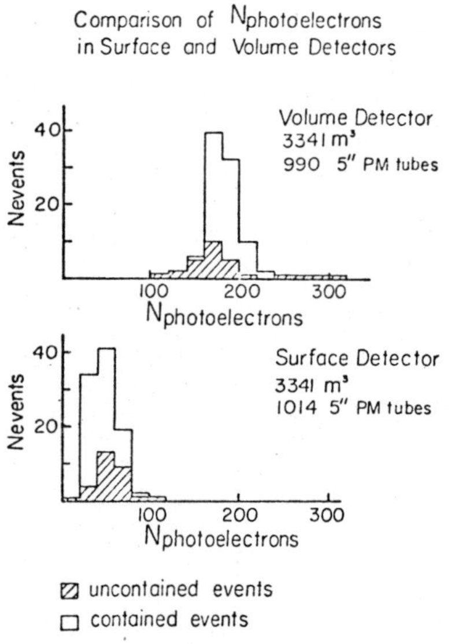

FIGURE 9.

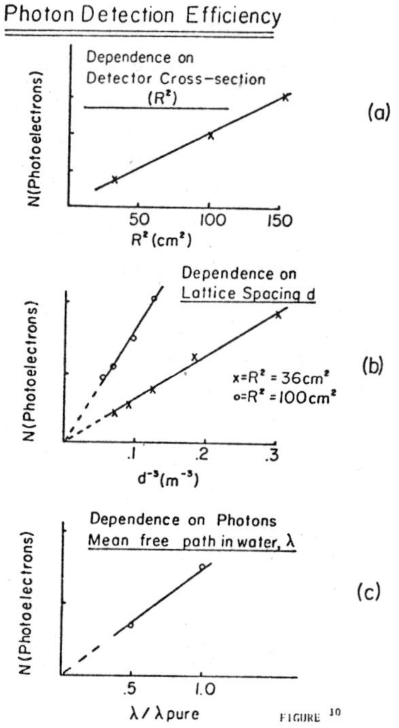

FIGURE 10

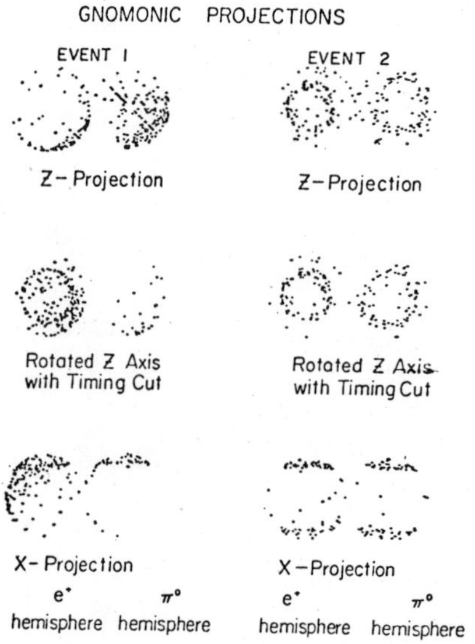

FIGURE 11

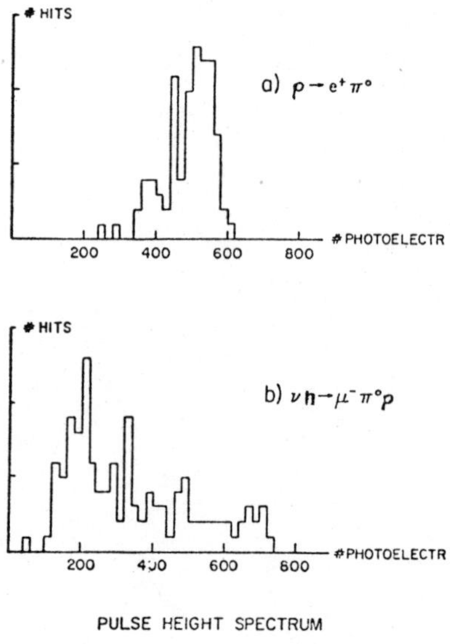

Figure 12

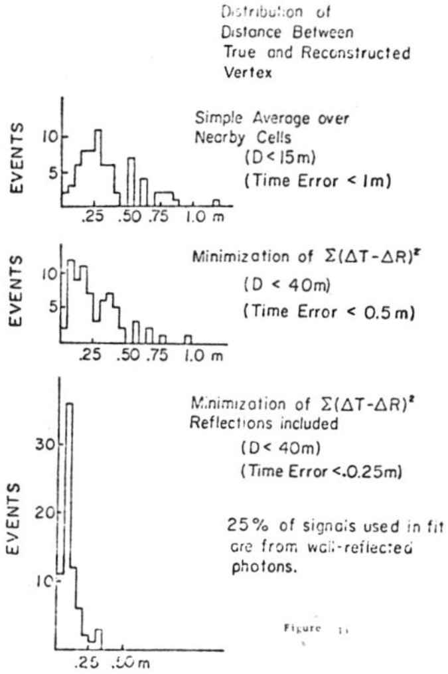

Figure 13

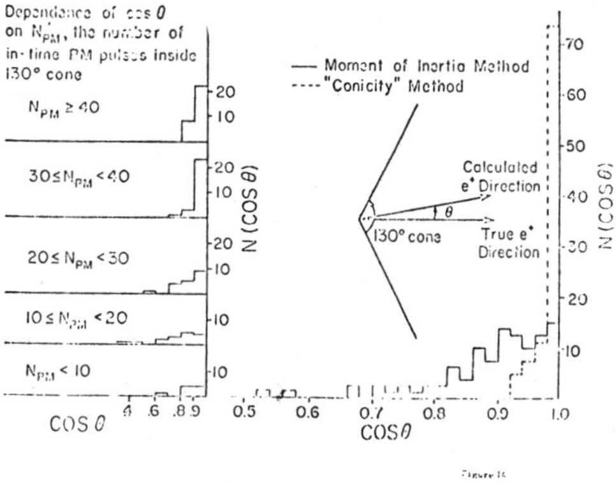

Figure 14

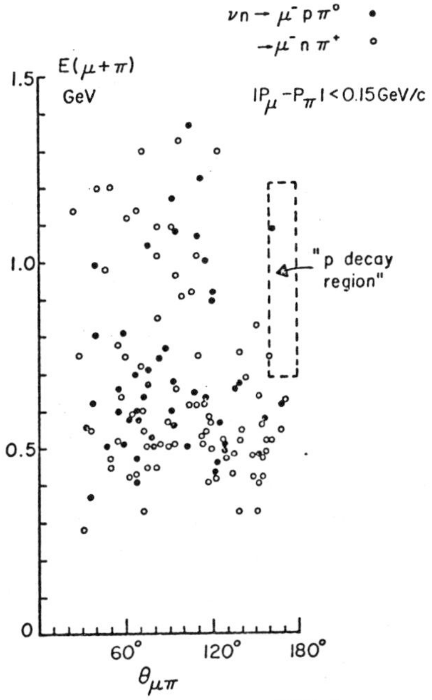

FIGURE 15

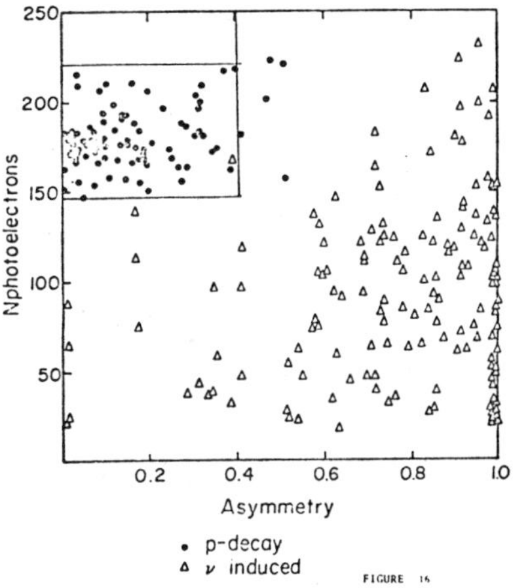

FIGURE 16

Figure 17 a.

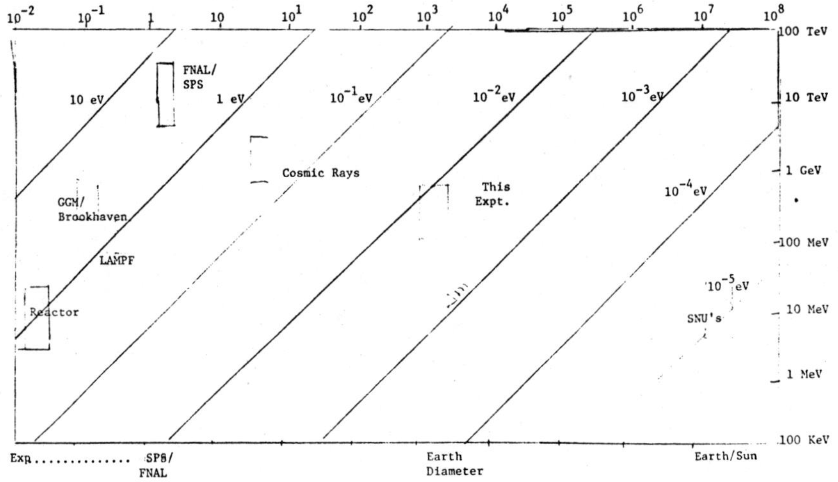

Figure 17b.

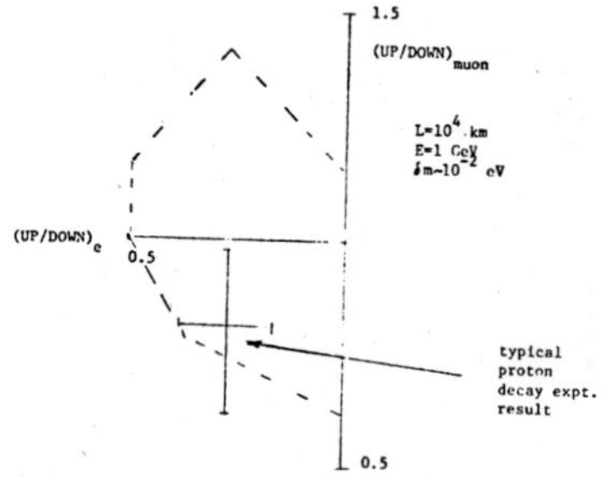

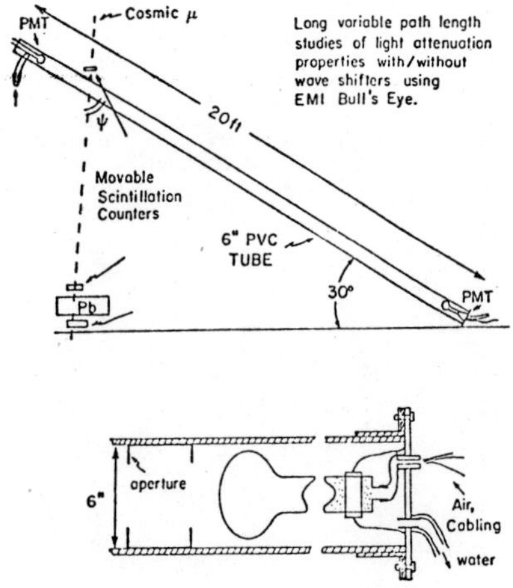

FIGURE 18

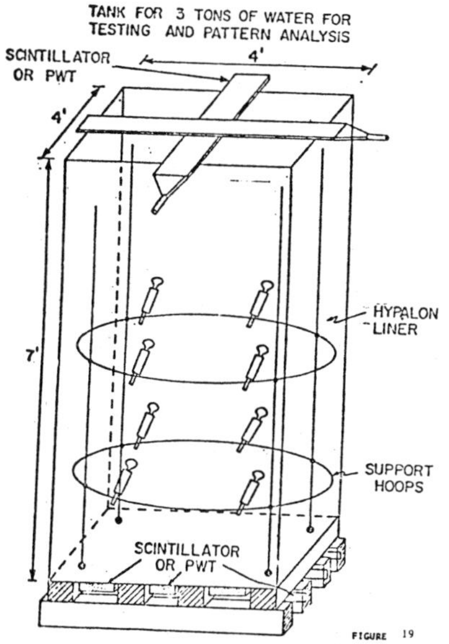

FIGURE 19

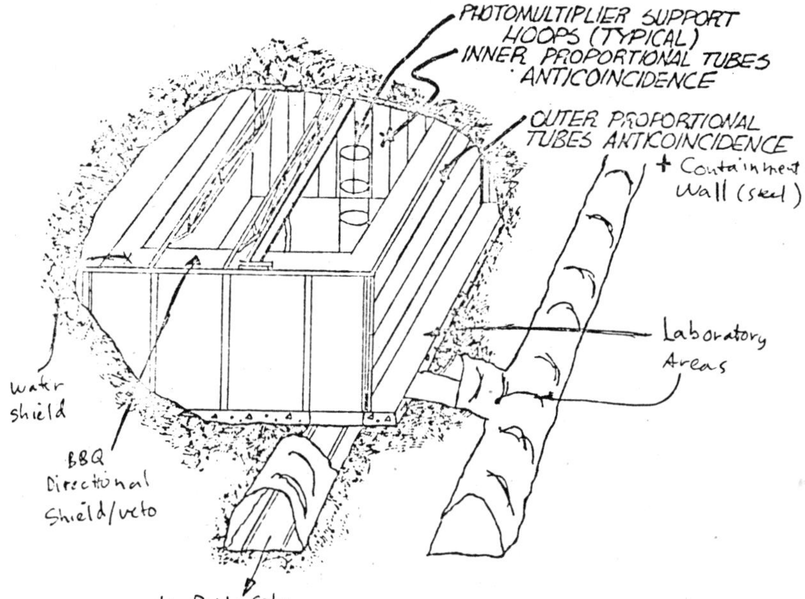

Site with P.M. Tube Support Figure 20

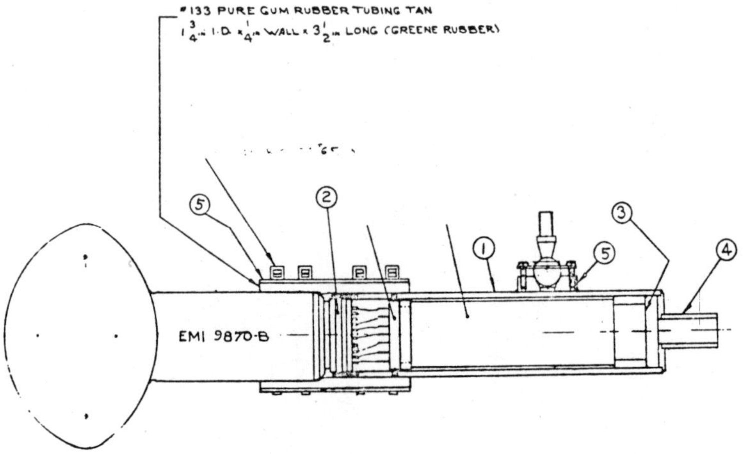

Figure 21

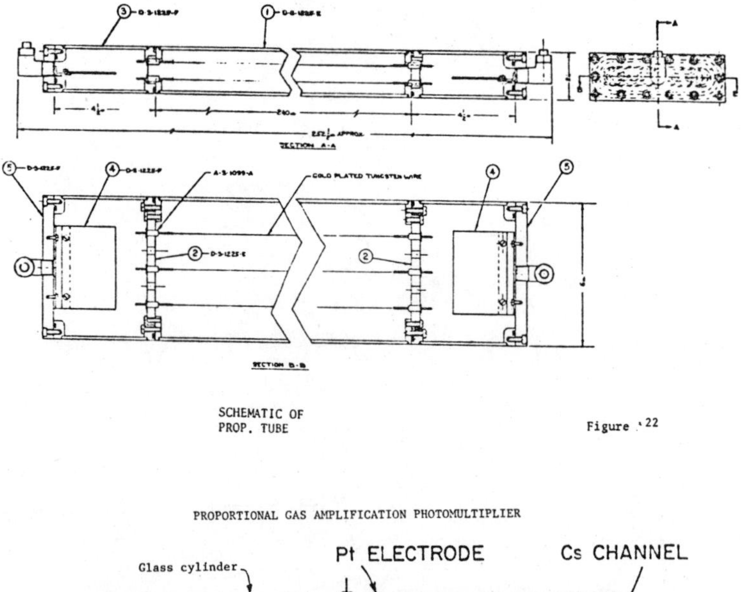

SCHEMATIC OF
PROP. TUBE

Figure 22

PROPORTIONAL GAS AMPLIFICATION PHOTOMULTIPLIER

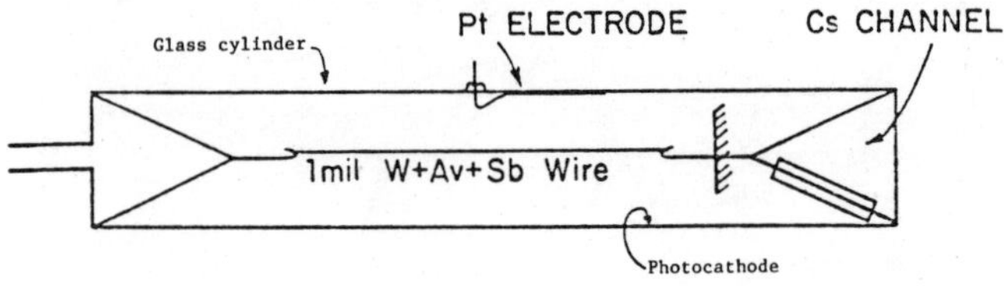

Figure 23

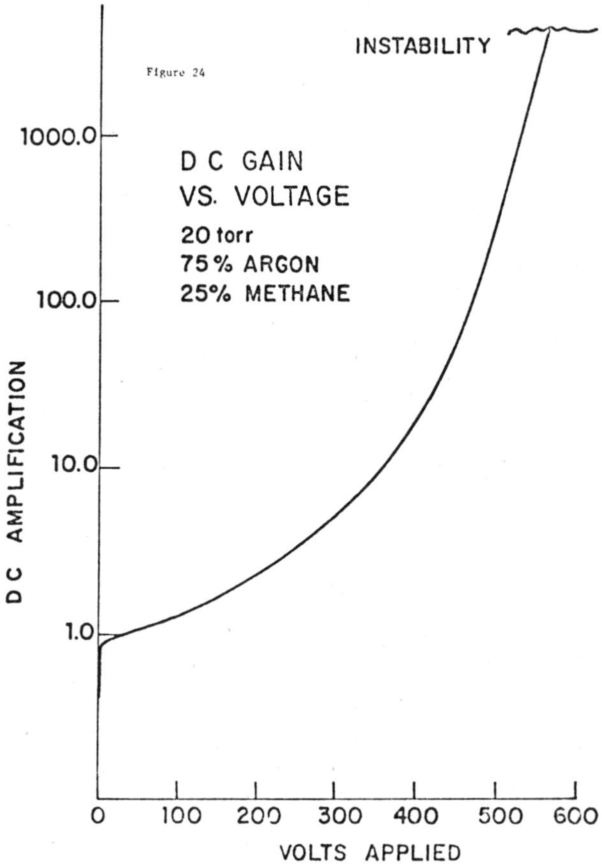

LARGE AREA LIGHT INTENSIFIER

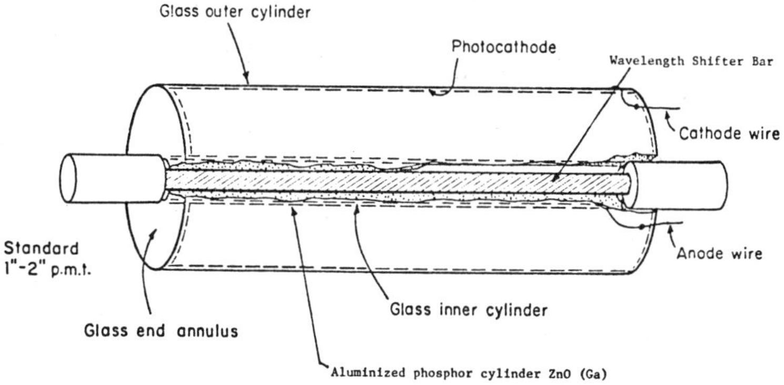

Figure 25

POSSIBILITIES OF EXPERIMENTS TO MEASURE NEUTRON-ANTINEUTRON MIXING

Richard Wilson

Harvard University

1. INTRODUCTION

The neutron and antineutron have quantum numbers which differ only by the baryon number. The masses are believed to be identical, but they can be separated by a magnetic field which will act in opposite ways on the neutron and antineutron magnetic moments and in ordinary matter by the differences between the neutron-nucleus potential and antineutron-nucleus potential.

Following the successes of the Weinberg-Salam theory, grand unification theories are becoming popular. These theories predict proton (or neutron) decay into mesons and leptons. This is a change of baryon number one, and while the theories neither predict nor deny appreciable neutron-antineutron mixing, our eyes have been opened and a change of baryon number two subject only to the constraints of existing experiments can be contemplated [1, 2].

However we shall see that these experiments are not as sensitive as experiments on proton or neutron decay. This study was stimulated by conversations with Sheldon Glashow, who discussed the problem in some summer school lectures at Cargese [3].

2. THEORY

I outline briefly the theory of mixing. This follows the well known K_o, $\bar{K}_o$ mixing theory. We start with a 2 component wave function

$$\psi = \begin{pmatrix} \bar{n} \\ n \end{pmatrix}$$

and the mass is given by the mass matrix

$$\begin{pmatrix} M + V_{\bar{N}} & \alpha \\ \alpha & M + V_N \end{pmatrix}$$

where α is the transition parameter (with units of mass) mixing the $\bar{n}$ and n states. $V_{\bar{N}}$ and V_N represent additional terms for the antineutron and neutron respectively. This can include the magnetic field interaction $V_{\bar{N}} = -\vec{\mu}_N \cdot \vec{B}$ and $V_N = +\vec{\mu}_N \cdot \vec{B}$ or it could represent neutron-nucleus potentials.

The mass eigenstates are

$$|N_1\rangle = |N\rangle + \theta|\bar{N}\rangle$$
$$|N_2\rangle = -\theta|N\rangle + |\bar{N}\rangle$$

where $\theta = \alpha/(V_{\bar{N}} - V_N)$.

Then if we start with a neutron ($|N>$) at time t = 0, the probability of finding an antineutron at time t = t is

$$P(t) = [|N(t)>]^2 = \frac{\theta^2}{2} \left[1 - \cos \frac{(V_{\bar{N}} - V_N)t}{\hbar}\right]$$

(where we ignore the free neutron, or antineutron, decay).

For small t, this becomes

$$P(t) = \frac{\alpha^2 t^2}{\hbar^2}$$

3. EXPERIMENT

If we imagine a beam of neutrons of intensity N/sec, after a time t, we expect

$$\frac{\alpha^2}{\hbar^2} (Nt^2)$$

antineutrons. The experiment, then, is to produce as many neutrons, N, as possible, allow them to be observed for as long a time, t, as possible and watch for antineutrons at the end. Figure 1 shows a schematic of such an experiment.

The condition of small t, for free neutrons in a magnetic field, is

$$\frac{2\mu Bt}{\hbar} \ll 1$$

Physically what this formula tells us is that unless N and $\bar{N}$ have the same mass, the amplitudes will get out of phase and the amplitudes will not add coherently to give an intensity proportional to t^2.

Since neutrons of velocity v will travel a distance vt in this time, this becomes

$$\frac{2\mu}{\hbar v} \int_0^L B\, dl \ll 1$$

Since μ/h = 762 cycles/sec gauss, and for thermal neutrons v = 2000 m/sec, this becomes

$$\int_0^L B\, dl < 1 \text{ gauss meter.}$$

The earth's field must be shielded, but a simple "conetic" shield can shield this by a factor of 10^3 which is more than adequate.

The criterion for sensitivity of an experiment is maximization of Nt^2.

Let us go through some ways of doing this.

If I have a reactor with a hole of area A in the shielding, through which neutrons may pass, then the solid angle varies as A/L^2. t^2 varies as L^2/v^2, so the sensitivity Nt^2 is independent of distance, L. However, as we increase L, and bring the detector further away from the reactor, two advantages arise:

(i) the background goes down

(ii) once L exceeds the shielding size, the area of the beam can be allowed to increase and Nt^2 increases with L^2.

If I moderate neutrons in a cold moderator of liquid hydrogen or deuterium (20°K), the velocity of the neutrons goes down. Then $Nt^2 = NL^2/v^2$ goes up inversely as T. (Nt^2 increases by a factor of 15 in going from thermal neutrons to neutrons of liquid hydrogen temperature.)

It is tempting to try to increase t almost indefinitely without a concomitant increase in detector size by using an internally reflecting neutron guide such as used at Institut Laüe-Langevin (ILL), or in the limit using a bottle with reflecting walls such as used at Leningrad and ILL. But this will not work. Neutrons reflect off the metal surfaces but the antineutron nucleus amplitude will be different, and, since antineutrons annihilate in matter, complex. Unless antineutrons also reflect off the walls, the "clock" must be restarted at every reflection. It must be noted that even if N nucleus and $\bar{N}$ nucleus amplitudes differ by as little as 10%, the neutrons penetrate the mirror very slightly and there will be a small phase change between the neutron and antineutron amplitudes, limiting the number of reflections to ten which would not be enough to compensate for the loss of intensity in producing the ultra cold neutrons.

Clearly we want a lot of neutrons, and we either find them bound in stable matter (which has problems) or else we get the next best intensity near a large nuclear reactor. We might contemplate a 3000 megawatt (thermal) nuclear reactor such as one of those built for producing electric power. But these are designed to contain the neutrons as much as possible and therefore the time, t, will be very short.

In contrast, there exist a number of high power experimental reactors (10 megawatt to 60 megawatt) where neutrons can be allowed out through various beam ports. In what follows, I will describe the experimental design for some of these.

The reactors chosen for these studies are the MIT 10 megawatt reactor, the 60 megawatt ILL reactor at Grenoble, and the 40 megawatt OR reactor at Oak Ridge. The MIT reactor has recently been rebuilt to have an undermoderated light water cooled core, surrounded by a heavy moderator. The flux peaks just below the core, and the beam tubes are arranged to look at this part of the reactor.

Since we want the highest possible intensity we want to have a large beam tube and we therefore would plan to use an existing beam tube which has a 12 inch (30 cm) diameter as it goes through the shield, but has two 10 cm diameter tubes as it starts at the reactor.

The effective flux at the end of these tubes is 5×10^{13} neutrons/cm^2sec or about 2000×10^{13} neutrons/sec over the 400 cm^2 area.

The distance to the end of the beam tube is about 3 meters and the area of the beam tube is $\frac{\pi}{4} \times 30^2 = 1000\ cm^2$, so that the solid angle is

$$\frac{1000}{300^2} \times \frac{1}{4} \approx \frac{1}{400}$$

so that out of the reactor shielding we will get about 5×10^{13} neutrons/sec. Outside the reactor we can arrange special shielding but are probably blocked at the edge of the containment vessel or 10 meters from the reactor; the neutrons are thermalized in the reactor with an average velocity 2000 m/sec, so that we will observe them for 1/200 sec leading to a value of

$$Nt^2 = \frac{5 \times 10^{13}}{4 \times 10^4} \simeq 10^9 \text{ sec.}$$

This is illustrated in figure 2.

We could open a 2 foot diameter hole in the reactor shielding and have double the drift length, L, by extension of a beam tube outside (figure 3). If we stop at the MIT reactor property line we can double the drift length and drift time, t. The solid angle would be reduced a little and we would get $Nt^2 = 3 \times 10^9$ secs. If we proceed to the edge of the empty lot we get $Nt^2 = 10^{10}$ secs. In the limit we might go to the main road (Massachusetts Ave.) but the detector gets ever larger.

The Oak Ridge research reactor has a smaller flux but has a large area looking straight at the core (figure 4). Although the flux in the center is 3×10^{14} n/cm^2sec, it is less in this hole (3×10^{13} n/cm^2sec). This hole has an area 2' x 3' = 60 cms x 90 cms = 5400 cm^2. We can continue this 20 meters or more to the road (figure 5). Let us assume we do this and keep an area of 1 meter square at the end of the detector.

Number of neutrons is

$$3 \times 10^{13}\ n/cm^2 sec \times 5400\ cm^2 \times \frac{(100\ cm)^2}{4\pi \times (2000\ cm)^2} = 3.2 \times 10^{13}\ n/sec$$

Neutrons travel 2000 m/sec, so the time to travel 20 m is 1/100 sec. $Nt^2 = 3 \times 10^9$ sec. If we go further back or make a bigger detector we get a bigger effect. A detector 3 meters x 3 meters gives $Nt^2 = 3 \times 10^{10}$ secs.

The nice advantage of the reactor at ILL is the cold source. There are several beam tubes looking at the cold source. One of these, H18, has 5×10^9 n/cm^{2}sec over an area 3 cm x 13 cm; N becomes 2×10^{11} n/sec. If we observe for 2 meters t^2 would be 10^{-6} secs for thermal neutrons, but becomes 1.5×10^{-5} secs for cold neutrons. For such an experiment, $Nt^2 = 3 \times 10^6$ secs. This has been formally proposed for ILL [4].

For beam H8 at Grenoble, a 2 m flight path is all that is conveniently available. But with more work, a similar beam could be used with a 20 m path; the detector size goes up to 1 meter x 1 meter and Nt^2 about 10^8.

In the following table I list various possible values for Nt^2.

TABLE 1

<u>Lifetime Experiments</u>

Existing	100 tons	$N'\tau \sim 2 \times 10^8$ secs
Proposed	1000 tons	$N'\tau \sim 2 \times 10^9$ secs

<u>Possible Reactors</u>

MIT, inside containment	$Nt^2 \sim 10^9$ secs
MIT, inside property line	$Nt^2 \sim 3 \times 10^9$ secs
MIT, including empty lot	$Nt^2 \sim 10^{10}$ secs
ORNL, ORR	$Nt^2 \sim 3 \times 10^{10}$ secs
ILL, H18, 2m length	$Nt^2 \sim 3 \times 10^6$ secs

If we find 1 count in 10^7 secs ($\approx$ 120 days) we have

$$\frac{1}{10^7 \text{ secs}} = Nt^2 \frac{\alpha^2}{\hbar^2}$$

For $Nt^2 = 3 \times 10^6$ secs then $\frac{\alpha}{\hbar} = 5 \times 10^6$ secs. For $Nt^2 = 3 \times 10^{10}$ secs, then $\frac{\alpha}{\hbar} = 5 \times 10^8$ secs.

4. EXISTING DATA

We can now ask what our present information is. It comes from lifetime experiments.

In a 100 ton detector we have 6×10^{31} nucleons or 3×10^{31} neutrons (N'). When these try to generate antineutrons the amplitudes get out of phase in a characteristic time

$$\tau = \hbar/(V_{\bar{N}} - V_N).$$

If we set $V_{\bar{N}} - V_N = 100$ mev

$$\tau = \frac{10^{-27}\ \text{erg secs}}{100\ \text{mev}} = 6 \times 10^{-24}\ \text{secs}.$$

The sensitivity of such an experiment to be compared to Nt^2 becomes

$$N'\tau = 3 \times 10^{31} \times 6 \times 10^{-24} = 2 \times 10^{8}\ \text{secs}$$

We note this is <u>linear</u> with τ; it would be quadratic, as for the beam experiment, except that we can imagine the experiment being repeated after every τ secs.

Mohaptra and Marshak [5] find a similar limit in a different way. They note that in a nucleus the $\Delta B = 2$ transition $N + N \rightarrow$ pions is possible. They estimate the effect of this by setting $\alpha^2 = \Gamma M$ where M is a typical hadronic mass and find $\alpha \leq 10^{-20}$ ev with a lifetime of the neutron of 10^{30} years or $\alpha \leq 3 \times 10^{-21}$ ev for a lifetime of 10^{31} years which then gives $\hbar/\alpha \geq 2 \times 10^{5}$ secs.

I am not sure whether these are two different limits or whether they are two different ways of looking at the same phenomena.

5. DETECTION

It seems best to detect the N-$\bar{N}$ mixing by observing the annihilation of antineutrons after passing through a region of free travel. The antineutrons would be stopped on lithium and give 2 Gev of energy. Neutrons by contrast would be captured on the lithium of isotope 6 without gamma ray emissions.

Of the 2 Gev energy of annihilation, isotopic spin considerations suggest that 1/3 of the energy will be in neutral particles--mostly π^0 mesons, and we can then expect 600 Mev of gamma rays. We can detect these in a bank of lead glass Cerenkov counters or lead scintillator sandwich counters. These will hopefully be insensitive to neutron capture gamma rays.

There will be 10^{13} neutrons/sec striking the lithium wall. We estimate (or perhaps guess) that less than 1 out of 10^{3} neutrons will produce a gamma ray. Then we will have less than 10^{10} γ/sec incident on perhaps 100 lead glass counters, or 10^{8} γ/sec per counter. The resolving time of a single counter can be as low as 10^{-8} secs (= 10 nsecs) so that there is some probability of getting two or three γ ray pulses piling up--but not enough to produce a pulse of 600 Mev.

We also expect to have a cosmic ray anticoincidence to identify cosmic rays passing through or near the detector bank. Alternatively we could use a very thin target of graphite so that the neutrons would pass and antineutrons stop. Then we might be able to set up charged particle detectors to identify the annihilation vertex and thereby reduce cosmic ray background [6].

6. CONCLUSION

It is possible to perform an experiment to measure the lifetime for mixing of free N and $\bar{N}$ states if it is 5×10^8 secs or less. At the present moment stability of nuclear matter gives a lower limit of 10^5 secs. This is much less than the 10^{31} years discussed for experiments for lifetime in matter. The intermediate mass involved would be of the order of 10 Gev [3] and might be a Higgs boson. But unlike proton decay, there is no prediction so it is not clear whether such an extension of the lower limit is worth the effort.

7. ACKNOWLEDGEMENTS

I have had very helpful conversations with N.F. Ramsey and M. Goodman about experimental possibilities. Prof. Harley of MIT and Dr. Koehler of ORNL spent some time informing me of details of their respective reactors. Prof. Luke Mo kindly brought references 1 and 2 to my attention.

8. REFERENCES

1. A.D. Sakharov, ZHETF Pis Red 6, 772 (1967) [JETP Lett. 6, 236 (1967)].
2. V.A. Kuz'min, ZHETF Pis Red 6, 335 (1970) [JETP Lett. 12, 228 (1970)].
3. S.L. Glashow, Harvard preprint HUTP-79/A059.
4. K. Green, G. Fidecaro and M. Baldo-Ceolin, proposal to ILL, 8 Feb. 1980.
5. R.N. Mohapatra and R.E. Marshak, "Phenomonology of Neutron Oscillations,"CCNY-VPSU preprint, 1978.
6. Suggested by Dr. Matthew Goodman.

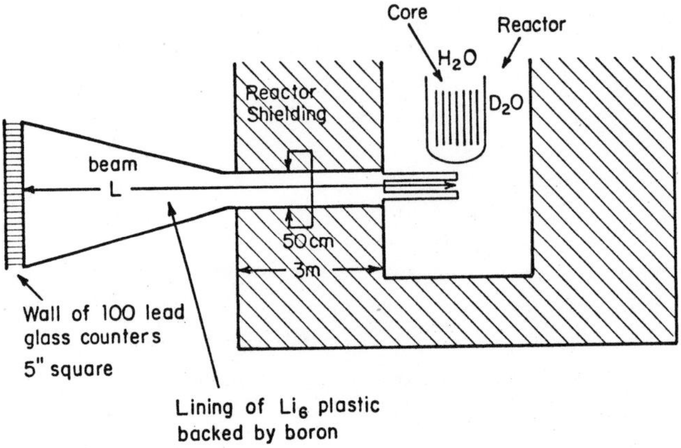

SCHEMATIC FOR MIT REACTOR

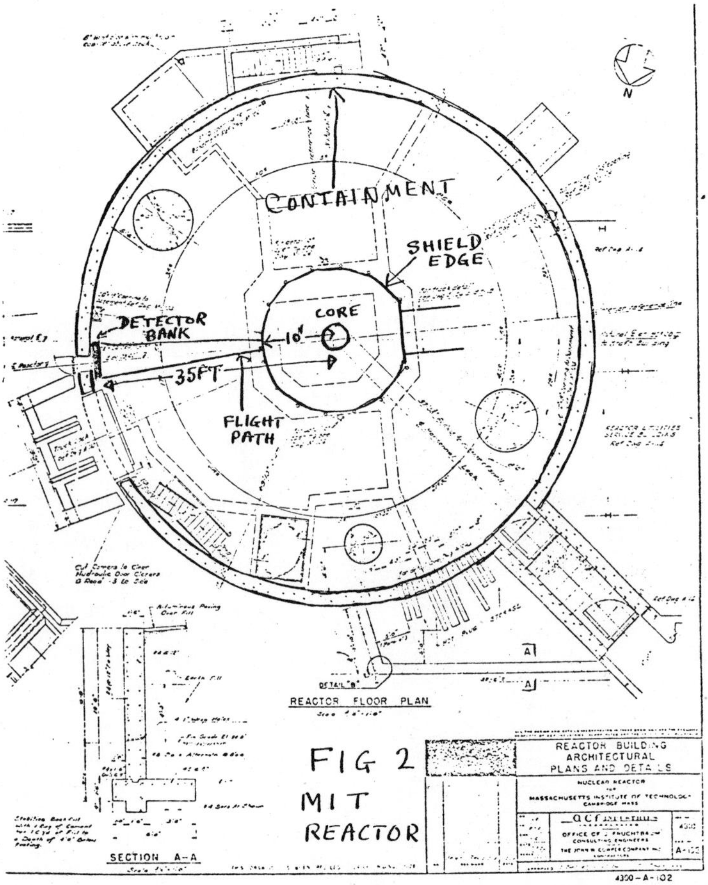

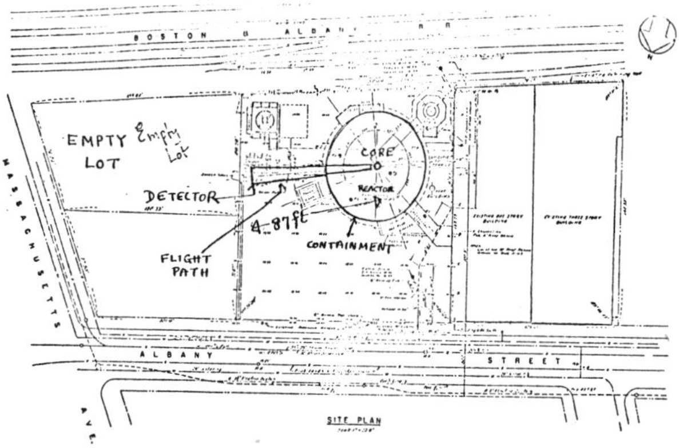

ARRANGEMENT FOR OUTSIDE FIG3

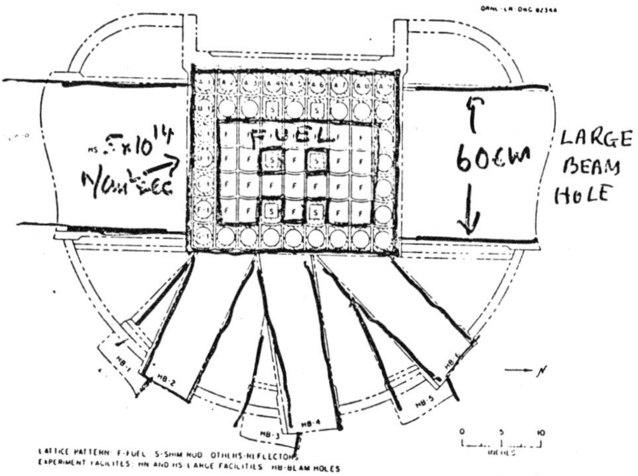

LATTICE PATTERN: F-FUEL S-SHIM ROD OTHERS-REFLECTORS
EXPERIMENT FACILITIES: HN AND HS LARGE FACILITIES HB-BEAM HOLES

Fig. 1.1. Horizontal Section Through Reactor Center Line. ORR

FIGURE 4.
SECTION THRO' REACTOR

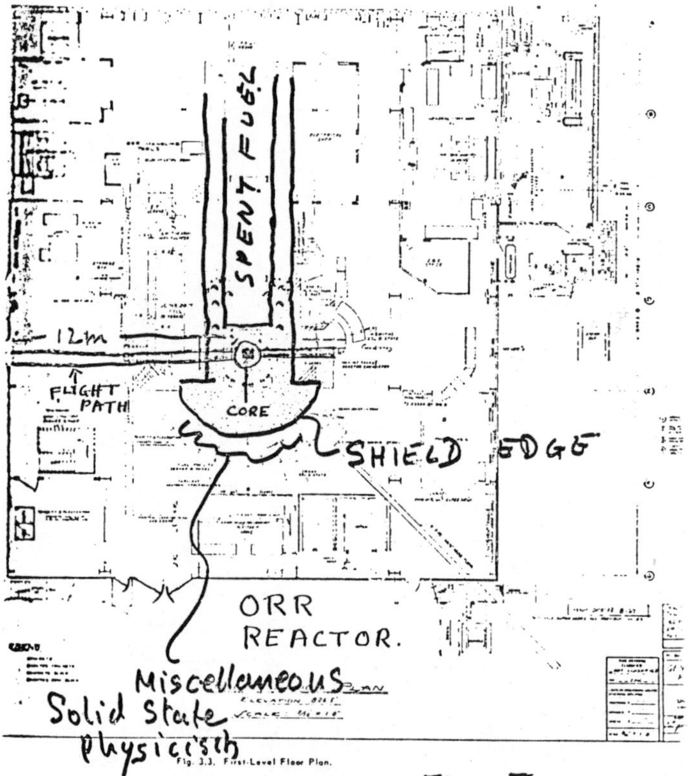
SPENT FUEL
12m
FLIGHT PATH
CORE
SHIELD EDGE
ORR REACTOR.
Miscellaneous Solid State physicists
Fig. 3.3. First-Level Floor Plan.

FIG. 5

THE SEARCH FOR NEUTRINO OSCILLATIONS: PRESENT EXPERIMENTAL DATA AND FUTURE EXPERIMENTS

David B. Cline

Fermilab
Batavia, IL

and

University of Wisconsin
Madison, WI

Abstract

We review the analysis of neutrino oscillations for the three neutrino system in present experimental data. The solar neutrino experiment, reactor experiments, and deep mine experimental data provide evidence for oscillations. Three broad classes of solutions are found. Using these clues we discuss definitive experiments to discover neutrino oscillations and to measure the parameters of the lepton mixing.

1. THEORY OF NEUTRINO OSCILLATIONS

The interesting possibility of neutrino oscillations has long been recognized,[1] but no clear signal has yet been established. Many excellent discussions of neutrino oscillation phenomenology exist in the literature.[2] Our goal here is to study the limitations on the neutrino mixing parameters from present data in the context of new experiments that might be carried out to search for oscillations.

Neutrino oscillations depend on differences in mass m_i between the neutrino mass eigenstates ν_i. The latter are related to the weak charged current eigenstates ν_α (distinguished by Greek suffices) through a unitary transformation

$$|\nu_\alpha\rangle = U_{\alpha i}|\nu_i\rangle \quad .$$

Starting with an initial neutrino ν_α of energy E, the probability for finding a neutrino ν_β after a path length L can be compactly written (for $E^2 \gg m_i^2$):

$$P(\nu_\alpha \to \nu_\beta) = \delta_{\alpha\beta} + \sum_{i<j} 2|U_{\alpha i}U_{\beta i}U_{\alpha j}U_{\beta j}|[(\cos(\Delta_{ij} - \phi_{\alpha\beta ij}) - \cos\phi_{\alpha\beta ij}] \quad ,$$

where

$$\phi_{\alpha\beta ij} = \arg(U_{\alpha i}U^*_{\beta i}U^*_{\alpha j}U_{\beta j})$$

and

$$\Delta_{ij} = \frac{1}{2}(m_i^2 - m_j^2)\frac{L}{E} \quad .$$

For a diagonal transition or an off-diagonal transition with CP conservation (U real), we obtain the simple formula

$$P(\nu_\alpha \to \nu_\beta) = \delta_{\alpha\beta} - \sum_{i<j} 4U_{\alpha i}U^*_{\beta i}U^*_{\alpha j}U_{\beta j} \sin^2(\tfrac{1}{2}\Delta_{ij}) \quad .$$

With L/E in m/MeV and m_i in eV units, the oscillation argument in radians is

$$\frac{1}{2}\Delta_{ij} = 1.27\ \delta m_{ij}^2 \frac{L}{E}$$

where

$$\delta m_{ij}^2 \equiv m_j^2 - m_j^2 \quad .$$

For antineutrinos replace U by U^* above.

The oscillations are periodic in L/E. Oscillations arising from a given δm_{ij}^2 can be most readily mapped out at L/E values of order $1/\delta m_{ij}^2$. The experimentally accessible ranges of L/E in m/MeV are $\sim 10^{10}$ (solar, $\sim 10 - 10^5$ (deep mine), 1-7 (low energy accelerators.). Figure 1 presents a summary of the available L/E for existing or imagined neutrino beams. After many cycles, detectors cannot measure L or E precisely enough to resolve individual oscillations and are sensitive only to average values. In the limit

$$\frac{L}{E} \gg (m_i^2 - m_j^2)^{-1} \quad ,$$

$$\text{for all } i = j \quad ,$$

the average asymptotic values are given by

$$\langle P(\nu_\alpha \to \nu_\beta)\rangle = \sum_i |U_{\alpha i}U^*_{\beta i}|^2 \quad .$$

Since only ν_e, ν_μ, and ν_τ neutrino types are known, we specialize to a three neutrino world. The matrix U can then be parameterized in the form introduced by Kobayashi and Maskawa,[5] in terms of angles θ_1, θ_2, θ_3 with ranges $(0, \pi/2)$ and phase δ with range $(-\pi, \pi)$. In our present analysis we neglect CP violation (thus $\delta = 0$ or $\pm\pi$).

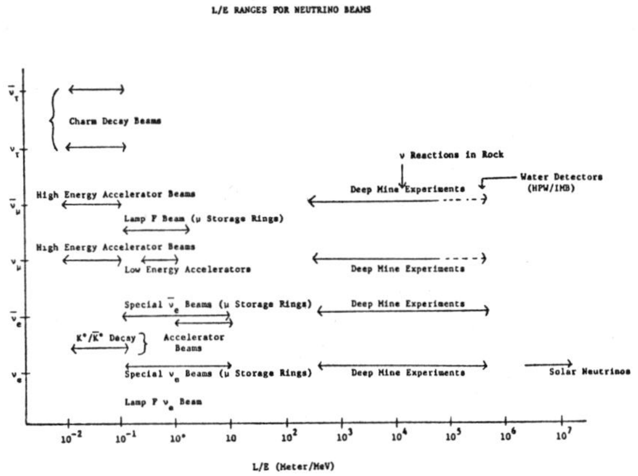

Figure 1. Available range of L/E in current and conceivable neutrino beams.

2. ANALYSIS OF PRESENT DATA ON NEUTRINO OSCILLATORS

To limit the regions of the θ_i, δm^2_{ij} parameter space, we consider first the constraints placed by solar, deep mine, and accelerator data. These data are taken at face value. The analysis has been carried out with V. Barger, R.J.N. Phillips, and K. Whisnant.

Solar neutrino observations and deep mine experiments: The solar neutrino data[6] suggest that

$$\langle P(\nu_e \to \nu_e) \rangle \simeq 0.3 - 0.5$$

at $L/E \simeq 10^{10}$ m/MeV. For three neutrinos, Eq. (4) gives

$$\langle P(\nu_e \to \nu_e) \rangle = c_1^4 + s_1^4 c_3^4 + s_1^4 s_3^4 \quad , \tag{5}$$

where $c_i = \cos\theta_i$ and $s_i = \sin\theta_i$. The minimum value of Eq. (5) is 1/3, and this requires

$$c_1 = \frac{1}{\sqrt{3}} \quad , \qquad c_3 = \frac{1}{\sqrt{2}} \quad .$$

For $\langle P(\nu_e \to \nu_e) \rangle$ to be near its minimum, all mass differences must satisfy

$$\delta m^2 \gg 10^{-10}\ \mathrm{eV}^2 \quad .$$

At this minimum all transition averages are specified, independent of θ_2; in particular,

$$P(\nu_\mu \to \nu_\mu) = \frac{1}{2} \quad .$$

In fact, there are indications from deep mine experiments[7-9] that

$$P(\nu_\mu \to \nu_\mu) \sim \frac{1}{2}$$

(see footnote f in Table 1). Since the ν_μ neutrinos detected in deep mines have traversed terrestrial distances, this measurement suggests that all

$$\delta m^2 \gtrsim 10^{-3}\ eV^2 \quad .$$

Based on these considerations, we may suppose that the true solution is not far from the above θ_1, θ_3 values. If we only require

$$P(\nu_e \to \mu_e) < 0.5 \quad ,$$

then θ_1 and θ_3 are constrained to a region approximated by the triangle $\theta_1 > 35°$, $\theta_3 > \theta_1 - 45°$, and $\theta_3 < 135° - \theta_1$.

$\nu_\mu \to \nu_e, \nu_\tau$ <u>oscillations</u>: Stringent experimental limits exist on these transitions[10-13] at L/E in the range 0.01 to 0.3 m/MeV (see Table 1). For $\delta m^2 \gg 1\ eV^2$, these oscillations do not appear until $L/E \gg 1$ m/MeV. With a single $\delta m^2 \gtrsim 1\ eV^2$, these oscillations can be suppressed by choice of θ_2 (if θ_1, θ_3 are taken as above).

Observable	Source Refs.	$\frac{L}{E}$ $\frac{m}{MeV}$	Present Limit	Solution A	Solution B	Solution C
$\langle P(\nu_e \to \nu_e)\rangle$	S 6	10^{10}	$\gtrsim \frac{1}{4}, \lesssim \frac{1}{2}$	0.41	0.33	0.41
$P(\bar\nu_e \to \bar\nu_e)$	R 3,4	1-3	> 0.5	0.6-1.0	0.8-1.0	0.8 mean
	R a	5-20		0.1-0.9	0.05-0.5	0.1-0.9
$P(\nu_e \to \nu_e)$	A	0.04	> 0.85 e	1.0	1.0	0.9
	M 12	0.3	1.1 ± 0.4	0.95	1.0	0.8 mean
	M b	1-3		0.6-1.0	0.8-1.0	0.8 mean
$P(\bar\nu_\mu \to \bar\nu_e)$	M 12	0.3	< 0.04	10^{-4}	10^{-3}	10^{-3}
	M b	3		0.03	0.11	0.03
$P(\nu_\mu \to \nu_e)/P(\nu_\mu \to \nu_\mu)$	A 10,11	0.04	$< 10^{-3}$	10^{-6}	10^{-5}	10^{-4}
	A 18 c	1-7		0-0.2	0-0.8	0-0.2
$P(\nu_e \to \nu_\tau)$	A d	0.04	< 0.2 e	10^{-3}	10^{-5}	0.1
$P(\nu_\mu \to \nu_\tau)/P(\nu_\mu \to \nu_\mu)$	A 13	0.04	$< 2.5\times10^{-2}$	10^{-5}	10^{-5}	10^{-3}
$\langle P(\nu_\mu \to \nu_\mu)\rangle$	D f	10^2-10^3	~ 0.5	0.51	0.51	0.51
$\langle P(\nu_c \to \nu_\mu)\rangle$	D g	10^3-10^5		0.48	0.44	0.48
$\langle P(\nu_c \to \nu_e)\rangle$	D g	10^3-10^5		0.42	0.33	0.42
$P(\nu_c \to \nu_\mu)$	D g	10-10^2		0.3-0.7	0.3-0.7	0.3-0.7
$P(\nu_c \to \nu_e)$	D g	10-10^2		0.2-0.6	0.2-0.6	0.2-0.6

Notation: S(solar), R(reactor), M(meson factory), A(accelerator), D(deep mine); $\nu_c = (2\nu_\mu + \nu_e)/3$.

Table 1. Experimental limits on neutrino oscillations and neutrino flux predictions. For references see page 243.

<u>Reactor $\bar{\nu}_e$-oscillations:</u> The $\bar{\nu}_e$ flux at distances L = 6m and 11.2 m from a reactor core center was measured by Reines, et al., using the known cross section for the inverse beta-decay reaction $\bar{\nu}_e p \to e^+ n$. The reactor $\bar{\nu}_e$ flux at the core has been calculated using semi-empirical methods.[14-17] The ratio of measured flux at L to the calculated flux measures $P(\bar{\nu}_e \to \bar{\nu}_e)$. Neutrino oscillation interpretations of the data thereby depend on the calculated spectrum about which there is some uncertainty.

Figure 2 shows a comparison of the measured $\bar{\nu}_e$ flux at L = 6 m and L = 11.2 m with calculated spectra. We note that the Avignone-1978 calculated flux[16] accommodates best the L = 6 m measurements for $E_{\bar{\nu}_e} > 6$ MeV. The data for $P(\bar{\nu}_e \to \bar{\nu}_e)$ obtained with Avignone-1978[16] and Davis et al.[17] calculated spectra are shown in Fig. 3. The horizontal error bars in Fig. 2 take into account the finite size of the reactor core source. We observe that $P(\bar{\nu}_e \to \bar{\nu}_e)$ seems to follow an oscillation pattern with one node in the range of L/E covered by the measurements. The possibility of such a solution in which a short wavelength occurs was not considered by Reines et al.[3] in their analysis of the 11.2 m data based on a similar calculated spectrum.[15]

The oscillation in Fig. 3 is well-described by the formula

$$P(\bar{\nu}_e \to \bar{\nu}_e) = 1 - 0.44 \sin^2(1.27\ L/E) \quad .$$

This corresponds to a mass difference

$$\delta m^2 = 1\ \text{eV}^2 \quad ,$$

which we can arbitrarily identify as δm^2_{13}. A nonzero δm^2_{12} with $\delta m^2_{12} \ll \delta m^2_{13}$ is required to bring $P(\nu_e \to \nu_e)$ down asymptotically to the solar neutrino result. The value of δm^2_{12} is not tightly constrained, other than the indication from deep mine measurements of $P(\nu_\mu \to \nu_\mu)$ that

$$\delta m^2_{12} \gtrsim 10^{-3}\ \text{eV}^2 \quad .$$

A solution which accomodates all known constraints is

	δm^2_{13}	δm^2_{12}	θ_1	θ_2	θ_3	δ	
SOLUTION A:	1.0 eV^2	0.05 eV^2	45°	25°	30°	0°	(5)

The predictions for subasymptotic transition probabilities are shown in Fig. 4.

A more conservative interpretation of the reactor $\bar{\nu}_e$ data could be that $P(\bar{\nu}_e \to \bar{\nu}_e)$ falls to around 0.7 - 0.8 in the range of L/E considered, but that oscillatory behavior is not established. If so, two other classes of solution are possible: Class B, where $\bar{\nu}_e \to \bar{\nu}_e$ is suppressed by a short

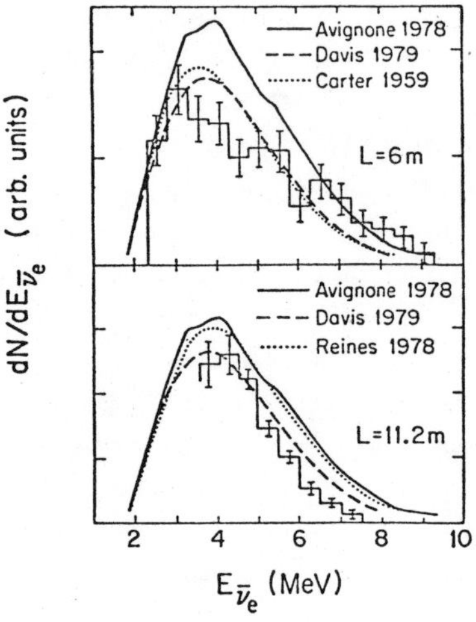

Figure 2. The $\bar{\nu}_e$ reactor flux measurements of Reines et al. at L = 11.2 m (Ref. 3) and L = 6 m (Ref. 4) compared with the calculated spectra of Refs. 3, 4, 14-17.

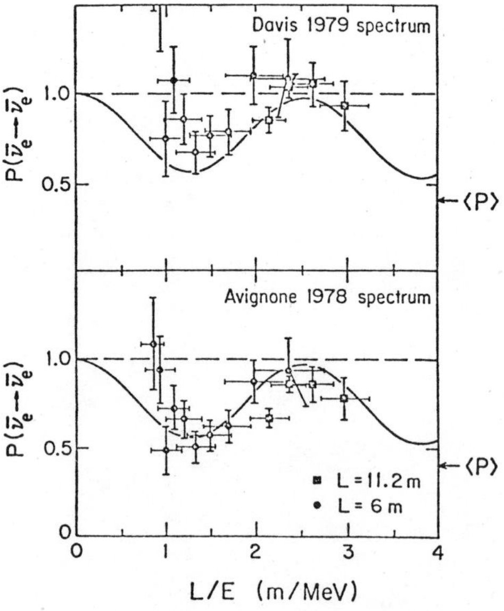

Figure 3. Transition probability $P(\bar{\nu}_e \rightarrow \bar{\nu}_e)$ versus L/E deduced from the ratio of the observed to the calculated $\bar{\nu}_e$ reactor flux from Refs. 16-17. The curve represents neutrino oscillations with an eigenmass difference squared of $\Delta m^2 = 1\ eV^2$ (Solution A of Eq. (5)).

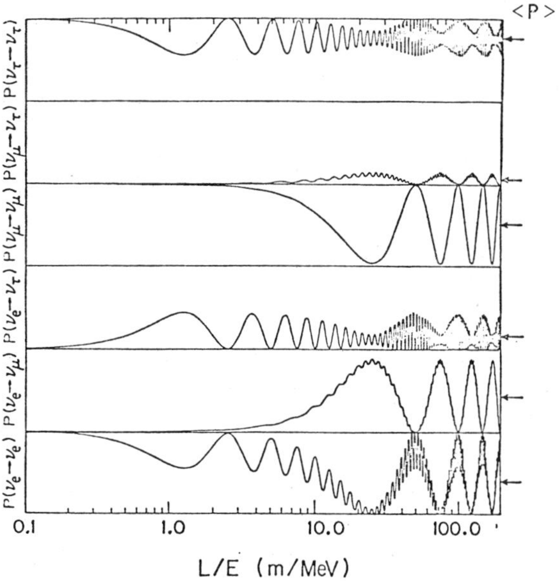

Figure 4. Subasymptotic neutrino oscillations for all channels based on Solution A in Eq. (5). Arrows on the right-hand side denote asymptotic mean values.

wavelength oscillation that may have many nodes below L/E = 1 m/MeV. Illustrative solutions of these classes are as follows (we emphasize that their parameters are less constrained than in Class A).

	δm^2_{13}	δm^2_{12}	θ_1	θ_2	θ_3	δ	
SOLUTION B:	0.15 eV^2	0.05 eV^2	55°	0°	45°	0°	(6)
SOLUTION C:	10 eV^2	0.05 eV^2	45°	25°	30°	0°	

We note that equivalent solutions to Eqs. (5) and (6) are obtained with

$$\delta m^2_{13} \leftrightarrow \delta m^2_{12} \quad , \qquad \delta = \pi \quad ,$$

and

$$\theta_3 \to \frac{\pi}{2} - \theta_3 \quad ,$$

with θ_1, θ_2 unchanged. Table 1 presents a capsule summary of the present experimental limits on oscillations and summarizes predictions of Solutions A, B and C for existing and planned experiments. For completeness, we present predictions for the $\bar{\nu}_e$ flux at 8.5 m in Fig. 5 to be compared with the Caltech-Grenoble-Munich experiment being carried out at the Grenoble reactor. We have only considered Solution A in this prediction.

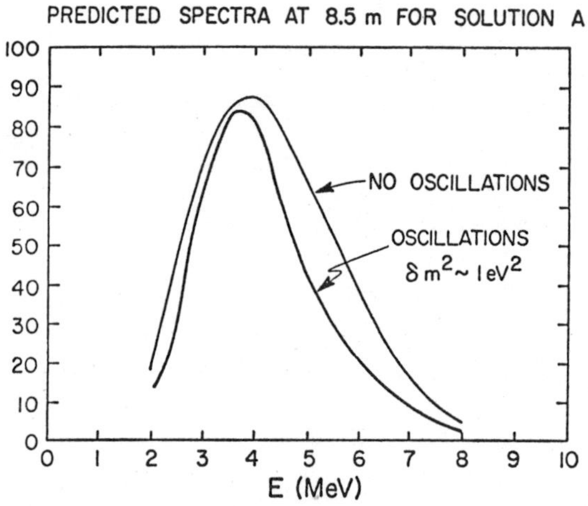

Figure 5. Expected $\bar{\nu}_e$ flux at 8.5 m from the center of the reactor.

3. NEW DEFINITIVE EXPERIMENTS TO ESTABLISH OSCILLATIONS

a.1 Ratio of Charged Current Neutral Current as a Flux Independent Measurements

Since leptonic charged current processes are sensitive to the neutrino flavor but semileptonic neutral currents are not, the ratio

$$\frac{R(\nu_\ell + N \rightarrow \ell + X)}{R(\nu_\ell + N \rightarrow \nu_\ell + X)}$$

should be less sensitive to the uncertainty in neutrino flux. A hybrid ratio that could be used is

$$\frac{R(\bar{\nu}_e + p \rightarrow e^+ + n)}{R(\bar{\nu} + e^- \rightarrow e^- + \bar{\nu})}$$

but the $\bar{\nu} + e$ cross section is extremely small.

As an example, consider

$$\bar{\nu}_e + d \rightarrow e^+ + n + n$$

and

$$\bar{\nu}_e + d \rightarrow \bar{\nu}_e + n + p \quad .$$

Both reactions have been observed experimentally by Reines et al. These cross sections are slightly sensitive to the $\bar{\nu}_e$ flux and rather sensitive to the neutrino energy threshold.[25] Figure 6 shows the calculated cross sections for these two reactions as a function of E_{neutrino}.[26] However, in a detector with very good neutron detector efficiency, these concerns would disappear.

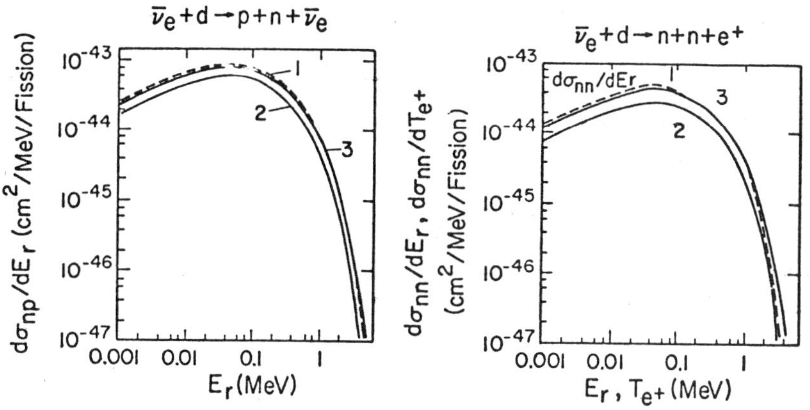

Figure 6. (a) Differential cross sections of the reaction $\bar{\nu}_e + d \rightarrow p + n + \bar{\nu}_e$ for incident antineutrinos from equilibrium fission versus neutron-proton relative motion energy E_r.

(b) Differential cross sections of the reaction $\bar{\nu}_e + d \rightarrow n + n + e^+$ for incident antineutrinos from equilibrium fission versus neutron-proton relative motion energy E_r and positron recoil energy T_{e^+}.

a.2 Ratio of Rates at Two Different Distances

The ratio of differential rates for $\bar{\nu}_e + p \rightarrow e^+ + n$ at reactor energies at two different distances:

$$\frac{\dfrac{dN_{e^+}}{dE_{e^+}}[L_1]}{\dfrac{dN_{e^+}}{dE_{e^+}}[L_2]}$$

is flux independent but very sensitive to neutrino oscillations. Figure 7 shows this ratio as an example only for the data of Reines et al.[3,4] at $L_1 = 6$ m and $L_2 = 11.2$ m. We also compare with the expected ratio for the three Solutions A, B, and C in the paper of Barger et al. Note that Solution A and C are favored.

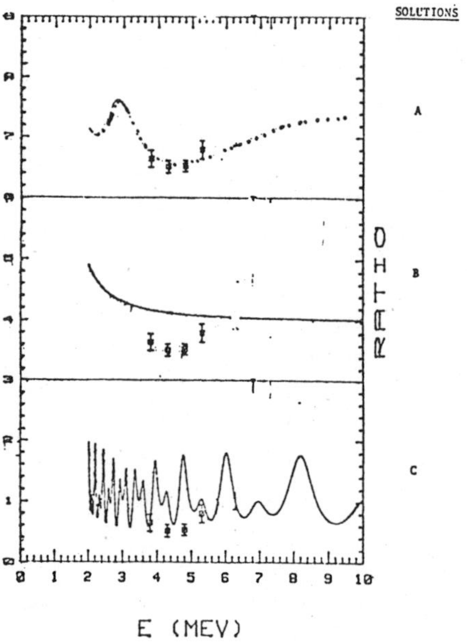

Figure 7. Flux independent ratio of the event rate at two L values. These data are obtained from Refs. 3 and 4.

b. Deep Mine Experiments

Consider a deep mine detector and the event rate of neutrino interactions that produce detected muons. Such an experiment is conceptually shown in Fig. 8. The rate for such muons is given by[9]

$$\frac{\partial P_\mu(X,E_\mu,\theta)}{\partial X} = \frac{\partial}{\partial E}\,[f(E_\mu)P_\mu(X,E,\theta)] + G_\mu(E_\mu,\theta)$$

where G_μ is the rate of production of muons in a distance dX and P_μ is the probability that μ's enter the detector. The exact solution (one dimensional solution) is given by

$$P_\mu(E_\mu,X) = \int_{X_0}^{X} G_\mu(E+\beta(X-X'))\,dX' + P_\mu(E_\mu+\beta(X-X_0);\ X_0)$$

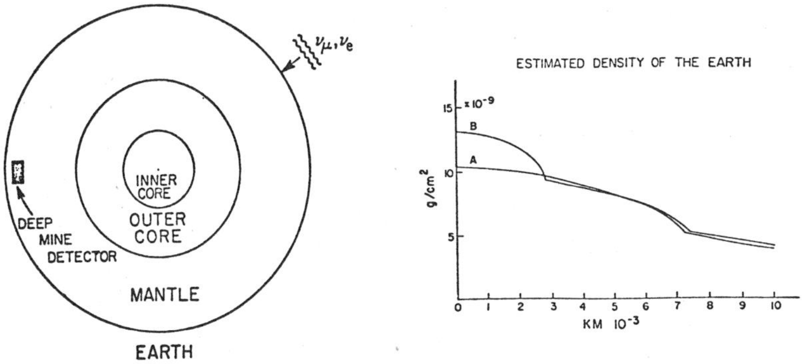

(a) (b)

Figure 8. Conceptual view of the neutrino flux through the earth and the estimated density of the earth as a function of radius.

ATMOSPHERIC NEUTRINOS

$(\nu_\mu + \bar{\nu}_\mu)$

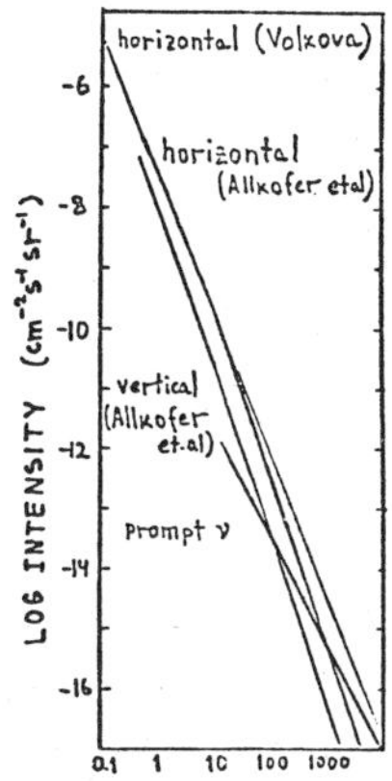

E(TeV)

Figure 9. The flux of atmospheric neutrinos, from papers presented to the 1979 Kyoto Cosmic Ray Conference and published in the proceedings.

where $\beta \equiv f(E_\mu)$ energy loss. The input neutrino spectrum is shown in Fig. 9 and assumes

$$\sigma_\nu \sim 3\sigma_{\bar{\nu}} = 0.7\, E \times 10^{-38}\ \mathrm{cm}^2/\mathrm{GeV} \quad .$$

The resulting integral rate of muon events is given by

$$\int_{P_\mu}^{\infty} P(E_\mu, X) = \tilde{P}(E_\mu)$$

$$\text{at} \quad x \quad \text{of detector} \quad .$$

Table 2 gives the values of this integral and Fig. 10 shows the expected μ flux as a function of the cut-off μ energy in the detector. Note that very high energy neutrino interactions are observed in such experiments and that the L/E for these processes is $\sim(10-1)$ m/MeV. Thus, these interactions have sufficient energy to excite heavy sequential leptons if the mass is less than $\sim$30 GeV and if the mixing parameters with ν_μ are appreciable. The present generation of detectors being constructed could be sensitive to these possibilities. However, it is likely that much larger underground detectors (> 100,000 ton water detectors, for example) would be necessary to isolate the effects of neutrino oscillations at the higher energies.

	Horizontal	Vertical	$P_\mu < E_\nu$
$\tilde{P}$ =	$9 \times 10^{-13}\ \mu/\mathrm{cm}^2$ sec sr	3.3×10^{-13}	0.3 GeV
	8×10^{-13}	2.77×10^{-13}	1.0 GeV
	5.6×10^{-13}	1.64×10^{-13}	10 GeV
	3.1×10^{-13}	0.64×10^{-13}	100 GeV
	1.1×10^{-13}	0.13×10^{-13}	1000 GeV

Table 2

c. Special ν_e, ν_τ Beams for Neutrino Oscillation Experiments

Consider a detector of radius R at a distance of L from the source of an electron neutrino beam of energy E. The only intense sources of energetic electron neutrino beams come from in-flight μ or K decay. The angular spread of the beam at the detector is set by the transverse momentum in the decay $\mu \to e + \bar{\nu}_e + \nu_\mu$ or $K \to \pi + e + \nu_e$, e.g.,

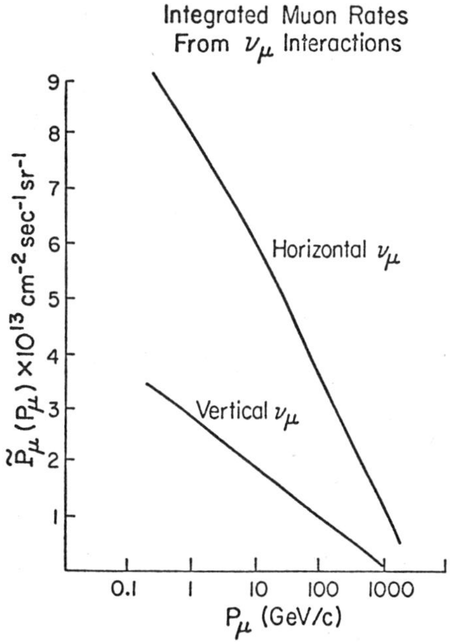

Figure 10. Calculated integral μ flux from neutrino interactions in the earth using the latest ν flux (Fig. 9).

$$\theta_{\mu \to \nu_e} \approx 30 \frac{\text{MeV/c}}{E}$$

$$\theta_{K \to \nu_e} \approx 150 \frac{\text{MeV/c}}{E} \quad .$$

If we match the detector radius and the beam radius, we find

$$R = L = \frac{L}{E} \cdot 30 \text{ MeV/c} \quad (\mu \to \nu_e)$$

$$= \frac{L}{E} \cdot 150 \text{ MeV/c} \quad (K \to \nu_e) \quad .$$

For existing neutrino detectors $R \lesssim 2$ m;

$$L/E)_{\mu \to \nu_e} \simeq \frac{1}{15} \text{ m/MeV}$$

$$L/E)_{K \to \nu_e} \simeq \frac{1}{75} \text{ m/MeV} \quad .$$

Thus the "natural" neutrino beam properties are matched to neutrino mass differences $> \text{few eV}^2$. The detection or study of smaller mass differences $(L/E \sim 1\ \text{m/MeV})$ can only be done with a large loss of neutrino interaction rate (this scales like $(L/E)^{-2}$; thus, for $(L/E) \sim 1\ \text{m/MeV}$ the $\mu \rightarrow \nu_e$ rate is depressed by $(1/15)^2$). However, the cross section for neutrino interactions increases like E, thus the rate scales like E/L^2* or

$$\frac{1}{E}\ \frac{1}{(L/E)^2} \quad .$$

Thus, in general, to increase (L/E) at constant target mass, the energy of the neutrino beam must be *decreased*. Since the probability for $\mu \rightarrow e\nu_e\bar{\nu}_\mu$ decay scales like $1/E$ the overall rate goes like

$$\frac{1}{E^2(L/E)} \quad :$$

low energy beams are required.

It is possible to construct special low energy μ beams to produce intense ν_e beams in two ways

A. Special long decay length beans--for ~ 1 GeV μ beams--such beams could be constructed at Fermilab using the booster synchrotron and the long $\bar{p}$ reverse injection line.

B. Special μ storage ringe--one example of such a ring would be to use the antiproton collector being constructed at Fermilab--to store π,μ beams--these storage rings will have a very large aperture and momentum spread. (See Fig. 11 for an example.)

Such beams should be used with large neutrino detectors (≥ 100 tons) (possibly water detectors) similar to the HPW proton decay detector.

In Table 3 we compare the characteristics of these ν_e, $\bar{\nu}_e$ beams with the beams available from reactors.

For ν_τ beams there is a minimum energy required to identify the beam through the charged current reaction

$$\nu_\tau + N \rightarrow \tau^- + X$$

which is

$$E_{\nu_\tau} \geq 10\ \text{GeV} \quad .$$

*This is only true for $E \gg 100$ MeV; at lower energy nuclear effects reduce the cross section.

FERMILAB ANTIPROTON PRECOOLER
RING USED AS A μ- STORAGE RING

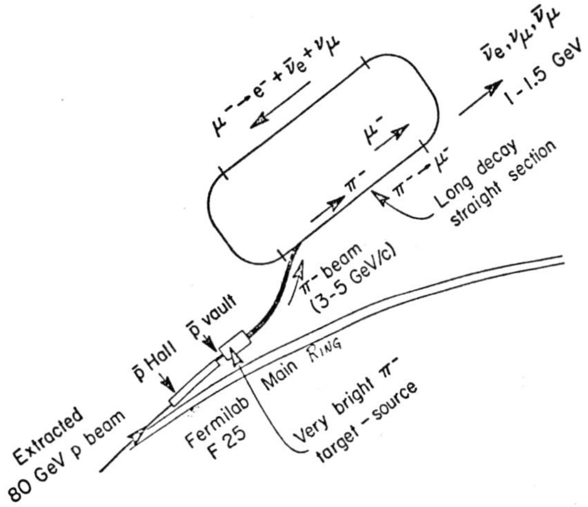

Figure 11. Conceptual view of a μ storage ring at Fermilab used to make $\nu_e/\bar{\nu}_e$ beams. In principle, the $\bar{p}$ precooler at Fermilab could be used for this purpose.

Requirements in New Experiments	Problems in Old Experiments
1. ν_e flux well known: Use $\mu^{\pm} \rightarrow e^{\pm} + \begin{Bmatrix} \nu_e \\ \bar{\nu}_e \end{Bmatrix} + \begin{Bmatrix} \bar{\nu}_\mu \\ \nu_\mu \end{Bmatrix}$ (decay in-flight)	Reactor flux not well known
2. Adequate ν_e event rate: (Use 1000 ton water detectors and $E_{\nu_e} > 100$ MeV.)	Cross sections are very small and threshold dependent $\sigma(\bar{\nu}_e + d \rightarrow nne^+) \approx 10^{-45}\ cm^2$ $\sigma(\bar{\nu}_e + p \rightarrow e^+ n) \approx 10^{-43}\ cm^2$
3. Adequate ν_e energy to produce $\nu_\tau \rightarrow \tau$ ($\rightarrow$ Use $K^\circ \rightarrow \nu_e/\bar{\nu}_e$).	Present high energy ν_e beams have small flux and very large ν_μ background.

Table 3: Definitive experiments to search for $\nu_e \leftrightarrows \nu_\tau$ (Solutions A or C)

For $\mu \to \nu_e \to \nu_\tau$ this requires $E_\mu > 30$ GeV. For $\mu \to \nu_e$ beams only μ storage rings would be suitable. A more likely source of energetic ν_e beams comes from $K^\pm$, K°, $\bar{K}^\circ$ decay (see Fig. 12). Charm decay neutrino beams have many disadvantages since the intrinsic transverse momentum is large and the production cross section is small. Ideally, K° beams of ~ 30 GeV would be used, and distances to the detector of $\simeq 30$ km for Solution A or B. However, present detectors are ≤ 1 km from the ν_e source and the typical ν_e energy is 40 GeV giving

$$\frac{L}{E} \simeq \frac{1000}{40,000} \simeq \frac{1}{40} \quad .$$

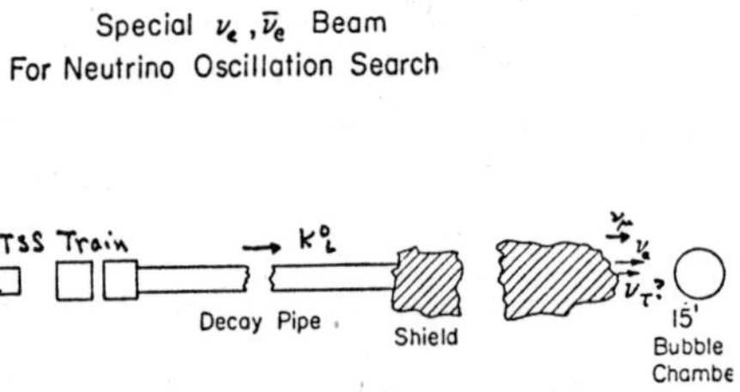

Figure 12. Special $\nu_e/\bar{\nu}_e$ beam at Fermilab for neutrino oscillations experiment.

In principle this ratio could be increased by a factor of four by running the SPS or Fermilab machines at lower energy and higher repetition rates. Additional increase in L/E must come from increasing L.

As an example, consider the δm_2^2 sensitivity for an experiment proposed at Fermilab using an enriched $K^\circ/\bar{K}^\circ$ beam to produce the $\nu_e/\bar{\nu}_e$ beam. Figure 13 shows the expected sensitivity of such an experiment. In Solution C the ν_τ rate will be ~ 0.1 of the ν_e rate. τ events can be identified by the $\tau^- \to e^-$ signature and the missing transverse momentum in the event as shown in Fig. 14 and Fig. 15. With high resolution operation of the bubble chamber, it may be possible to identify some of the τ^- events by observing the flight path. Such an experiment is being proposed for Fermilab.

d. Special Neutrino Detectors for Oscillations Experiments

The large water detectors being constructed for proton decay experiments could be used for neutrino oscillation experiments. Such detectors have excellent directional properties. In the HPW experimental detector the number

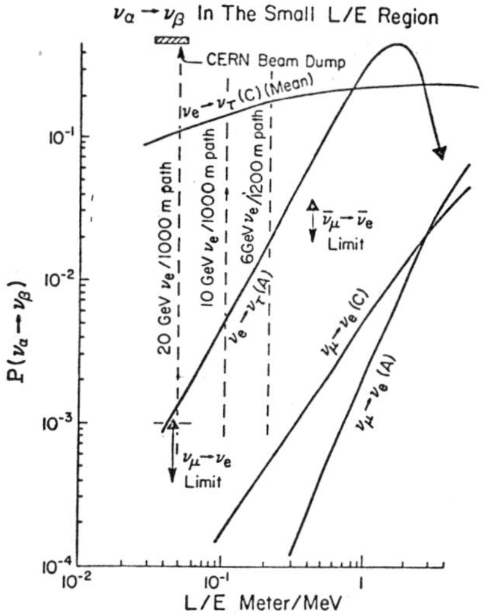

Figure 13. $P(\nu_\alpha \to \nu_\beta)$ in the small L/E region.

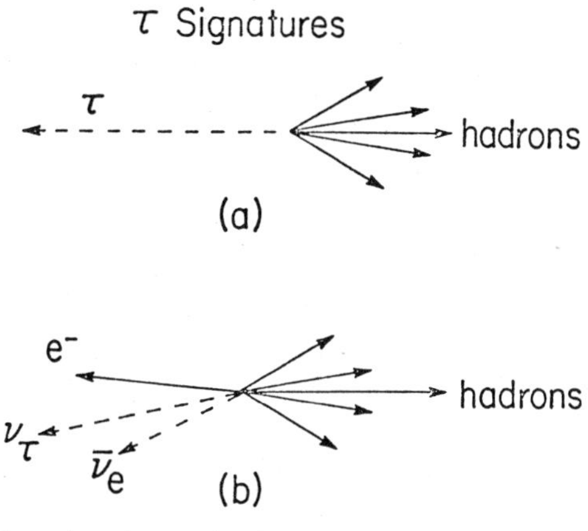

Figure 14. Schematic of the missing $P_\perp$ from τ decay.

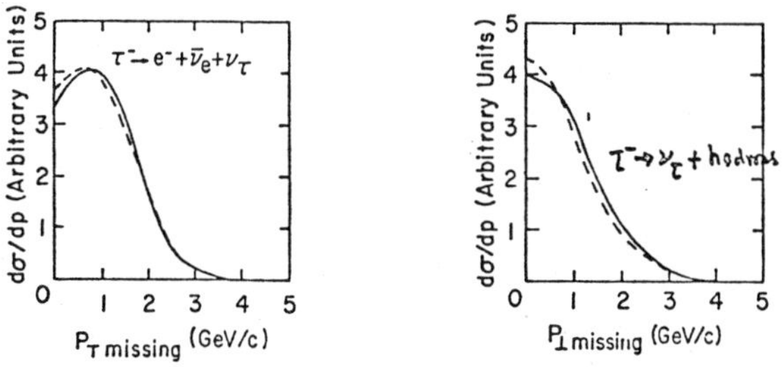

Figure 15. Estimated $P_\perp$ for τ^- production and decay (taken from C. Albright et al., "Signatures for Tau Leptons from Beam Dump Experiments".

of collected photoelectrons is $\simeq$ 0.7/MeV of the proton mass. Thus for 1 GeV ν_e interactions approximately 700 photoelectron would be collected. The mass of such detectors would be in the few 1000 ton range. e/μ separation might be obtained by range and shower properties.

As another example of the use of such a detector, consider a $\nu_e/\bar{\nu}_e$ beam produced by a K_L^o beam (Fig. 11) and a water detector placed at a distance of 30 km, such a detector could be used to measure the ratio

$$R^{E_{vis}}_{(\mu/e)} = \frac{\mu \text{ events}}{\text{non } \mu \text{ events}}$$

$$= \frac{\underline{P}(\nu_\mu \to \mu) + \underline{P}(\nu_\tau \to \tau \to \mu)}{\text{nuetral current events} + e^{\pm} \text{ events}}$$

This ratio is very sensitive to the existence of neutrino oscillations.

4. CONCLUSION

Our central conclusion is that no definitive neutrino oscillation experiments have yet been performed.* There are very interesting hints of oscillation effects in solar neutrino, reactor, and deep mine experiments and possibly the beam dump experiments at CERN. The allowed range of mixing angles and (mass differences)^2 is large. Experimentally, the neutrino flux uncertainties as well as the experimental difficulty of studying low energy neutrino interaction suggest a new generation of neutrino oscillation experiments at accelerators with well known neutrino fluxes and large signal to background ratios. Deep mine experiments being constructed are of crucial importance to establish lower limits on mixing angles and differences especially if one mass difference is smaller than 1 eV^2.

The new accelerator experiments will be modest in cost compared to the current generation of colliding beam machines and detectors, and considering the extreme importance of the question of neutrino masses, these possibilities should be seriously considered at Fermilab, BNL, and CERN. Of special importance would be the construction of μ storage rings for intense $\nu_e/\bar{\nu}_e$ beams. It appears that the $\bar{p}$ precooler being constructed at Fermilab could be used for such a μ storage ring if suitably configured with two long straight sections.

---

*After this talk was given, the new experimental results on the ratio of $\bar{\nu}_e \to e^+/\bar{\nu} \to$ neutral currents were reported by Reines et al. at the April Washington APS meeting.

Acknowledgments. I wish to thank A.K. Mann, F. Reines, and C. Rubbia for discussions and to thank V. Barger, Y. Chon, J. Ellis, R.J.N. Phillips, and K. Whisnant for help in the calculations reported here.

NOTES FOR TABLE 1

a. San Onofre reactor experiment by Reines et al.[3] in progress, with $L = 25 - 100$ m.

b. Possible LAMPF experiment with

$$E_{\nu_e}, E_{\bar{\nu}_\mu} = 30 - 50 \text{ MeV}$$

and $L = 30 - 100$ m.

c. Brookhaven experiment[18] in data analysis stage.

d. $\nu_e \rightarrow \nu_\tau$ oscillations can lead to an e/μ ratio different from unity in beam dump experiments (see, e.g., de Rujula et al., Ref. 2 and data of Ref. 19).

e. The excellent agreement of observed and calculated $\nu_e/\bar{\nu}_e$ flux at CERN and Fermilab indicates that most of the ν_e does not oscillate into ν_τ.

f. Deep mine experiments[7,8] have detected about 130 neutrino events ($E \sim 10^4 - 10^6$ MeV, $L \sim 10^6 - 10^7$ m). An unaccountably large number of multitrack events were observed in the Kolar gold field experiment;[7] assuming that these are not attributed to ν_μ, the event rate is about half the expected rate. In the Johannesburg mine experiment[8] a ratio 1.6 ± 0.4 of expected to observed ν_μ events was found. The analysis in Ref. 9 of these experiments is consistent with $\langle P(\nu_\mu \rightarrow \nu_\mu) \rangle \sim 0.5$. A new deep mine experiment is operating at Baksan, USSR which is sensitive to ν_μ flux through the earth.[20]

g. Deep mine experiments in construction[21] will detect neutrinos of energies $E \sim 10^2 - 10^3$ MeV using very large water detectors placed in deep mines. At these energies the composition[22] of the ν-flux from π, K, and μ decays of the secondary cosmic ray component in the atmosphere is roughly $(2\nu_\mu + \nu_e)/3$. Upward events in the detector will have $L \simeq 10^6 - 10^7$ m and downward events will have $L \simeq 10^4$ m. For events in which distances $L \gtrsim 10^7$ m are traversed in the earth, our vacuum oscillation results can be significantly modified by ν_e charged-current scattering on electrons (see Ref. 23).

REFERENCES

1. B. Pontecorvo, *Soviet Phys. JETP* 53, 1717 (1967); V. Gribov and B. Pontecorvo, *Phys. Lett.* 28B, 493 (1969).

2. S.M. Bilenky and B. Pontecorvo, *Phys. Reports* 41, 225 (1978); A. de Rujula et al., CERN TH-2788 (1979); S. Eliezer and D.A. Ross, *Phys. Rev.* D10, 3088 (1974); S.M. Bilenky and B. Pontecorvo, *Phys. Lett.* 61B, 248 (1976); S. Eliezer and A. Swift, *Nucl. Phys.* B105, 45 (1976); B. Pontecorvo, *JETP Lett. (Pisma JETP)* 13, 281 (1971); H. Fritzsch and P. Minkovsky, *Phys. Lett.* 62B, 72 (1976); A.K. Mann and H. Primakoff, *Phys. Rev.* D15, 655 (1977); J.N. Bahcall and R. Davis, *Science* 191, 264 (1976).

3. F. Reines, *Unification of Elementary Forces and Gauge Theories*, D.B. Cline and F.E. Mills (eds.), Harwood Academic Pub., 1978, p. 103; S.Y. Nakamura

et al., *Proc. Inter. Neutrino Conf.*, Aachen, H. Faissner, et al. (eds.), Vieweg (1977).

4. F. Nezrick and E. Reines, *Phys. Rev.* 142, 852 (1966).

5. M. Kobayashi and T. Maskawa, *Prog. Theor. Phys.* 49, 652 (1973). We follow the convention in V. Barger and S. Pakvasa, *Phys. Rev. Lett.* 42, 1585 (1979).

6. R. Davis, Jr., J.C. Evans, and B.T. Cleveland, *Proc. Conf. on Neutrino Physics*, E.C. Fowler (ed.), Purdue University Press, 1978.

7. M.R. Krishnaswamy et al., *Proc. Phys. Soc. Lond.* A323, 489 (1971).

8. M.F. Crouch et al., *Phys. Rev.* D18, 2239 (1978).

9. L.V. Volkova and G.T. Zatsepin, *Sov. J. Nuvl. Phys.* 14, 117 (1972).

10. E. Belloti et al., *Lett. Nuovo Cim.* 17, 553 (1976).

11. J. Blietschau et al., *Nucl. Phys.* B133, 205 (1978).

12. S.E. Willis et al., *Phys. Rev. Lett.* 44, 522 (1980).

13. A.M. Cnops et al., *Phys. Rev. Lett.* 40, 144 (1978).

14. R.E. Carter, F. Reines, R. Wagner, and M.E. Wyman, *Phys. Rev.* 113, 280 (1959).

15. F.T. Avignone, III, *Phys. Rev.* D2, 2609 (1970).

16. F.T. Avignone, III, and L.P. Hopkins, in *Proc. Conf. on Neutrino Physics*, E.C. Fowler (ed.), Purdue Univ. Press, 1978.

17. B.R. David, P. Vogel, F.M. Mann, and R.E. Schenter, *Phys. Rev.* C19, 2259 (1979).

18. L.R. Sulak et al., in *Proc. Int. Conf. on Neutrino Physics and Astrophysics*, Elbrus, USSR (1977).

19. H. Wachsmuth, CERN-EP/79-115C (1979).

20. A. Chudakov and G. Zatsepin (private communication).

21. Irvine-Michigan-Brookhaven collaboration (F. Reines et al.); Harvard-Purdue-Wisconsin collaboration (J. Blandino et al.).

22. J.L. Osborne and E.C.M. Young, in *Cosmic Rays at Ground Level*, A.W. Wolfendale (ed.), Institute of Physics, London (1973).

23. L. Wolfenstein, *Phys. Rev.* D17, 2369 (1978).

24. Considerations are in progress by D. Cline and B. Burman for such a neutrino oscillation experiment at LAMPF.

25. F. Reines et al., *Phys. Rev. Lett.* 43, 96 (1979).

26. S.A. Fayans, L.A. Mikaelyan, and Y.L. Dobryin, *J. Phys. G.: Nucl. Phys.* 5, 209 (1979); T. Ahrens and T.P. Lang, *Phys. Rev.* C3, 979 (1971).

27. U. Camerini, C. Canada, D. Cline, et al., *An Experiment to Search for* $\left\{\begin{matrix}\nu_\mu \\ \nu_e\end{matrix}\right\} \to \nu_\tau$ *Neutrino Oscillations Using an Enriched* ν_e *Beam*, Wisconsin, Athens, Fermilab, Padova Proposal to Fermilab, April 1980.

NEUTRINOS IN COSMOLOGY--GOOD NUS FROM THE BIG BANG

Gary Steigman

Bartol Research Foundation of the Franklin Institute
University of Delaware
Newark, DE 19711

1. INTRODUCTION

According to the standard hot big bang model, the early evolution of the universe provided a cosmic accelerator in which elementary particles, those already known as well as those yet to be discovered, were produced copiously. During the expansion and cooling of the universe from its early, high density, high temperature state, the environmental impact was severe and very few particles survived. Fortunately, some hardy "relics" from earlier epochs do survive to influence the subsequent evolution of the universe. Indeed, the present universe is the debris of earlier epochs.

The neutrino is among the most elusive of the particles produced in the cosmic accelerator. Still, relic neutrinos may have played an important role during early epochs, they may dominate the present evolution, and their presence may determine the ultimate fate of the universe. Although direct detection of relic neutrinos is unfeasible (microwave neutrinos!), they may call attention to themselves through their gravitational interactions. Indeed, it is the gravitational effects of relic neutrinos which may thrust them onto center stage in the past, present, and future evolution of the universe.

The intent here is to focus on the basic physics relating to the role played by neutrinos in cosmology. This topic has been the subject of several recent reviews (Steigman, 1979; Steigman, 1980; Schramm and Steigman, 1980 a,b), the results of which are summarized here. For details and further references, the reader is urged to consult these papers. The discussion here will concentrate on "light" neutrinos ($m_\nu \ll 1$ MeV); for the contributions from, and the constraints on, "heavy" neutrinos ($m_\nu \gg 1$ MeV), see Lee and Weinberg (1977), Dicus et al. (1977), Gunn et al. (1978), Steigman et al. (1978). Light neutrinos were and are comparable in abundance to photons. During the early evolution when photons contribute significantly to the total energy density, so do the neutrinos. Since, at those times, the density determines the expansion rate, neutrinos influence the early evolution of the universe; their effect on primordial nucleosynthesis is particularly important. During the recent past and at present, non-relativistic particles play a dominant role. Since there are, roughly, a billion photons for each nucleon in the universe, there are, as well, a billion times as many neutrinos as nucleons. It follows, therefore, that if neutrinos are more than a billionth as massive as nucleons, they dominate the recent, present, and future evolution of the universe. These cosmic connections will be explored.

2. RELIC NEUTRINOS

At any stage in the evolution of the universe, non-relativistic (NR) as well as extremely relativistic (ER) particles contribute to the total density. With increasing temperature (T), the ER contribution dominates, since $\rho_{ER} \propto T^4$ and $\rho_{NR} \propto T^3$; the early, hot universe is "radiation dominated" (RD):

$$\rho \approx \rho_{ER} \propto T^4 \quad .$$

During the early evolution of the universe, the age (or its inverse, the expansion rate) and the density are related by

$$t^{-1} \propto \rho^{1/2} \quad . \tag{1}$$

For RD epochs, $\rho \propto T^4$, so that $t^{-1} \propto T^2$. Very roughly, the universe is ≈ 1 second old when $T \approx 1$ MeV (for a more precise result, see Steigman, 1979).

Ordinary electromagnetic interactions (e.g., $e^+ + e^- \leftrightarrow \gamma + \gamma$) maintain the photons in an equilibrium (black body) distribution whose number density and energy density are

$$n_\gamma = \frac{2.404}{\pi^2}\left(\frac{kT}{\hbar c}\right)^3 \quad , \qquad \rho_\gamma c^2 = \left(\frac{\pi^2}{15}\right) kT \left(\frac{kT}{\hbar c}\right)^3 \quad . \tag{2}$$

At the same time, light neutrinos are created and destroyed by neutral current weak interactions such as

$$e^+ + e^- \leftrightarrow \nu_i + \bar{\nu}_i \quad , \qquad i = e, \mu, \tau, \ldots \tag{3}$$

At high temperatures the rate of reaction (3) is fast:

$$\Gamma(T) = n(T) < \sigma v >_T \propto T^3 \times T^2 \tag{4}$$

By comparing the expansion rate (1) with the reaction rate (4), it follows that the neutrinos can be maintained in equilibrium for $T \gtrsim T_*$ (≈ 1 MeV). T_* is the decoupling temperature below which reactions (3) become too slow (compared to the expansion rate) to ensure equilibrium.

At high temperatures $(T > T_* \gg m_\nu)$, the neutrinos are a non-degenerate, relativistic fermi gas. Their densities differ little from those of the photon (relativistic boson) gas:

$$\frac{n_\nu}{n_\gamma} = \frac{3}{4}\left(\frac{g_\nu}{g_\gamma}\right)\left(\frac{T_\nu}{T_\gamma}\right)^3 \quad ; \quad \frac{\rho_\nu}{\rho_\gamma} = \frac{7}{8}\left(\frac{g_\nu}{g_\gamma}\right)\left(\frac{T_\nu}{T_\gamma}\right)^4 \quad . \tag{5}$$

In Eqs. (5) g_ν and g_γ $(=2)$ are the number of neutrino and photon spin states, respectively. For Majorana neutrinos, $g_\nu = 2$; for Dirac neutrinos $g_\nu = 4$. In equilibrium $(T > T_*)$,

$$T_\nu = T_\gamma \equiv T \quad .$$

As the universe expands and cools below T_*, few new neutrino pairs are produced and those already present fail to annihilate. The neutrino temperature is red-shifted by the expansion and:

$$n_\nu \propto T^3 \quad , \qquad \rho_\nu \propto T^4 \qquad (\text{for } T_\nu > m_\nu)$$

or

$$\rho_\nu \propto m_\nu T_\nu^3 \qquad (\text{for } T_\nu < m_\nu) \quad .$$

When the temperature drops below the electron mass, $e^\pm$ pairs disappear, their energy going to heat the photons relative to the decoupled neutrinos $(T_* > m_e)$. Thus, for $T \equiv T < m_e$,

$$\frac{T_\nu}{T} = \left(\frac{4}{11}\right)^{1/3} \quad , \tag{6a}$$

$$\frac{n_\nu}{n_\gamma} = \frac{3}{22}\, g_\nu \quad . \tag{6b}$$

The relative energy densities depend on whether the neutrinos are extremely relativistic (ER; $T > m$) or non-relativistic (NR; $T < m$):

$$\left(\frac{\rho_\nu}{\rho_\gamma}\right)_{ER} = \frac{7}{16}\left(\frac{4}{11}\right)^{4/3} g_\nu \quad , \tag{7a}$$

$$\left(\frac{\rho_\nu}{\rho_\gamma}\right)_{NR} = \frac{m_\nu n_\nu}{\rho_\gamma} \approx \frac{m_\nu n_\nu}{2.7\,kT\,n_\gamma} \approx \frac{g_\nu m_\nu}{20\,kT} \quad . \tag{7b}$$

Relic neutrinos, therefore, are always comparable to or dominate photons in the universe: $n_\nu \approx n_\gamma$, $\rho_\nu \gtrsim \rho_\gamma$.

Indeed, during RD epochs, neutrinos are an important contributor to the early expansion rate. Their influence on the outcome of primordial nucleosynthesis is reviewed in the next section.

During more recent "matter dominated" (MD) epochs, photons are relatively unimportant, but relic neutrinos with small masses could dominate the mass in the universe. This aspect, most exciting at present, will also be discussed.

3. RELIC NEUTRINOS AND PRIMORDIAL NUCLEOSYNTHESIS

As the early universe cools below $T \approx 10$ MeV, neutrons and protons are engaged in two very different sorts of interactions. Neutrons are turned into protons and vice versa through the familiar charged current weak interactions:

$$p + e^- \leftrightarrow n + \nu_e \quad ,$$
$$n + e^+ \leftrightarrow p + \bar{\nu}_e \quad , \qquad (8)$$
$$n \to p + e^- + \bar{\nu}_e \quad .$$

At the same time, deuterons are formed via nuclear interactions.

$$n + p \leftrightarrow d + \gamma \quad . \qquad (9)$$

At high temperatures $(T \gtrsim 0.1$ MeV), photodissociation of deuterium is so rapid that the deuterium abundance is negligible. There is no platform, then, on which to build the heavier elements. All the while, neutrons and protons are interconverting via reactions (8). For $T \gtrsim 1$ MeV, the weak interactions are fast enough to maintain the neutron-to-proton ratio in equilibrium.

$$\frac{n}{p} = \exp\left(-\frac{\Delta m}{T}\right) \quad . \qquad (10)$$

As the temperature decreases, the neutron abundance decreases, leaving fewer neutrons available for the build-up of the heavier elements during nucleosynthesis.

However, with decreasing temperature, the weak interaction rate $(\Gamma_{wk} \propto T^5)$ decreases more rapidly than the expansion rate $(t^{-1} \propto \rho^{1/2} \propto T^2)$ and, for $T \lesssim T_f$, the neutron-to-proton ratio "freezes-out". That is, below T_f, the rate at which the neutron abundance decreases is much less than would be the case (Eq. (10)) if equilibrium could be maintained. Notice that since the expansion rate $t^{-1}(T)$ increases with the total density $\rho(T)$, the more types of light neutrinos there are, the faster the universe expands and the earlier the reactions (8) drop out of equilibrium $(T_f \propto g_\nu^{1/6})$, leaving more neutrons available for nucleosynthesis.

For $T \lesssim 0.1$ MeV, photodissociation of deuterium decreases in importance and the deuteron abundance increases. Now there is a basis on which to build the heavier elements. In a sequence of very rapid, two body nuclear reactions, ^{3}H, ^{3}He, and ^{4}He are quickly synthesized (for details and references, see Schramm and Wagoner, 1977). There is, however, a gap at mass-5 (no stable nucleus), which prevents the build-up of significant amounts of heavier nuclei. As a result, virtually all the neutrons available find themselves incorporated in the most tightly bound of the light nuclei: ^{4}He.

The abundance of primordial ^{4}He is, then, a sensitive probe of the neutron abundance at nucleosynthesis. The neutron abundance is a probe of the early expansion rate leading to a constraint on the number of types of light neutrinos (Steigman et al., 1977; Yang et al., 1979; Steigman et al., 1979). For example, for each extra two component neutrino ($2N_L = \Sigma g_\nu$), one percent (by mass) more ^{4}He is synthesized. There are several interesting implications of this constraint. They are outlined below in order of increasing uncertainty.

i) Limit to the nucleon-to-photon ratio

The higher the nucleon density, the earlier the photo-dissociation of the deuteron is overcome as an obstacle to nucleosynthesis. The earlier nucleosynthesis begins, the higher the neutron abundance and, therefore, the more ^{4}He produced. In Yang et al. (1979) the observations are reviewed that lead to an estimate of the primordial abundance of ^{4}He (by mass): $Y_p \lesssim 0.25$. For at least three types (ν_e, ν_μ, ν_τ) of two component neutrinos ($\Sigma g_\nu \geq 6$, $N_L \geq 3$), this abundance would be exceeded unless the nucleon-to-proton ratio is sufficiently small. At present,

$$\eta \equiv \left(\frac{n_N}{n_\gamma}\right)_0 \lesssim 4.2 \times 10^{-10} \quad . \qquad (11)$$

All our estimates here depend on the beta decay rate (8) and have been calculated with an assumed neutron half-life $\tau_{1/2}(n) = 10.6$ minutes. If the true $\tau_{1/2}(n)$ is shorter (longer), the weak interactions remain in equilibrium to lower (higher) temperatures at which point there are fewer (more) neutrons leading to less (more) ^{4}He. It is important for particle physics and for cosmology that the uncertainty in $\tau_{1/2}(n)$ be reduced.

ii) Constraint on N_L

If a lower limit to η is known, then, combined with the requirement that $Y_p \lesssim 0.25$, an upper limit to N_L may be derived. There is, however, a problem here. Until very recently it has been natural to assume that nucleons dominate the mass in the universe today. Were this the case, estimates of the total mass are equivalent to estimates of the nucleon abundance. If, however, neutrinos have a small but finite rest mass ($m_\nu \gtrsim 1$ eV), relic neutrinos may dominate the total mass in the universe today. Given this possibility, it is very difficult to obtain a reliable *lower* limit to η. Of course, the luminous (including x-rays) matter in the universe must be traceable to nucleons (stars, hot gas). This leads (Schramm and Steigman, 1980a,b) to an estimate of a lower limit to η

$$\eta \gtrsim 10^{-10} \quad . \tag{12}$$

For $\eta \gtrsim 10^{-10}$, $Y_p \lesssim 0.25$ and $\tau_{1/2}(n) \approx 10.6$ min., there can be at most four types of two component neutrinos.

$$N_L \lesssim 4 \quad . \tag{13}$$

This limit excludes right handed counterparts to the usual left handed neutrinos unless the right handed interactions are so weak that decoupling occurred above 200 MeV (Steigman et al., 1979); this would correspond to

$$M_{W_R} \gtrsim 50\, M_{W_L} \quad .$$

An extreme lower bound to η could, in principle, be quite small:

$$\eta \gtrsim 10^{-11}$$

(Schramm and Steigman, 1980a,b). Were the nucleon abundance this low, much less helium and much more deuterium would have been produced primordially:

$$Y_p \lesssim 0.14 \ , \qquad X_D \gtrsim 10^{-2}$$

(Yang et al., 1979). One might be forced to such a low nucleon abundance if more lepton types were discovered ($N_L \gtrsim 4$) and/or if convincing data became available suggesting that the abundance of primordial helium is less than twenty-five percent by mass.

4. A NEUTRINO DOMINATED UNIVERSE

Observations of the microwave background suggest the present temperature of the relic photons is in the range

$$2.7 \lesssim T_0 \lesssim 3.0°\,K$$

(Thaddeus, 1972; Hegyi et al., 1974; Woody et al., 1975; Danese and DeZotti, 1978). The density of relic photons is

$$n_{\gamma_0} \approx 400\left(\frac{T_0}{2.7}\right)^3 \ cm^{-3} \quad . \tag{14}$$

The present neutrino temperature is lower (see Eq. (6a)):

$$T_{\nu_0} = 0.71 \ T_0$$

and the density of relic neutrinos is (see Eq. (6b))

$$n_{\nu_0} \approx 109\left(\frac{g_\nu}{2}\right)\left(\frac{T_0}{2.7}\right)^3 \ cm^{-3} \quad . \tag{15}$$

Although virtually impossible to detect directly, relic neutrinos are as abundant as microwave photons. They are, in particular, much more abundant than ordinary nucleons (compare Eqs. (6b) and (11))

$$\left(\frac{n_\nu}{n_N}\right)_0 \gtrsim 6.5 \times 10^8 \left(\frac{g_\nu}{2}\right) \quad . \tag{16}$$

Relic neutrinos with a finite mass could dominate the mass in the universe today.

$$\left(\frac{\rho_\nu}{\rho_N}\right) = \left(\frac{m_\nu}{m_N}\right)\left(\frac{n_\nu}{n_N}\right) \gtrsim \left(\frac{g_\nu}{2}\right)\left(\frac{m_\nu}{1.4 \ eV}\right) \quad . \tag{17}$$

This estimate, that the universe would be dominated by relic neutrinos with $m_\nu \gtrsim 1.4$ eV, is free from uncertainties in T_0 and in other cosmological quantities such as the present value of the Hubble parameter. To pursue further the cosmological constraints on m_ν, it is, unfortunately, necessary to introduce and deal with these quantities.

For the simplest isotropic and homogeneous cosmological models, there is a critical density separating those models which expand forever $(\rho_0 \leq \rho_c)$ from those which eventually stop expanding and collapse $(\rho_0 > \rho_c)$.

$$\rho_c = \frac{3H_0}{8\pi G}$$

$$\approx 2 \times 10^{-29} \ h_0^2 \ (gcm^{-3}) \quad . \tag{18}$$

In Eq. (18), G is Newton's gravitational constant and H_0 is the present value of the Hubble parameter. To allow for the range of uncertainty in H_0 (Sandage and Tammann, 1976; deVaucouleurs and Bollinger, 1979; Branch, 1979; Aaronson et al., 1979), h_0 has been introduced where $0.4 \lesssim h_0 \lesssim 1$ and

$$H_0 = 100\ h_0\ (\text{km s}^{-1}\ \text{Mpc}^{-1}) \quad ,$$

$$H_0^{-1} = 10\ h_0^{-1} \times 10^9\ \text{yr}. \tag{19}$$

It is convenient to express the density as a fraction of the critical density.

$$\Omega \equiv \frac{\rho}{\rho_c} \quad . \tag{20}$$

For massive relic neutrinos,

$$\Omega_\nu = \left(\frac{1}{100\ h_0^2}\right)\left(\frac{T_0}{2.7}\right)^3 \sum\left(\frac{g_\nu}{2}\right) m_\nu \quad . \tag{21}$$

In Eq. (21) and throughout, m_ν is in eV. In the following, for convenience, consideration will be given to one, two component neutrino $(\Sigma g_\nu m_\nu = 2m_\nu)$; remember, though, that Ω_ν is related to a sum over all neutrinos. Rewriting Eq. (21), then, the neutrino mass mas be related to a combination of cosmological parameters:

$$m_\nu = 100(\Omega_\nu h_0^2)\left(\frac{2.7}{T_0}\right)^3 \text{eV} \quad . \tag{22}$$

In a sequence that becomes increasingly uncertain, more and more constraining upper limits to m_ν will be obtained.

Since $T_0 \gtrsim 2.7°$K,

$$m_\nu \lesssim 100\ \Omega_\nu h_0^2 \quad ;$$

what are the upper limits to Ω_ν and h_0?

Clearly, $\rho_\nu < \rho_{TOT}$, so that $\Omega_\nu < \Omega_0$ (there are, after all, some nucleons in the universe). A very conservative upper limit to Ω_0 (which is probably too high by at least a factor two) is $\Omega_0 \lesssim 2$; this would correspond to a deceleration parameter $q_0 \lesssim 1$, and

$$m_\nu \lesssim 200\ h_0^2\ \text{eV} \quad . \tag{23}$$

Published data on H_0 would permit $h_0 \approx 1$ (de Vaucouleurs and Bollinger, 1979; Aaronson et al., 1979). It is likely, however, that H_0 is smaller.

The expansion of the universe, in the standard cosmological model, is always decelerating due to the gravitational effect of the matter in the universe.

$$t_0 = H_0^{-1} f(\Omega_0) \quad , \qquad f(\Omega_0) \leq 1 \quad . \tag{24}$$

Since the universe must be older than the oldest stars, it is suggested that $t_0 \gtrsim 13 \times 10^9$ yr (Iben, 1974) and $h_0^2 \lesssim \frac{1}{2}$, so that

$$m_\nu \lesssim 100 \text{ eV} \quad . \tag{25}$$

A neutrino mass in excess of 100 eV would result in relic neutrinos contributing too much to the mass in the universe today; with little difficulty this limit probably could be reduced to $m_\nu \lesssim 50$ eV. With somewhat greater uncertainty, a more severe constraint can be obtained.

There is the long-standing puzzle of the missing light in the universe (for an excellent recent review, see Faber and Gallagher, 1979). On increasing scales from galaxies to pairs and groups of galaxies to large clusters of galaxies, more and more of the mass is nonluminous. Neutrinos, of course, might provide an ideal solution to this puzzle (Cowsik and McClelland, 1972). If sufficiently heavy (Tremaine and Gunn, 1979), they would cluster along with ordinary matter (nucleons), contributing to the mass, but not the light. In any astrophysical system (galaxy, group, cluster) formed by the collapse of neutrinos and nucleons, the ratio of mass in each component is limited to

$$\frac{M}{M_N} \lesssim \frac{M_{TOT}}{M_N} \quad . \tag{26}$$

Comparing Eqs. (17) and (26), it follows (for $g_\nu = 2$) that

$$m_\nu \lesssim 1.4 \left(\frac{M_{TOT}}{M_N} \right) \quad . \tag{27}$$

As mentioned earlier, it is difficult to separate the nuclear contribution from the total.

In going from the inner parts of galaxies (~ 10 kpc) to the scale of binary galaxies (~ 100 kpc), there is the suggestion that $M_{TOT} \lesssim 10\, M_N$ (see Faber and Gallagher, 1979, for details and references). It should be noted that the nonluminous mass *could* be due to nucleons. There is, however, evidence suggesting that this ratio does not increase much in going to the scale of large clusters (~ 1 Mpc). Many clusters are x-ray sources, emitting thermal bremsstrahlung radiation from a hot intracluster gas. Although estimates of the amount of gas are uncertain, it is suggested (Lea et al., 1973; Cavaliere and Fusco-Femiano, 1976; Malina et al., 1976) that

$$\left(\frac{M_{TOT}}{M_N} \right)_{\text{Clusters}} \lesssim 28\, h_0^{3/2} \quad . \tag{28}$$

For $h_0 \lesssim 0.7$, this corresponds to $m_\nu \lesssim 23$ eV.

Although neutrino masses in excess of 20-30 eV may be excluded by this argument, it should not go unnoticed that for $m_\nu \gtrsim$ a few eV and $m_\nu \lesssim 20$ eV, most of the mass in the universe could be in relic neutrinos.

For $m_\nu \lesssim$ a few eV, neutrinos are too light to cluster at all (Tremaine and Gunn, 1979). Such relic neutrinos would not contribute to the dark mass on scales of clusters; they would be more homogeneously distributed. Until H_0 and T_0 (as well as g_ν and m_ν) are better constrained, it is difficult to estimate the full contribution of all relic neutrinos to the total mass in the universe (Ω_0). Consider, however, the following illustrative example.

For $\Omega_N \approx 0.1$ and $\Omega_\nu \gtrsim 0.9$, the total density exceeds the critical density ($\Omega_0 > 1$); the expansion will ultimately stop and the universe will eventually collapse. For $T_0 \approx 3.0°\,\mathrm{K}$ and $H_0 \approx 50\,\mathrm{km\,s^{-1}Mpc^{-1}}$ ($h_0 \approx \frac{1}{2}$) the condition for this to occur is

$$\sum \left(\frac{g_\nu}{2}\right) m_\nu \gtrsim 16.4 \text{ eV} \quad . \tag{29}$$

Suppose there is one, heaviest Majorana neutrino with $m_\nu \approx 14$ eV and two lighter neutrinos with

$$\sum g_\nu m_\nu \gtrsim 3 \text{ eV} \quad .$$

The universe, then, could be closed by relic neutrinos.

There is, of course, no direct evidence that this example is realized in the real world. More likely estimates are

$$0.04 \lesssim \Omega_N \lesssim 0.08 \quad , \qquad \Omega_\nu \lesssim \frac{1}{2} \quad . \tag{30}$$

5. SUMMARY

In the cosmic accelerator of the early universe, large numbers of neutrinos were produced. Emerging from the big bang, light relic neutrinos are as abundant as photons. The weakness of the weak interaction ensures that direct detection of relic neutrinos, in contrast to relic photons, will be difficult. Through their gravitational interactions, however, relic neutrinos may have played an important role during the earlier evolution of the universe, and they may dominate the present and future evolution.

The presence of three kinds of relic neutrinos (ν_e, ν_μ, ν_τ) at the time of nucleosynthesis provides a lower limit to the expansion rate during that epoch. If too much ^{4}He is not to be produced primordially ($Y_p \lesssim 0.25$), this leads to an upper limit to the nucleon-to-photon ratio (11), which corresponds to $\Omega_N \lesssim 0.12$ (for $h_0 \gtrsim 0.4$, $T_0 \lesssim 3.0°\,\mathrm{K}$). The universe cannot be closed ($\Omega_0 > 1$) by nucleons. Furthermore, if the nucleon density is not too small ($\eta \gtrsim 0.01$), extra neutrino types would lead to the primordial production of too much ^{4}He: $N_L \lesssim 4$.

Relic neutrinos with a small but finite mass ($m_\nu \gtrsim 1$ eV) could dominate the mass in the universe at present. The astrophysical implications of a neutrino mass in excess of a few eV will be enormous. Stable ($\tau_\nu \gtrsim t_0$) neutrinos cannot be too heavy ($m_\nu \lesssim 50$-100 eV), but relatively light neutrinos ($3 \lesssim m_\nu \lesssim 30$ eV) could solve the puzzle of the missing light in clusters of galaxies. Lighter neutrinos ($m_\nu \lesssim 1$ eV) would be unclustered and they would make a relatively small contribution to the total mass in the universe.

Acknowledgments. I would like to thank P. Frampton and A. Yildiz for having organized an exciting workshop. I have learned a great deal from many colleagues in particle physics and astrophysics. I especially wish to acknowledge David Schramm, John Ellis, and Ed Witten for valuable discussions relating to the content of this paper.

REFERENCES

M. Aaronson, J. Mould, J. Huchra, W.T. Sullivan, R.A. Schommer, and G.D. Bothun, *Ap. J.* (In press, 1980).

D. Branch, *MNRAS* 186, 609 (1979).

A. Cavaliere and R. Fusco-Femiano, *Astron. and Astrophys.* 49, 137 (1976).

L. Danese and G. DeZotti, *Astron. and Astrophys.* 68, 157 (1978).

D. Dicus, E.N. Kolb, and V. Teplitz, *Phys. Rev. Lett.* 39, 168 (1977).

S.M. Faber and J.S. Gallagher, *Ann. Rev. Astron. Astrophys.* 17, 135 (1979).

J.E. Gunn, B.W. Lee, I. Lerche, D.N. Schramm, and G. Steigman, *Ap. J.* 223, 1015 (1978).

D.J. Hegyi, W.A. Traub, and N.P. Carleton, *Ap. J.* 190, 543 (1974).

S.M. Lea, J. Silk, E. Kellogg, and S. Murray, *Ap. J. (Lett.)* 184, L105 (1973).

B.W. Lee and S. Weinberg, *Phys. Rev. Lett.* 39, 165 (1977).

R. Malina, M. Lampton, and S. Bowyer, *Ap. J.* 209, 678 (1976).

A. Sandage and G.A. Tammann, *Ap. J.* 210, 7 (1976).

D.N. Schramm and R.V. Wagoner, *Ann. Rev. Nucl. Part. Sci.* 27, 37 (1977).

D.N. Schramm and G. Steigman, First Prize Essay, Gravity Research Foundation (to appear in GRG 1980)(1980a).

D.N. Schramm and G. Steigman, submitted to *Ap. J.* (1980b).

G. Steigman, *Ann. Rev. Nucl. Part. Sci.* 29, 313 (1979).

G. Steigman, Astrophysical constraints on neutrino physics, to appear in *Proc. Int. Meeting on Astrophysics and Elementary Particles, Common Problems*, Rome, 21-23 February, 1980, G. Salvini (ed.).

G. Steigman, C.L. Sarazin, H. Quintana, and J. Faulkner, *Astron. J.* 83, 1050 (1978).

G. Steigman, K.A. Olive, and D.N. Schramm, *Phys. Rev. Lett.* 43, 239 (1979).

P. Thaddeus, *Ann. Rev. Astron. Astrophys.* 10, 305 (1972).

S. Tremaine and J.E. Gunn, *Phys. Rev. Lett.* 42, 407 (1979).

G. deVaucouleurs and G. Bollinger, *Ap. J.* 233, 433 (1979).

D.P. Woody, J.C. Mather, N. Nishioka, and P.L. Richards, *Phys. Rev. Lett.* 34, 1036 (1975).

J. Yang, D.N. Schramm, G. Steigman, and R.T. Rood, *Ap. J.* 227, 697 (1979).

BIG BANG BARYOSYNTHESIS

Michael S. Turner

Astronomy and Astrophysics Center
Enrico Fermi Institute
The University of Chicago
5640 S. Ellis Ave.
Chicago, IL 60637

Abstract

Grand Unified Theories predict the instability of the proton, and therefore the eventual demise of matter in an open universe; however, they may also explain the existence of matter in the first place. It has been suggested that B, C, and CP violating interactions allow an initially baryon symmetrical universe to evolve a baryon asymmetry which, after most of the baryons and essentially all the antibaryons annihilate, leaves only matter in the universe. Detailed calculations are discussed which show that the observed baryon-to-photon ratio of $10^{-9.4 \pm 0.3}$ could have been produced by the actions of a superheavy gauge boson with mass $\sim 3 \times 10^{14}$ GeV if C and CP are violated in its decays by $\sim 10^{-5.3}$, or by a superheavy Higgs boson with mass $\gtrsim 3 \times 10^{13}$ GeV if C and CP are violated in its decays by $\sim 10^{-8}$. The effects of many superheavy species on the evolution of the baryon asymmetry are also discussed. At present, progress in the understanding of the C and CP violations in the superheavy boson system is sorely needed.

1. INTRODUCTION

One of the most startling predictions of Grand Unified Theories (GUTs) is the instability of the proton. As we have heard at this workshop, the estimate of the proton lifetime, $10^{31 \pm 1}$ years in SU_5, is tantalizingly close to the present lower limit, 10^{30} years,[1] and within reach of the experiments currently being undertaken.[2] In an open universe, the instability of the proton leads to the ultimate demise of matter.

One unexpected consequence of GUT is the possible explanation of the existence of matter in the universe in the first place. If this idea proves to be correct, it will solve a long-standing astrophysical puzzle which involves the following two observations: i) the absence of antimatter in our galaxy (anti/matter/matter $\lesssim 10^{-4}$) and perhaps out to the Virgo cluster[3] in a universe where the laws of physics are nearly particle-antiparticle symmetrical; and ii) the observed baryon-to-photon ratio of $10^{-9.4 \pm 0.3}$, almost all the photons being in the 3K background and the constraint on the baryon density resulting from Big Bang nucleosynthesis.[4] A more useful quantity is kn_B/s, the net baryon number to specific entropy ratio, which remains constant when baryon number is conserved and the expansion is isentropic. The species contributing most to the specific entropy today are: the 3K microwave photons, and the 2K background of neutrinos (ν_e, ν_μ, and ν_τ), resulting in

$$\frac{kn_B}{s} = 10^{-10 \pm 0.3} \quad .^5$$

A conventional Big Bang cosmological model which is initially baryon symmetrical and in which baryon number is conserved fails to account for either of these observations.[3] The number of baryons and antibaryons is always equal, and a baryon-to-photon ratio of $\sim 10^{-18}$ evolves due to the incompleteness of annihilations.

Many authors (including S. Hawking, L. Parker, A. Sakharov, S. Weinberg, and Ya. B. Zeldovich[6]) have suggested that if baryons were not conserved it might be possible for the universe to evolve a slight excess of baryons, leaving the matter we see today after most of the matter and all the antimatter annihilate. Yoshimura and Ignatiev et al. first discussed how the B violating interactions predicted by GUTs could allow the generation of a baryon excess.[7,8]

Three ingredients are necessary for a baryon excess to evolve: i) baryon violating interactions; ii) both C and CP violations--an arrow to specify the direction of the excess; and iii) a departure from thermal equilibrium. This final point is a subtle one (which Yoshimura overlooked) and was pointed out by Dimopoulos and Susskind,[9] Toussaint et al.,[10] and Weinberg,[11] and even before GUTs by Sakharov.[6] With CPT and unitarity alone one can show that when all particle species assume their equilibrium distributions, the rate of baryon and antibaryon production is equal, regardless of B, C, and CP violations, so that no excess evolves (e.g., see Appendix A of Ref. 5). Departures from thermal equilibrium occur frequently in the expanding universe. As the universe expands its temperature falls as $R(t)^{-1}$. Equilibrium distributions are changing and unless reactions which allow particle distributions to adjust are occurring rapidly (rates $\Gamma >$ expansion rate $H \equiv \dot{R}/R$), particle distributions will not be equilibrium distributions. For example, today the matter and radiation temperatures are very different because interactions between matter and the 3K background are occurring much less rapidly than the expansion.

A variety of scenarios have been suggested for the generation of a baryon excess, including production by primordial black holes, quantum gravity effects, and semi-classical quantum gravity effects.[6-16] The mechanism for generating a baryon asymmetry by GUT processes alone which appears to work is the out-of-equilibrium decay scenario of Weinberg[11] and Wilczek.[10] I will briefly outline it. Throughout, I will take

$$\hbar = k_B = c = 1 \quad .$$

The superheavy boson in this scenario, denoted by S with mass M_S ($\sim 10^{15}$ GeV) and coupling α_S, might be either a gauge ($\alpha_S \sim 10^{-2}$) or Higgs ($\alpha_S \sim 10^{-4} - 10^{-6}$) boson. At the Planck time ($\sim 10^{-43}$ s) the universe begins as a baryon symmetrical hot soup of all the fundamental particles (quarks, leptons, gauge and Higgs bosons). The temperature $\sim 10^{19}$ GeV is $\gg M_S$, so that S bosons are about as abundant as any other species (e.g., photons)--all species are ultrarelativistic. When the temperature of the universe falls below $\sim M_S$, then the equilibrium abundance of S bosons relative to photons is $\sim \exp(-M_S/T)$. In order for S bosons to achieve their equilibrium distribution, they must diminish in number, e.g., by decays or annihilations--decays are the more important of the two.

If the decay rate Γ_D ($\sim \alpha_S M_S$) is $< H$ ($\sim T^2/10^{19}$ GeV) when $T \sim M_S$ (lifetime of $S >$ age of universe), then S bosons *will not* diminish, but will reamin as abundant as photons. Eventually, Γ_D becomes $> H$; however, when it does, T is $\ll M_S$, and so S and $\bar{S}$ bosons decay freely, inverse decays not occurring since typical fermion pairs have energy $\sim T \ll M_S$. If C and CP are both violated, then the average net baryon number of the decay products of the S and $\bar{S}$ need not be equal and opposite. Let $\varepsilon \equiv$ the average baryon excess produced by the decay of an $S,\bar{S}$ pair $\sim$ size of the C and CP violations. Since before they freely decayed S bosons were as abundant as photons, the baryon excess produced is $\sim \varepsilon n_\gamma$ (n_γ = number of photons). The specific entropy, s/k, is $\sim$ number of relativistic particles $\sim N n_\gamma$ (N = number of fundamental species ~ 100). Thus, the baryon asymmetry generated is

$$\frac{kn_B}{s} \sim 10^{-2}\,\varepsilon \quad .$$

On the other hand, if Γ_D is $> H$ when $T \sim M_S$, we expect the equilibrium abundance of S bosons to be maintained to some extent by decays and inverse decays, and a lesser asymmetry should evolve (decreasing as Γ_D/H increases).

In the most general theory which breaks down to

$$SU_3^C \times SU_2^L \times U_1$$

there are three generic species of superheavy bosons (S) which mediate B and L nonconserving interactions (and hence could produce a baryon excess).[17] There are two vector (gauge) types: i) XY --an isodoublet, color triplet (charge $\pm 4/3$, $\pm 1/3$); and ii) X'Y' --an isodoublet, color triplet (charge $\pm 2/3$, $\pm 1/3$). The third type of superheavy boson is a scalar (Higgs): H --isosinglet, color triplet (charge, $\pm 1/3$).

In the remainder of this paper I will discuss in detail the evolution of the baryon asymmetry of the universe due to the actions of these superheavy bosons. In Section 2 numerical calculations of the effect of each of the three generic species acting alone will be discussed. In general, a species can damp preexisting asymmetries, produce new asymmetries, or both, depending upon $K \simeq \Gamma_D/H$ for $T \simeq M_S$ ($\equiv 2.9 \times 10^{17}$ GeV α_S/M_S). If K is $\ll 1$, preexisting asymmetries are not damped and an asymmetry ($\sim 10^{-2}\,\varepsilon$) is produced due to delayed free decays. If K is $\gg 1$ preexisting asymmetries are damped and a lesser asymmetry is produced.

Even in the minimal SU_5 theory, there are two species of superheavy bosons (XY and H) and so to understand the evolution of kn_B/s one must consider their combined effects. In a more complex theory there may be many species of superheavy bosons. In Section 3 the effects of many species ("hierarchical effects") are discussed. In most situations the effects of many species can be analyzed sequentially, each species modifying the asymmetry produced by the next heavier species, and so on. Any attempt to explain the baryon asymmetry observed today ($kn_B/s \approx 10^{-10.3}$) must, of course, take the hierarchy of effects into account. A summary and concluding remarks are contained in Section 4.

2. EVOLUTION OF kn_B/s DUE TO ONE SUPERHEAVY SPECIES

One of the great triumphs of the Big Bang theory and theoretical astrophysics is the detailed calculation of Big Bang nucleosynthesis (see, e.g., Ref. 18). These calculations show that about 25% of the mass in the universe should have been converted into ^{4}He when the universe was $\sim$ 3 minutes old (the exact amount depends upon the baryon-to-photon ratio and the number of neutrino types; see, e.g., Steigman's contribution to these proceedings). The concordance of these calculations with the observed mass fraction of ^{4}He ($\approx$25%) provides strong support for the Big Bang model; in fact, one can (and does) use Big Bang nucleosynthesis as a probe of the early universe and to put constraints on the properties of hypothetical particles.

A dynamical explanation of the observed baryon asymmetry would be an equally impressive achievement. The goal of the work I am about to describe was to imitate the Big Bang nucleosynthesis calculations wherever possible in order either to establish more firmly or to refute the idea of Big Bang baryosynthesis. This work was carried out by James Fry, Keith Olive, and myself.[5,19,20] Similar calculations have also been done by Kolb and Wolfram[21] and their results are in agreement with outs wherever comparison is possible.

a. Assumptions

In the Big Bang nucleosynthesis calculations, the assumptions which must be made are rather basic and straightforward: i) validity of general relativity, ii) isotropy and homogeneity of the universe, and iii) no particle species was degenerate.[18] The situation with regard to cosmological baryon generation is much more complicated. At present we have no GUT, nor do we even know what the fundamental particles of the theory are. We do know that the theory must be larger than SU_5 and hope that it is not too much larger. Very little is known about one of the key ingredients-- C and CP violation in the superheavy system. Therefore, we make the following assumptions in order to make the problem tractable: i) the universe is described by the Friedmann-Robertson-Walker metric, and during the times of interest ($t \lesssim 10^{-35}$ s) the universe is dominated by ultrarelativistic particle species. The comoving time coordinate (age of the universe) and temperature are related by

$$t = 0.301\, m_p T^{-2} g_*^{-1/2} \quad , \tag{1}$$

where m_p is the Planck mass ($\equiv G^{-1/2} \approx 1.22 \times 10^{19}$ GeV) and g_* is the total effective number of degrees of freedom of all relativistic species ($\equiv$ total number of Bose degrees of freedom + (7/8) total number of Fermi degrees of freedom). The specific entropy density is

$$\frac{s}{k} = g_* \left(\frac{2\pi^2}{45}\right) T^3 \quad . \tag{2}$$

ii) The fundamental particle species are those of the minimal SU_5 model:[22] three generations of fermions, 24 gauge bosons, and a $\underset{\sim}{24}$ and $\underset{\sim}{5}$ of Higgs bosons. At temperatures greater than any particle mass $g_* = 160$. Additional particle species (beyond minimal SU_5) will affect the expansion rate as $(g_*/160)^{1/2}$ and the specific entropy as $(g_*/160)$ and therefore enlarging the particle content of the theory by a factor of 2 or even more should not greatly change our results.

iii) The interactions included all involve the superheavy boson and are: a) Decays (D) and inverse decays (ID) of the superheavy bosons into fermions. We find that these are the most important interactions and to a good approximation all other interactions can be neglected. b) Annihilations of the superheavy bosons. For gauge bosons these processes are $O(\alpha^2)$, while D and ID are $O(\alpha)$, and so are not as important. For Higgs bosons, D and ID are $O(\alpha_H)$ ($\alpha_H \sim 10^{-4} - 10^{-6}$), while annihilations are $O(\alpha^2)$ ($\alpha^2 \sim 10^{-3}$), and so annihilations are potentially very important. But because annihilations tend to "turn themselves off" and because of the quasiequilibrium which exists during the epoch of baryosynthesis, they are not significant.[19] c) Baryon nonconserving processes mediated by superheavy bosons. These processes are all $O(\alpha^2)$ or $O(\alpha_H^2)$, and include two fermion to two fermion scatterings, fermion + superheavy boson to fermion + superheavy (Compton-like processes), and two superheavy bosons to two fermions (annihilation-like processes). These processes turn out to be unimportant compared to D and ID. Baryon conserving interactions are assumed to be happening rapidly ($\Gamma > H$) and thus maintain kinetic equilibrium. This assumption is *not* crucial--we have relaxed it and it affects our results by $\lesssim 50\%$.[5]

iv) The C and CP violations in the superheavy boson system are parameterized by $\varepsilon \equiv$ the average baryon excess produced by the decay of an $S,\bar{S}$ pair. CPT and unitarity require that ε be $< \alpha \sim 10^{-2}$ (the Born graph conserves CP). The issue of C and CP violation in the superheavy system has been addressed in detail by Nanopoulos and Weinberg[19] and by Barr et al.[23]

v) The Boltzmann equations are linearized in the baryon and lepton asymmetries. This certainly seems justifiable, since the observed baryon asymmetry $kn_B/s \simeq 10^{-10.3}$. The resulting set of coupled differential equations is then numerically integrated.

b. Results

The evolution of the baryon asymmetry due to actions of one superheavy species can be well described in terms of just one parameter, K. K is $\simeq \Gamma_D/H$ (decay rate of the superheavy boson/expansion rate) for $T \simeq M_S$; specifically,

$$K \equiv 2.9 \times 10^{17} \alpha_S \ \mathrm{GeV}/M_S \quad , \tag{3}$$

where α_S is the coupling strength of the superheavy boson (for a gauge boson $\sim 1/45$, and for a Higgs boson $\sim 10^{-4} - 10^{-6}$) and M_S is its mass ($\sim 10^{15}$ GeV). When K is < 1, then $\Gamma_D < H$ for $T = M_S$ and one expects a departure from equilibrium (S bosons remain too abundant for $T < M_S$) and baryon production. For $K \gtrsim 1$, Γ_D is $> H$ for $T = M_S$ and one expects equilibrium to be maintained to *some* degree (however, not perfectly).

i) C or CP conserved ($\varepsilon = 0$)

If the interactions of the superheavy boson conserve C or CP, then $\varepsilon = 0$ and no baryon generation is possible. However, a preexisting baryon asymmetry may be reduced. The preexisting asymmetry might be an initial condition, or might be due to some other superheavy boson or non-GUT processes such as quantum gravity or primordial black hole evaporation.[12-15]

The damping occurs primarily by the two-step process: ID, D; e.g.,

$$u + u \to \bar{X} \,, \qquad \bar{X} \to \bar{d} + e^+$$

($\Delta B = 1$). Most of the damping takes place when $T \simeq M_S$. This is easy to understand: for $T \gg M_S$, the decay and inverse decay rates are equal and are suppressed by time dilation since the superheavy boson is very relativistic (Γ_D, $\Gamma_{ID} < H$). For $T \ll M_S$, inverse decays are suppressed since typical fermion pairs are not energetic enough to produce a superheavy boson ($E \sim T \ll M_S$) and so $\Gamma_{ID} < H$. Only for $T \approx M_S$ can Γ_{ID} and Γ_D be $> H$, and then

$$\frac{\Gamma_{ID}}{H} \approx \frac{\Gamma_D}{H} \approx K \quad .$$

Gauge bosons damp pre-existing asymmetries by $\exp(-5.5\,K)$, that is,

$$\begin{aligned}(kn_B/s)\Big|_{T \ll M_S} &= (kn_B/s)\Big|_{T \gg M_S} \exp(-5.5\,K) \\ &= (kn_B/s)\Big|_{T \gg M_S} \exp(-3.5 \times 10^{16}\ \text{GeV}/M_S)\end{aligned} \tag{4}$$

The interactions of the gauge bosons conserve Q (total charge), I_3 (total weak isospin), $B - L$ (total baryon number - total lepton number), and 5-ness. "5-ness" is the property of being in the $\bar{\underset{\sim}{5}}$ of SU(5): ν_L, L_L, and $\bar{d}_L$ have 5-ness $= +1$ and their antiparticles have 5-ness $= -1$, where ν, L, and d are the neutrino, electron-like lepton, and d-like quark of any generation.[24] Baryon asymmetries accompanied by nonzero values of any of these conserved quantities are *not* damped. Of course, one expects both Q and I_3 to be identically zero, since they are gauged quantities. Asymmetries with $B - L \neq 0$ cannot be produced by superheavy bosons; however, it is possible for a superheavy Higgs boson to produce a baryon asymmetry with net 5-ness. (Note: Although the gauge bosons of interest (XY, X'Y') conserve 5-ness, other gauge bosons in the theory may not, e.g., the gauge boson of a SU_2^R symmetry. Together, the gauge boson which does not conserve 5-ness and the gauge boson which mediates B violations (XY or X'Y') could damp baryon asymmetries with net 5-ness.)

It is interesting to note that if a superheavy boson with mass $< 10^{15}$ GeV exists, then it will reduce any pre-existing asymmetry by more than a factor of 10^{11} (unless that asymmetry has $B - L \neq 0$ or 5-ness $\neq 0$), ruling out the possibility of explaining the observed $kn_B/s \approx 10^{-10.3 \pm 0.3}$ as merely being an initial condition, and thereby *necessitating* cosmological baryon generation. In SU_5, gauge boson masses have been estimated to be $\sim 3 \times 10^{14}$ GeV.[1] In addition, $M \lesssim 10^{15}$ GeV is interesting because it implies a proton lifetime of $\lesssim 10^{34}$ years.

Because the superheavy Higgs species of interest are isosinglets and have only one spin degree of freedom, they are slightly less efficient at damping pre-existing asymmetries. Their damping is given by $\exp(-K)$, or

$$(kn_B/s)\Big|_{T \ll M_H} = (kn_B/s)\Big|_{T \gg M_H} \exp(-K) \quad . \tag{5}$$

For

$$\alpha_H \gtrsim 10^{-4} \quad \text{and} \quad M_H > 3 \times 10^{13}\ \text{GeV}$$

damping by Higgs bosons is not significant (less than a factor of e). The interactions of the Higgs bosons conserve only Q, I_3, and $B-L$. Only baryon asymmetries accompanied by nonzero values of $B-L$ cannot be reduced by Higgs bosons.

ii) Both C and CP violated ($\varepsilon \neq 0$)

If the decays of the superheavy boson violate both C and CP, then generation of a net baryon number is possible. For $K \ll 1$ ($\Gamma_D \ll H$ for $T \sim M_S$), we find the qualitative picture of Weinberg and Wilczek to be correct; S bosons remain as abundant as photons even after $T < M_S$ and eventually ($T \ll M_S$) they decay freely, producing a baryon excess. For gauge bosons this excess is

$$(kn_B/s) = 1.5 \times 10^{-2}\, \varepsilon \quad , \tag{6}$$

and for Higgs bosons it is

$$(kn_B/s) = 0.5 \times 10^{-2}\, \varepsilon \quad ; \tag{7}$$

the factor of three difference results because gauge bosons have three spin states while Higgs bosons have only one.

For $K \gg 1$ the situation is rather surprising. Although one might expect equilibrium to be maintained, since $\Gamma_D, \Gamma_{ID} \gg H$, and therefore no baryon production; this is *not* the case. A non-negligible baryon asymmetry can be produced. Let me briefly explain why this occurs.

The relevant Boltzmann equations (when simplified) are:

$$\begin{aligned} \Delta' &= -zK\Delta - S'_{EQ} \\ &\approx -zK\Delta + \exp(-z) \quad , \end{aligned} \tag{8}$$

$$\begin{aligned} B' &= zk(\varepsilon\Delta - \gamma_{ID}B) \\ &\approx zK[\varepsilon\Delta - \exp(-z)B] \quad , \end{aligned} \tag{9}$$

where

$$\Delta = S - S_{EQ} \quad ,$$

S is the number density of S bosons relative to pbotons, S_{EQ} is their equilibrium density $\approx \exp(-z)$, B is the baryon number density relative to photons ($kn_B/s \sim 10^{-2}$ B), $z \equiv M_S/T$, and prime denotes d/dz. Without the S'_{EQ} term, Eq. (8) would imply that any deviation from equilibrium relaxes to zero exponentially. However, since the universe is expanding, S_{EQ} is changing $(T \propto R^{-1})$ and S is "chasing" an equilibrium value which is constantly decreasing (hence, the S'_{EQ} term). For $z \gtrsim 1$ $(T \lesssim M_S)$ both rates in Eq. (8) are large (and of opposite sign) and Δ adjusts itself so that $\Delta' \approx 0$,

$$\Delta \approx \exp(-z)/zK \quad ; \tag{10}$$

the relative departure from equilibrium, Δ/S_{EQ}, only decreases as $(zK)^{-1}$.

In the equation governing B, $\varepsilon\Delta$ is the source term, and $\gamma_{ID}B$ represents damping due to inverse decays and decays $(\gamma_{ID} \sim \exp(-z))$. As long as ID are occurring rapidly $(\Gamma_{ID} \gtrsim H)$ both rates in Eq. (9) are large (and of opposite sign) so that B takes on a value such that $B' \approx 0$,

$$B \approx \frac{\varepsilon\Delta}{\exp(-z)}$$

$$B \approx \frac{\varepsilon}{zK} \quad . \tag{11}$$

Therefore, B decreases as z^{-1} as long as $\Gamma_{ID} \gtrsim H$. Eventually ID freeze out $(\Gamma_{ID} \approx H)$; this occurs for

$$z = z_f \quad ,$$

$$K z_f^{7/2} e^{-z_f} = 1 \quad .$$

From that time forward, B does not decrease and

$$B(z > z_f) \approx \frac{\varepsilon}{z_f K}$$

$$\frac{kn_B}{s} \approx \frac{10^{-2}\,\varepsilon}{z_f K} \quad . \tag{12}$$

Asymptotically, $z_f \to \ln K$; however, this is a poor approximation for most values of K (e.g., for $K = 10$, $z_f = 10.5$). The evolution of kn_B/s is shown in Fig. 1.

When K is $\gg 1$, baryon production by gauge bosons is

$$(kn_B/s) \sim 1.5 \times 10^{-2}\,\varepsilon(8.6\,K)^{-1.5}$$

and by Higgs bosons is

$$(kn_B/s) \sim 0.5 \times 10^{-2}\,\varepsilon(3\,K)^{-1.2} \quad ,$$

decreasing only as a small power of K. Over the whole range of K our

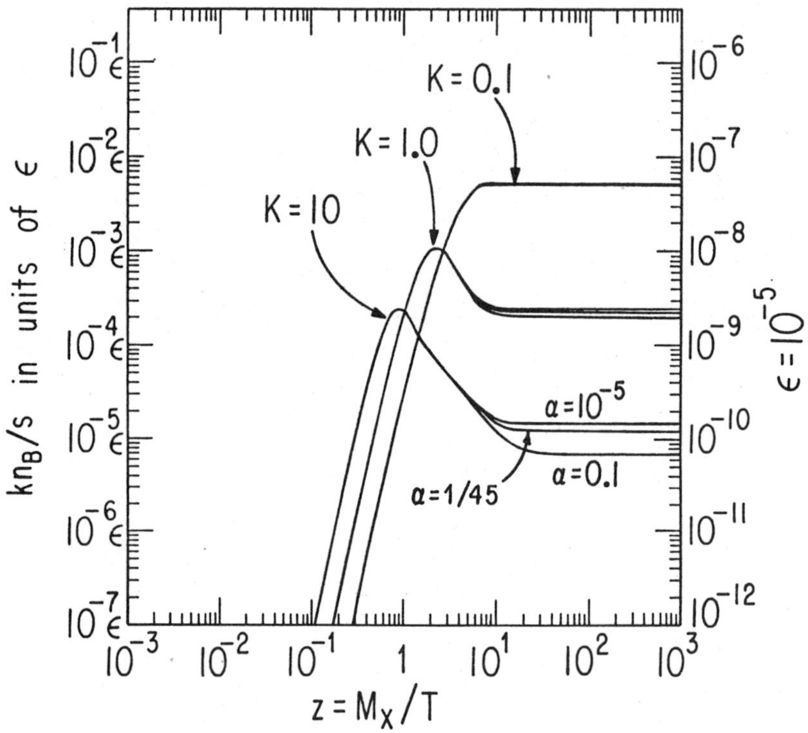

Figure 1. The time evolution $(t \sim z^2)$ of kn_B/s in units of ε due to a superheavy gauge boson for $K = 0.1$, 1, and 10 and $\alpha = 0.1$, 1/45, and 10^{-5}. The slight dependence upon α for fixed K shows the small effect of $O(\alpha^2)$ processes. For $K = 0.1$, the baryon asymmetry is produced by delayed free decays $(z \sim 5)$. For $K = 10$, there is a period of initial nonequilibrium growth (for $z \lesssim 1$ both Γ_D and $\Gamma_{ID} < H$) followed by a period of quasiequilibrium $(1 \lesssim z \lesssim z_f \approx 10.5)$ where $kn_B/s \propto z^{-1}$, and finally, "freeze out" at $z = z_f$. The final value of $kn_B/s \sim 10^{-2}\varepsilon/Kz_f$,

$$Kz_f^{7/2} e^{-z_f} = 1 \quad .$$

<u>Note</u>. For this figure, $\varepsilon/2 \equiv$ the average baryon number produced by the decay of a $S,\bar{S}$ pair, while in the text $\varepsilon \equiv$ the average baryon number produced. (From Ref. 5.)

numerical results are well represented by

$$\frac{kn_B}{s} \simeq \frac{1.5 \times 10^{-2}\ \varepsilon(8.6\,K)^{-1.5}}{1+(8.6\,K)^{-1.5}} \tag{13a}$$

$$\frac{kn_B}{s} \simeq \frac{0.5 \times 10^{-2}\ \varepsilon(3\,K)^{-1.2}}{1+(3\,K)^{-1.2}}\ , \tag{13b}$$

for gauge and Higgs bosons, respectively. Our numerical results are shown in Figs. 2 and 3.

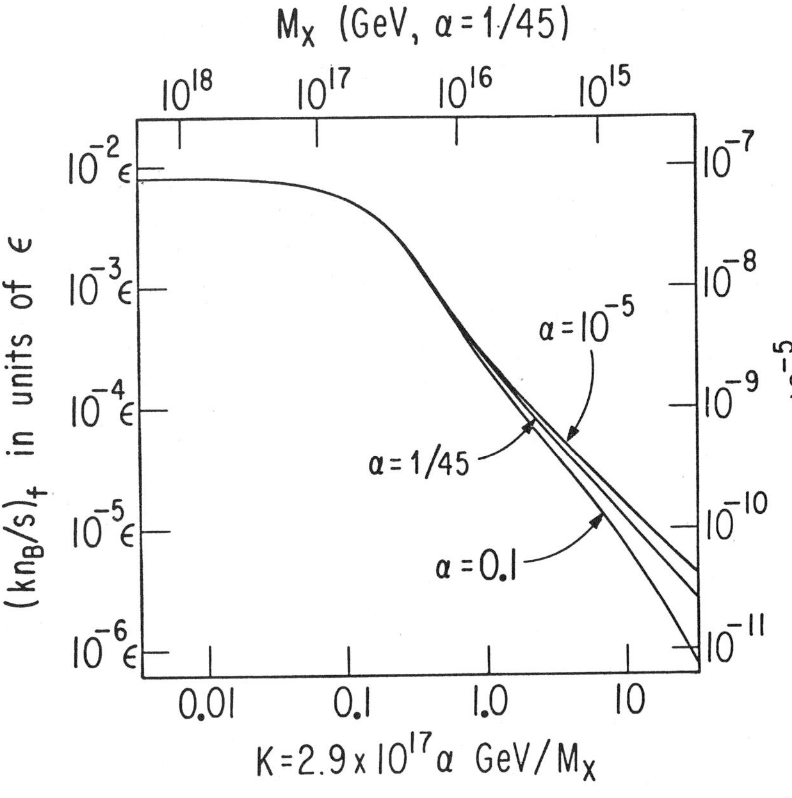

Figure 2. The final value of the baryon asymmetry produced by a gauge boson in units of ε (see note in Fig. 1) as a function of K and α. For fixed K there is negligible dependence on α. A mass scale is shown for $\alpha = 1/45$. (From Ref. 5.)

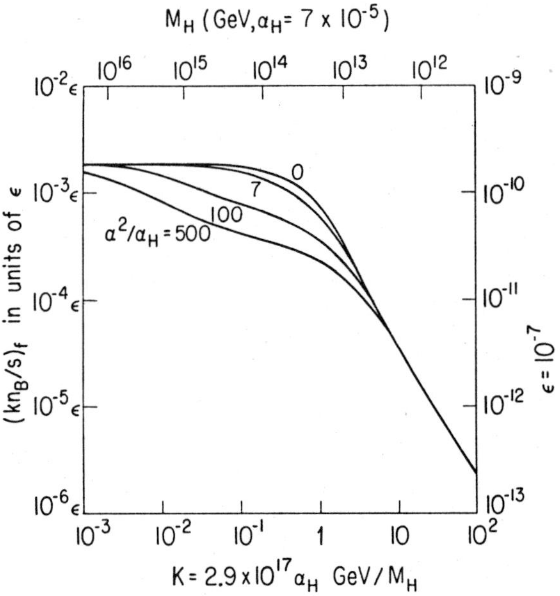

Figure 3. The final value of the baryon asymmetry produced by a Higgs boson in units of ε (see note in Fig. 1) as a function of K and α^2/α_H. The dependence upon α^2/α_H shows the effect of Higgs annihilations on baryon production, which is slight. (Note. The rate of annihilations ~ decay rate $\times\, \alpha^2/\alpha_H$.) A mass scale is shown for $\alpha_H = 7\times10^{-5}$. (From Ref. 19.)

The baryon asymmetry produced by a 3×10^{14} GeV gauge boson alone (α = 1/45, K = 21.5) is $(kn_B/s) \approx 10^{-5}\ \varepsilon$, requiring $\varepsilon \approx 10^{-5.3}$ to produce the observed $10^{-10.3}$. For a Higgs boson with $\alpha_H \lesssim 10^{-4}$ and $M_H \gtrsim 3\times10^{13}$ GeV (K < 1) the asymmetry produced is the saturation value $(kn_B/s) \approx 0.5\times10^{-2}$, requiring $\varepsilon \approx 10^{-8}$ to produce the observed $10^{-10.3}$.

3. HIERARCHY OF BARYON GENERATION

a. General Remarks

Even the simplest GUT, the minimal SU_5 theory, has two species of superheavy bosons (XY and H) and more complex theories will have many species of superheavy bosons. Each species will affect the evolution of the baryon asymmetry. In principle, the calculation of the evolution of kn_B/s due to all species is a formidable task.

However, because the effects of a given species occur for $T \sim M_S$, the effects can be treated approximately sequentially. For purposes of baryon generation, each superheavy boson is described by its mass M_i, its value of Γ_D/H, K_i, and the CP violation in its decays, ε_i. We shall assume that different species have different masses; if two species have the same mass, $M_i \approx M_j$, then they can be replaced by one "effective species" with $K = K_i + K_j$ and $\varepsilon = \varepsilon_i + \varepsilon_j$.

The time development of the baryon asymmetry can be found as follows: The universe is assumed to begin with an initial asymmetry, $(kn_B/s)_0$ (either an initial condition or produced by pre-GUT processes); then the interactions of the most massive superheavy boson (Species 1) cause that asymmetry to change (when $T \sim M_1$), so that when its effect on kn_B/s is complete ($T \ll M_1$), the baryon asymmetry is $(kn_B/s)_1$. Next, $(kn_B/s)_1$ is used as the initial asymmetry for the "baryon asymmetry changing" epoch ($T \sim M_2$) of the next most massive superheavy boson, Species 2, and after its effects have ceased ($T \ll M_2$), the asymmetry is $(kn_B/s)_2$. This procedure continues until the effect of the lightest species has been taken into account. Thus, the problem of computing the evolution of kn_B/s reduces to calculating the asymmetry $(kn_B/s)_i$ produced by the species i, in the presence of a pre-existing ($T > M_i$) asymmetry $(kn_B/s)_{i-1}$.

Since we have linearized the Boltzmann equations, the solution to the above problem is simply obtained by adding the damping of $(kn_B/s)_{i-1}$ to the asymmetry produced by species i; for gauge bosons

$$\left(\frac{kn_B}{s}\right)_i = \left(\frac{kn_B}{s}\right)_{i-1} \exp(-5.5\,K_i) + \frac{1.5 \times 10^{-2}\, \varepsilon_i (8.6\,K_i)^{-1.5}}{1 + (8.6\,K_i)^{-1.5}} , \tag{14}$$

and for Higgs bosons

$$\left(\frac{kn_B}{s}\right)_i = \left(\frac{kn_B}{s}\right)_{i-1} \exp(-K_i) + \frac{0.5 \times 10^{-2}\, \varepsilon_i (3\,K_i)^{-1.2}}{1 + (3\,K_i)^{-1.2}} . \tag{15}$$

(Of course we should remember that gauge bosons *cannot* damp pre-existing asymmetries with nonzero $B - L$ or 5-ness, and Higgs bosons cannot damp ones with $B - L \neq 0$.)

Equations (14) and (15) have simple consequences in the two limiting cases, $K \ll 1$ and $K \gg 1$. For $K \ll 1$, Eqs. (14) and (15) reduce to

$$\left(\frac{kn_B}{s}\right)_i = \left(\frac{kn_B}{s}\right)_{i-1} + 1.5 \times 10^{-2}\, \varepsilon_i \tag{14a}$$

$$\left(\frac{kn_B}{s}\right)_i = \left(\frac{kn_B}{s}\right)_{i-1} + 0.5 \times 10^{-2}\, \varepsilon_i \tag{15a}$$

that is, Species i just algebraically adds its saturation asymmetry to the pre-existing asymmetry. For $K \gg 1$, Eqs. (14) and (15) reduce to

$$\left(\frac{kn_B}{s}\right)_i = 1.5 \times 10^{-2}\, \varepsilon_i (8.6\, K_i)^{-1.5} \tag{14b}$$

$$\left(\frac{kn_B}{s}\right)_i = 0.5 \times 10^{-2}\, \varepsilon_i (3\, K_i)^{-1.2} \quad , \tag{15b}$$

that is, species i erases the pre-existing asymmetry, and the resulting asymmetry depends *only* upon its parameters (ε_i and K_i) --the previous evolution of (kn_B/s) becomes irrelevant.

b. Two Plausible Scenarios

i) Suppose that the lightest gauge boson is less massive than 10^{15} GeV and that there is one Higgs boson less massive than this, but more massive than 3×10^{13} GeV. In this case, K for the gauge boson is 6.5, so that its action erases any pre-existing asymmetry and kn_B/s after its epoch is just:

$$\frac{kn_B}{s} \approx 1.5 \times 10^{-2}\, \varepsilon_{gauge} (8.6\, K_{gauge})^{-1.5} \quad .$$

For the Higgs boson K is < 1 (for $\alpha_H < 10^{-4}$), so that it will not reduce the asymmetry produced by the lightest gauge boson, but will only add to it, resulting in a final asymmetry

$$\frac{kn_B}{s} = 1.5 \times 10^{-2}\, \varepsilon_{gauge} (8.6\, K_{gauge})^{-1.5} + 0.5 \times 10^{-2}\, \varepsilon_{Higgs} \tag{16}$$

so that the post-GUT epoch asymmetry depends *only* upon the lightest gauge boson and Higgs boson. Various combinations of ε_{gauge} and ε_{Higgs} can produce the observed

$$\frac{kn_B}{s} = 10^{-10.3} \quad .$$

ii) Suppose that the lightest superheavy boson is a Higgs boson and is light enough or couples strongly enough so that K_{Higgs} is $\gtrsim 10$ (for $\alpha_H \approx 10^{-4}$, $M_H < 3 \times 10^{12}$ GeV). In this case, the Higgs boson erases any pre-existing asymmetry, and then produces an asymmetry which depends only upon its parameters ε_{Higgs} and K_{Higgs},

$$\frac{kn_B}{s} = 0.5 \times 10^{-2}\, \varepsilon_{Higgs} (3\, K_{Higgs})^{-1.2} \quad ; \tag{17}$$

thus the post-GUT epoch value of the asymmetry depends *only* on the lightest Higgs boson.

The evolution of the baryon asymmetry due to an arbitrary set of superheavy bosons can, of course, be found from Eqs. (14) and (15) by using the procedure outlined in (a). (This procedure will be described in greater detail in a paper which is to be submitted to *Phys. Rev. Lett.*[20])

4. SUMMARY AND CONCLUDING REMARKS

Today, two years after Yoshimura and Ignatiev[7,8] suggested the idea that the baryon asymmetry, $kn_B/s \approx 10^{-10.3}$, might have evolved due to GUT processes and over a year after Weinberg and Wilczek[10,11] suggested the out-of-equilibrium mechanism, detailed numerical calculations[5,19,20,21] have demonstrated that this is a viable mechanism. The existence of a superheavy gauge boson with mass $\lesssim 10^{15}$ GeV (which could be inferred by observation of proton decay) would almost necessitate cosmological baryon generation, as any initial asymmetry put in *ad hoc* to explain $kn_B/s \approx 10^{-10.3}$ would be damped by more than a factor of 10^{11} (unless it is accompanied by a nonzero value of $B-L$ or 5-ness). At the very least, our thinking about explaining $10^{-10.3}$ as an initial condition would have to be modified.

The GUT scenario for baryon generation works even better than one might have expected. Even when $\Gamma_D > H$ for $T \sim M_S$ $(K > 1)$, a sizeable asymmetry can be produced, as baryon production falls off for large K only as a power law, with exponent ~ -1.2 to -1.5. The observed asymmetry could be produced by the actions of a 3×10^{14} GeV gauge boson if

$$\varepsilon_{gauge} \approx 10^{-5.3}$$

or by a Higgs boson $(\alpha_H \leq 10^{-4})$ more massive than 3×10^{13} GeV if

$$\varepsilon_{Higgs} \approx 10^{-8} \quad .$$

As discussed in Section 3, the existence of a light gauge boson $(\leq 10^{15}$ GeV) or a very light Higgs boson $(\leq 3 \times 10^{12}$ GeV) greatly simplifies the analysis of the evolution of kn_B/s due to all the superheavy species that may exist. In these situations, the final value of kn_B/s depends only upon the lightest few species (and possible pre-existing asymmetries with $B-L \neq 0$ or net 5-ness).

The question which looms bigger than even which GUT is correct (if any), is that of the C and CP violations in the superheavy boson system. CPT and unitarity restrict ε to be smaller than the largest superheavy coupling $\sim 10^{-2}$.[17] However, beyond that, little can be said. In the minimal SU_5 model (one $\underset{\sim}{5}$ and one $\underset{\sim}{24}$ of Higgs), it has been shown that ε is $< 10^{-10}$, ruling out baryon generation in the most economical theory. The addition of only one more $\underset{\sim}{5}$ of Higgs to the minimal model once again allows ε to be as large as 10^{-2}. (The minimal model also has difficulties explaining fermion mass ratios.) There is always the hope (probably unrealistic) that ε can be related to the CP violation we observe in the $K^0-\bar{K}^0$ system.[23] If this hope were realized, two fundamental puzzles, the matter-antimatter asymmetry in the universe and the $K^0-\bar{K}^0$ system, would be united by one equation,

$$10^{-10.3} \approx \frac{kn_B}{s}$$

$$\approx \eta\varepsilon \quad ,$$

a possibility suggested by J. Cronin shortly after the $K^0-\bar{K}^0$ CP violation was discovered. (Here $\varepsilon \sim 10^{-3}$, the CP asymmetry in the $K^0-\bar{K}^0$ system, and $\eta \sim 10^{-7.3}$ would be due to calculable GUT effects in the early universe.)

This work was supported by NSF AST 78-20402 and an Enrico Fermi Fellowship, both at the University of Chicago.

REFERENCES

1. T.J. Goldman and D.A. Ross, *Phys. Lett.* 84B, 208 (1979); J. Learned, F. Reines, and A. Soni, *Phys. Rev. Lett.* 43, 907 (1979).

2. The four U.S. experiments, BNL-Irvine-Michigan, Harvard-Purdue-Wisconsin, Minnesota, and Homestake (Pennsylvania) are discussed elsewhere in these proceedings.

3. G. Steigman, *Ann. Rev. Astr. Astrophys.* 14, 339 (1976).

4. K.A. Olive, D.N. Schramm, G. Steigman, M.S. Turner, and J. Yang, "The neutron half-life and constraints on neutrino types", in preparation.

5. J.N. Fry, K.A. Olive, and M.S. Turner, EFI Preprint 80-07, to be published in *Phys. Rev.* D (1980).

6. L. Parker, in *Asymptotic Structure of Space-Time*, F.P. Esposito and L. Witten (eds.), Plenum, New York, 1977; A. Sakharov, *Zh. Eksp. Teor. Fiz. Pis'ma Red.* 5, 32 (1967) (*JETP Lett.* 5, 24 (1967)); S. Weinberg, in *Lectures on Particles and Field Theory*, S. Deser and K. Ford (eds.), Prentice-Hall, New Jersey, 1964.

7. M. Yoshimura, *Phys. Rev. Lett.* 41, 281 (1978).

8. A. Ignatiev, N. Krasnikov, V. Kuzmin, and A. Taukhelidze, *Phys. Lett.* 76B, 436 (1978).

9. S. Dimopoulos and L. Susskind, *Phys. Rev.* D18, 4500 (1979).

10. D. Toussaint, S.B. Treiman, F. Wilczek, and A. Zee, *Phys. Rev.* D19, 1036 (1979).

11. S. Weinberg, *Phys. Rev. Lett.* 42, 850 (1979).

12. M.S. Turner, *Phys. Lett.* 89B, 155 (1979).

13. J. Barrow, *Mon. Not. Roy. Astron. Soc.* 192 (1980).

14. B.J. Carr and M.S. Turner, "Does explaining S need more than GUTs", EFI Preprint 80-09 (1980).

15. N.J. Papastamatiou and L. Parker, *Phys. Rev.* D19, 2283 (1979).

16. J. Ellis, M.K. Gaillard, and D.V. Nanopoulos, *Phys. Lett.* 80B, 360 (1978).

17. D.V. Nanopoulos and S. Weinberg, *Phys. Rev.* D20, 2484 (1979).

18. R.V. Wagoner, W.A. Fowler, and F. Hoyle, *Ap. J.* 148, 3 (1967).

19. J.N. Fry, K.A. Olive, and M.S. Turner, EFI Preprint 80-21 (1980).

20. J.N. Fry, K.A. Olive, and M.S. Turner, "Hierarchy of baryon generation", in preparation.

21. E.W. Kolb and S. Wolfram, "Baryon number generation in the early universe", Caltech Preprint OAP-579 (revised), 1980.

22. H. Georgi and S.L. Glashow, *Phys. Rev. Lett.* 32, 438 (1974).

23. S. Barr, G. Segre, and H.A. Weldon, *Phys. Rev.* D20, 2494 (1979).

24. S.B. Treiman and F. Wilczek, "Thermalization of baryon asymmetry", Princeton Preprint, April 1980.

SOME COMMENTS ON MASSIVE NEUTRINOS*

Edward Witten
Lyman Laboratory of Physics, Harvard University, Cambridge, MA 02138

Abstract

In this talk, the concept of massive Majorana neutrinos is explained; one particular scenario is pointed out by which grand unified theories may lead to "large" neutrino masses; and some astrophysical evidence for a neutrino weighing several tens of electron volts is reviewed.

At this conference we have had ample discussion of the possibility that neutrinos may have small, non-zero rest masses, which are likely to be so-called "Majorana masses". I would like to begin by reviewing the subject of what is a "Majorana mass".

First, let us recall why it has been conventionally believed that the neutrinos are massless. While experiment has long provided good upper bounds on neutrino masses, there is also a standard theoretical argument that the neutrino mass should be zero. This argument is based on the two component theory of the neutrino. It is argued that the neutrino has only one helicity state (left-handed), but a massive spin 1/2 fermion would have two helicity states, so the neutrino must be massless. The claim, in other words, is that the neutrino must be massless if the right-handed neutrino ν_R does not exist.

It has long been recognized that these arguments contain in principle a fallacy, although until recently most physicists doubted that nature really makes use of the fallacy. The fallacy is that in the two component theory of the neutrino, we have actually two helicity states, not one. There are *left-handed* neutrinos and *right-handed* antineutrinos. Two helicity states are the right number for a massive spin 1/2 fermion, so why can't we combine the negative helicity ν and positive helicity $\bar{\nu}$ into a massive fermion?

The answer is that lepton number conservation makes it impossible to combine ν and $\bar{\nu}$ into a massive fermion. The neutrino is a lepton, with $L = +1$; the antineutrino is an anti-lepton, with $L = -1$. The two helicity states of a massive fermion must have the same lepton number (*if lepton number is a symmetry*!), because rotations and boosts exchange the two helicity states. So the different lepton numbers of ν and $\bar{\nu}$ (and only that) prevent us from making a massive fermion out of ν and $\bar{\nu}$.

If lepton number is not conserved, we *can* combine them.

To see how this works mathematically, let us recall[1] that the Lie algebra of the Lorentz group O(3,1) can be decomposed as SU(2) x SU(2):

$$O(3,1) \simeq SU(2) \times SU(2). \tag{1}$$

Since the representations of SU(2) are labeled by an integer or half-integer, the representations of SU(2) x SU(2) or of O(3,1) are labeled by a pair of numbers (p,q) which are each integers or half-integers.

The usual neutrino field ν_L and its Dirac adjoint $\bar{\nu}_L$ transform as follows:

$$\nu_L \quad \left(\frac{1}{2},0\right) \qquad \bar{\nu}_L \quad \left(0,\frac{1}{2}\right) \tag{2}$$

$\bar{\nu}_L$ transforms oppositely to ν_L because the complex conjugation which is involved in going from ν_L to $\bar{\nu}_L$ exchanges the two factors of SU(2) in Eq. (11) and so exchanges p and q. (This in turn is because of some factors of i which must be introduced in relating the O(3,1) Lie algebra to SU(2) x SU(2).)

If the opposite helicity fields ν_R and $\bar{\nu}_R$ existed (they apparently don't, at least not in ordinary particle phenomenology), they would transform as follows:

$$\nu_R \quad \left(0,\frac{1}{2}\right) \qquad \bar{\nu}_R \quad \left(\frac{1}{2},0\right). \tag{3}$$

Now, what is a fermion mass term? Fermions (of spin 1/2) always transform as (1/2,0) or (0,1/2), and a mass term always combines two fermi fields *of the same type*. If one multiplies two fermi fields of the same type, let us say both of type (1/2,0), the decomposition is

$$\left(\frac{1}{2},0\right) \times \left(\frac{1}{2},0\right) = \left(0,0\right) + \left(1,0\right), \tag{4}$$

and the (0,0) piece is a Lorentz invariant which can appear in the Lagrangian. (By contrast, combining fermi fields of opposite type gives

$$\left(\frac{1}{2},0\right) \times \left(0,\frac{1}{2}\right) = \left(\frac{1}{2},\frac{1}{2}\right);$$

there is no Lorentz invariant component which could be a mass term. So a mass term always combines fields of the same type.)

Let us now return to our neutrino fields which transform as indicated in Eqs (2) and (3). If the right-handed neutrino existed, we could take ν_L and $\bar{\nu}_R$, both of them transforming as (1/2,0), and form the usual Dirac mass term

$$\bar{\nu}_R \, \nu_L, \tag{5}$$

which combines the two fields as

$$\left(\frac{1}{2},0\right) \times \left(\frac{1}{2},0\right) = \left(0,0\right) + \ldots$$

This Dirac mass term is obviously invariant under the "lepton number" transformation

$$\begin{aligned} \nu_L &\to e^{i\alpha} \, \nu_L \\ \bar{\nu}_R &\to e^{-i\alpha} \bar{\nu}_R. \end{aligned} \tag{6}$$

Since ν_R does not exist, the only (1/2,0) field at our disposal is ν_L, and to write a mass term we must write something bilinear in ν_L:

$$\nu_L \, \nu_L. \tag{7}$$

This is obviously not invariant under the lepton number transformation $\nu_L \to e^{i\alpha} \nu_L$. It leads to the lepton number violating Lagrangian

$$\mathcal{L} = \int dx \quad \bar{\nu}_L \, i\not{\partial} \, \nu_L - \left(\frac{m}{2} \nu_L \nu_L + \text{h.c.}\right) , \tag{8}$$

which is the Lagrangian for a massive Majorana neutrino. (The Dirac algebra in Eqs. (7) and (8) will be commented on later.)

The mass term (7) is known as a Majorana mass term; a Majorana mass is simply a mass which violates a lepton or fermion number conservation law.

To clarify the physical content of Lagrangian (8), this Lagrangian describes a massive *neutral* fermion--neutral in the sense that the particle is its own antiparticle. It is obvious that the particle described by (8) must be identical with its own antiparticle, because otherwise we would need four helicity states, two for the particle and two for the antiparticle, but if ν_R does not exist we have only two helicity states at our disposal.

We are quite acquainted in physics with particles which are identical with their own antiparticles--for example, the neutral pi meson π^0. The only novelty is that we do not usually deal with *fermions* which are their own antiparticles (or which, differently put, do not carry conserved additive quantum numbers).

Roughly speaking, the Majorana neutrino is to the Dirac electron as the neutral π^0 is to the charged π^+. The Majorana neutrino is described by a two component field and the electron by a four component field. The electron field has twice as many components as the neutrino field, and correspondingly it describes *two* massive spin 1/2 particles, e^+ and e^-, while the Majorana neutrino describes a single neutral massive fermion. Likewise, the neutral π^0 can be described by a real scalar field while the charged π^+ requires a complex scalar field. The complex field has twice as many degrees of freedom as the real field, and describes two states, $\pi^{\pm}$, while the real field describes only one, the π^0.

If from the beginning of the development of quantum field theory, fermions that do not carry conserved additive quantum numbers had been known, then Majorana fermions would probably be as familiar to us as neutral π^0's.

(Some further points about the Majorana mass term (7) should be clarified. To form (0,0) from (1/2,0) x (1/2,0), one combines the two fields *antisymmetrically* since only the *antisymmetric* combination of spin 1/2 and spin 1/2 makes spin 0. Therefore the two fields ν_L in (7) or (8) are combined antisymmetrically with respect to their spinor indices. But fermi fields anticommute and should be combined antisymmetrically, so the Majorana mass term is in fact consistent with Fermi statistics. By "$\nu_L \nu_L$" in Eq. (7) is meant the following. The field ν_L is a two component spinor field ν_L^{α}, $\alpha = 1,2$. By "$\nu_L \nu_L$" we mean the antisymmetric combination of the two fields, which transforms as (0,0). Introducing the two index antisymmetric tensor $\varepsilon_{\alpha\beta}$, $\varepsilon_{12} = +1$, the antisymmetric combination can be written more explicitly as

$\nu_L \nu_L = 1/2\, \varepsilon_{\alpha\beta}\, \nu_L^{\alpha}\, \nu_L^{\beta}$.)

The conclusion is that, despite the two component nature of the neutrino, the neutrino may have a mass--*if* lepton number is not conserved.

As is well known, in grand unified theories[2], lepton number is generally not conserved, for the same reason that baryon number is generally not conserved. Grand unified theories generally combine quarks and antiquarks, leptons and antileptons, into the same representation of a gauge group. The bosons of the unified group mediate transitions among the various states, and thus mediate violations of the various quantum numbers. Because of the lepton number violation that is introduced by grand unification, unified theories will generically have non-zero neutrino masses. (The major exception is the minimal SU(5) theory, in which neutrino masses are prevented by B-L conservation.) Moreover, this subject is lent some importance by the fact that neutrino masses are by far the most sensitive way to search for lepton number violation of the sort that unified theories suggest.

On the scale of grand unification, SU(2) x U(1) is a very good symmetry[3,4], and this leads to a simple estimate of the scale of neutrino masses that should be expected. (Much of the discussion below follows Ref. (4).) The simple Majorana mass term $\nu_L \nu_L$ that we have discussed above is not SU(2) x U(1) invariant, because the neutrino field ν_L is an SU(2) x U(1) nonsinglet. To form a gauge invariant expression, one must introduce the Weinberg-Salam doublet

$$\begin{pmatrix} \phi^+ \\ \phi^0 \end{pmatrix}$$

and replace the neutrino field ν_L by the gauge invariant form $(\phi^0 \nu_L - \phi^+ e^-_L)$. This is a gauge invariant version of ν_L because, after symmetry breaking, $(\phi^0 \nu_L - \phi^+ e^-_L) = \langle\phi\rangle \nu_L + \ldots$ The gauge invariant version of $\nu_L \nu_L$ is then[4]

$$(\phi^0 \nu_L - \phi^+ e^-_L)(\phi^0 \nu_L - \phi^+ e^-_L) = \langle\phi^0\rangle^2\, \nu_L \nu_L + \ldots \tag{9}$$

This operator has dimension five, so it is a nonrenormalizable interaction and will not be present in the fundamental Lagrangian. However, it may be induced as an effective interaction by the exchange of very massive particles with lepton-number violating couplings. ((9) will in fact then describe the dominant lepton number violation at low energies, because it is the lowest dimension lepton number nonconserving operator that can be formed from the usual particle fields.)

If the operator (9) does appear in the effective Lagrangian, then, on dimensional grounds, it will appear with a coupling constant that has dimensions of inverse mass. The mass in question will presumably be related to the mass scale of grand unification since we expect that the effective interaction (9) will be induced by diagrams with exchange of superheavy particles. So let us parametrize the coefficient of the operator (9) as a dimensionless constant f, which will depend on the model, divided by the grand unified mass scale M:

$$\mathcal{L}_{eff} = \frac{f}{M}(\phi^0\nu_L - \phi^+e_L^-)(\phi^0\nu_L - \phi^+e_L^-).$$

$$= \frac{f}{M}\langle\phi^0\rangle^2\,\nu_L\,\nu_L + \ldots \tag{10}$$

With this definition the neutrino mass is

$$m_\nu = \frac{f}{M}\langle\phi^0\rangle^2. \tag{11}$$

With $\langle\phi^0\rangle$ = 300 GeV and M equal to the usual unification scale of 10^{15} GeV, this is approximately

$$m_\nu = (.1\text{ eV})\,f. \tag{12}$$

The value of f is extremely model dependent and depends on the nature of diagrams which are assumed to generate the effective action (10). A particularly simple possibility is that this effective interaction may be generated by a tree diagram with an exchange of a superheavy fermion (figure (1)). This possibility was considered by Gell-Mann, Ramond, and Slansky in work that stimulated much of the current interest in neutrino masses.[5] (A similar mechanism in an SU(2) x SU(2) x U(1) model has been given by Mohapatra and Senjanovic.[6]) If no suitable heavy fermion exists or if the Yukawa couplings in figure (1) are extremely small, one might consider instead a loop diagram, such as the diagram of figure (2) (suggested by Weinberg).

Although f is extremely model dependent, in many models f will be much less than one, perhaps of order 10^{-3} or 10^{-4}. For example, from figure (1), we would get $f = \lambda^2$, the square of the Yukawa coupling constant. Yukawa couplings, of course, seem to be rather small. From figure (2), we would get f of order α^2. If f is as small as one might guess from looking at figures (1) and (2), then neutrino masses are $\lesssim 10^{-4}$ eV and are too small to be detected except in mixing experiments in which the path length is the radius of the earth (as in an experiment described at this conference by Lo Secco) or the earth-sun separation (as in the Davis solar neutrino experiment, which, of course, may have already provided evidence for neutrino mixing). Also--to anticipate ourselves a bit--if f is equal to or less than one, then neutrino masses are much too small to play a role in cosmology.

Many scenarios might lead to neutrino masses larger than the above pessimistic estimates. I will here just point out one simple possibility.

If the heavy fermion which is exchanged in figure (1) is much lighter than the other superheavy particles, then the neutrino masses will be enhanced, since the neutrino mass from figure (1) is $\lambda^2\langle\phi\rangle^2$ divided by the mass of the heavy fermion. In models in which the heavy fermion mass is a free parameter, we can make the neutrino masses as large as we wish by choosing the heavy fermion light enough. However, the procedure is not very natural and there is no predictive power.

It may happen, however, that the heavy fermion of figure (1) is naturally massless at the tree level and receives its mass from a loop correction, proportional to the other superheavy masses. In this case the heavy fermion of figure (1) will be naturally much lighter than the other superheavy particles, by a calculable factor. The neutrino masses will then be automatically "large".

This actually happens[7] in the minimal form of the O(10) model. In that model, the heavy fermion is massless at the tree level but gets mass from a two loop diagram, which is shown in figure (3). The mass is therefore (roughly) of order $\alpha^2 M$, M being the typical grand unified mass (for a more accurate discussion see reference (7)). The neutrino masses are then naturally *enhanced* by a factor $1/\alpha^2$.

After a certain amount of analysis, in which one must use relations among Yukawa couplings provided by O(10), one finds in this model the following formula for neutrino masses:

$$m_\nu = m_Q \frac{M_W}{\alpha^2 M} . \tag{13}$$

Here M is the grand unified mass, M_W is the W boson mass, and m_Q is the mass of the up quark in the same generation as whatever neutrino we are considering. With $M_W \approx 100$ GeV and $M \approx 10^{15}$ GeV (but actually the right M to use here is quite uncertain) this becomes

$$m_\nu = 10^{-9}\, m_Q . \tag{14}$$

Quantitatively, this means

$$\begin{aligned} m_{\nu_e} &= 10^{-9}\, m_u \approx 5 \times 10^{-3}\ \text{eV} \\ m_{\nu_\mu} &= 10^{-9}\, m_c \approx 1\ \text{eV} \\ m_{\nu_\tau} &= 10^{-9}\, m_t \approx 30\ \text{eV} \end{aligned} \tag{15}$$

(if, for instance, $m_t = 30$ GeV).

Also, it should be noted that the proportionality in equation (14) between neutrino masses and up quark masses is a proportionality not just between masses but between mass *matrices*. This means that, in this model, the neutrino mixing angles are equal to the Cabibbo-Kobayashi-Maskawa angles.

These results should not be taken too literally, for two reasons. First, although it is true that in this model the neutrino masses are proportional to the up quark masses, the constant of proportionality, quoted in equation (14) as 10^{-9}, is actually uncertain in a quite wide range because the superheavy masses that enter the diagram of figure (3) are not really known. Second, the model in question also predicts $m_d = m_e$, and so definitely requires modification. However, it may be that the mechanism considered here could have applications in other models or in a modified form of this model.

As the final subject in this talk, I would like to turn attention to astrophysics and point out that there actually are two interesting astrophysical arguments that suggest the existence of a neutrino with a mass of several tens of electron volts. These arguments are not new, but do not seem to be well known among particle physicists. One argument involves the mean mass density of the universe; the second concerns galactic halos, a subject about which my knowledge comes mainly from conversations with M. Davis and M. Lecar.

Considering first the average mass density of the universe, we know that according to general relativity, if the average density ρ of the universe is less than a certain critical density, ρ_c, the universe will expand forever, but if it is greater than ρ_c, the universe will recollapse. Because it depends on the uncertain Hubble constant, ρ_c isn't known accurately, but it is roughly 10^{-29} gm/cm^3.

Moreover, if ρ/ρ_c is less than 1, it goes to zero in time; if ρ/ρ_c is greater than 1, it diverges in time (as the universe recollapses to a singularity).

Experimentally, ρ/ρ_c isn't known reliably. The baryon density of the universe (which is estimated by measuring the total starlight from galaxies, and taking into account the average number of baryons per star and the average amount of starlight per star) seems to correspond to ρ/ρ_c of a few percent. However, indirect measurements (such as the observation of galactic halos, discussed below) suggest that ρ/ρ_c might be a few tenths. Thus, experiment suggests that ρ/ρ_c is less than one, but by a fairly modest factor, not by many orders of magnitude.

From the point of view of particle physics, there is something surprising about this. According to general relativity, if ρ/ρ_c is less than one, it goes to zero as a function of time for large time (in fact $(\rho/\rho_c) \sim 1/t$ for large t). But our universe, with an age of order 10^{10} years, is extremely old by elementary particle physics standards. The age of the universe is about 10^{40} in units of 1/GeV, or about 10^{60} in units of 1/(Planck mass).

If ρ/ρ_c is really going to zero in time, why, such a long time after the big bang, does ρ/ρ_c still differ from one only by one order of magnitude or so? Why is ρ/ρ_c not equal to, say, 10^{-20} or 10^{-40}?

The fact that in such an old universe ρ/ρ_c is fairly close to one suggests that ρ/ρ_c is not diverging in time either to zero or to infinity (as occurs if $\rho/\rho_c > 1$), but that ρ/ρ_c is exactly equal to one.

That ρ/ρ_c might equal exactly one is an old speculation, which goes back at least to Dirac. Until recently it was just a speculation, in the sense that no rational reason was ever given for $\rho/\rho_c = 1$. Recently, for the first time, a possible reason has been given. Guth[8] has described a class of theories in which it is possible to show, for dynamical reasons, that ρ/ρ_c *must* equal one.

Although there are many unanswered questions about Guth's theory, this theory is an important development because it is the first theory that has put the value of ρ/ρ_c on a scientific basis, rather than leaving it as a matter for speculation. Whether or not Guth's theory proves to be correct, it should encourage us to believe that the value of ρ/ρ_c is capable of being rationally understood.

In any event, if in our present universe the mass density ρ is equal to ρ_c, where is this mass density to be found? Since, as mentioned above, the baryon density seems to give a ρ that is only a few percent of ρ_c, perhaps most of the mass is in some form other than baryons. This reasoning led Cowsik and McLelland[9] and Lee and Weinberg[10] to consider the possibility that the neutrino might have a mass and that most of the mass of the universe might consist of massive neutrinos.

If this is true, it is easy to determine[9,10] what neutrino mass is required. The neutrino *number* density is easy to calculate from the standard big bang theory because at a high temperature (of order 1 MeV) the neutrinos were in thermal equilibrium, and since then the neutrino number has been conserved. In fact, the neutrino number density is expected to be approximately $100/\text{cm}^3$ for each species, ν_e, ν_μ, or ν_τ. (This number density is not affected by the neutrino mass, which, for the range of masses of interest, was a negligible perturbation when the neutrinos were last in thermal equilibrium. Likewise, the neutrino mass was a negligible perturbation with respect to nucleosynthesis.)

With a neutrino number density of $100/\text{cm}^3$, the mass density is simply $100/\text{cm}^3$ times the mass, summed over species:

$$\rho(\text{neutrinos}) = (100/\text{cm}^3) \sum_i m_i. \tag{16}$$

The condition that this neutrino mass density equals the critical density turns out to be

$$\sum_i m_i = (50 \text{ eV}) \, (h/75)^2, \tag{17}$$

where h is the Hubble constant in units of km/sec/M_{pc}. (h = 75 is currently favored). For $50 < h < 100$, the sum of the neutrino masses required to give $\rho = \rho_c$ ranges from 25 eV to 100 eV.

It is very interesting that there is also a second astrophysical argument which suggests a neutrino mass in roughly the same range. This argument, which is due originally to Gunn and Tremaine[11], involves the existence of dark galactic halos.

Galaxies are observed to be surrounded by dark matter. The dark matter is detected by its gravitational field; the gravitational fields of galaxies, including ours, are stronger than expected on the basis of the stars making up the galaxy.

The parameters characterizing the dark matter are observed to be roughly as follows. The radius and mass of the dark matter are about five or ten times those of the visible part of the galaxy (figure (4)). These quantities are measured by observing the orbits of particles (either neutral hydrogen atoms, which are detected by the radiation they emit, or large objects such as globular clusters) which are in orbit around the galaxy. From the velocity and orbital radius of an orbiting particle one can, using Newton's laws, determine the galactic mass. The galactic masses determined in this way exceed by a factor of five or ten the masses expected based on the stars contained in galaxies. Even more convincingly, the orbital velocity of a particle in orbit at radius R is determined, according to Newton's laws, by the total mass contained within the orbit. By studying the "rotation curve" of orbital velocity as a function of R one can determine the mass distribution of the galaxy as a function of R (for a review, see ref. (12)). In this way, it is determined that eighty or ninety percent of the mass of a galaxy lies outside the visible region.

The density of the dark matter is not accurately known, but the maximum observed density of dark matter seems to be about 10^{-24} gm/cm^3.

What does the dark matter consist of? Could it be a cloud of massive neutrinos, gravitationally bound to the galaxy? If so, then, as Gunn and Tremaine pointed out[11], there is an interesting *lower* bound for the neutrino mass.

The velocity of a neutrino which is gravitationally bound to a galaxy cannot exceed the escape velocity from the galaxy, V_{escape}, which is generally roughly 300 km/sec. For these nonrelativistic neutrinos, the momentum is simply $p = mV$. So the momentum of a neutrino which is gravitationally bound to the galaxy cannot exceed a maximum momentum $p_{max} = m\, V_{escape}$.

For neutrinos of momentum less than p_{max}, fermi statistics do not permit a number density greater than

$$2 \int^{p_{max}} \frac{d^3 p}{(2\pi\hbar)^3} = \frac{p_{max}^3}{3\pi^2\hbar^3} . \tag{18}$$

Actually, although it will not significantly affect the conclusion, we should note that, as Gunn and Tremaine showed, the maximum *plausible* neutrino number density based on big bang cosmology is one half of the maximum allowed by fermi statistics, or $p_{max}^3/6\pi^2\hbar^3$.

The neutrino mass density is now simply the mass times the number density, so the maximum possible neutrino mass density is

$$\rho_{max} = \frac{m\, p_{max}^3}{6\pi^2\hbar^3} = \frac{m^4 (300\ \text{km/sec})^3}{6\pi^2\hbar^3} \tag{19}$$

The condition that the observed density of dark matter of about 10^{-24} gm/cm^3 should be less than ρ_{max} now gives a *lower* bound on the neutrino mass. With $m^4(300 \text{ km/sec})^3/6\pi^2\hbar^3 \gtrsim 10^{-24}$ gm/cm^3, we find

$$m \gtrsim 20 \text{ eV}, \tag{20}$$

which is the bound first derived by Gunn and Tremaine.

If there are several species of massive neutrinos, then ρ_{max} involves a sum over neutrino species. Since ρ_{max} is proportional to the fourth power of the neutrino mass, the inequality (20) becomes

$$\sum_i m_i^4 \gtrsim (20 \text{ eV})^4. \tag{21}$$

(It should be noted that, on the basis of additional assumptions, Gunn and Tremaine derived additional inequalities that were inconsistent with (20), and concluded that a neutrino in this mass range could not exist. It is probably for this reason that their paper was not widely noticed among particle physicists. The additional inequalities of Gunn and Tremaine involved assumptions about the process of galaxy formation, and in my opinion are not nearly as reliable as (20).)

It is very interesting that two separate arguments lead to estimates of neutrino masses, (17) and (20), which, within the uncertainties of astrophysical quantities, substantially coincide. Both estimates indicate the existence of a neutrino weighing several tens of electron volts. Given that the two estimates of such different nature coincide, it is nautural to suspect that a neutrino in this mass range really exists.

Acknowledgement

I would like to thank M. Davis, M. Lecar, and G. Steigman for discussions.

References

(1) For example, see S. Weinberg, in *Lectures on Particles and Field Theory* (Prentice-Hall, Inc., Englewood Cliff, N.J., 1964), Vol. II, p. 439.

(2) H. Georgi and S. L. Glashow, Phys. Rev. Lett. 32 (1974) 438; J. C. Pati and A. Salam, Phys. Rev. D 8 (1973) 1240.

(3) F. Wilczek and A. Zee, Phys. Rev. Lett. 43 (1979) 1571.

(4) S. Weinberg, Phys. Rev. Lett. 43 (1979) 1566.

(5) M. Gell-Mann, P. Ramond, and R. Slansky, unpublished.

(6) R. N. Mohapatra and G. Senjanovic, Phys. Rev. Lett. 44 (1980) 912.

(7) E. Witten, Phys. Lett. 91B (1980) 81.

(8) A. Guth, to appear.

(9) R. Cowsik, J. McLelland, Phys. Rev. Lett. 29 (1972) 669.

(10) B. W. Lee and S. Weinberg, Phys. Rev. Lett. 39 (1977) 165.

(11) S. Tremaine and J. E. Gunn, Phys. Rev. Lett. 42 (1979) 407. For discussions of additional astrophysical aspects of massive neutrinos, see J. E. Gunn, B. W. Lee, I. Lerche, D. N. Schramm, and G. Steigman, Ap. J. 223 (1978) 1015; G. Steigman, C. L. Sarazin, H. Quintana, and J. Faulkner, Astron. J. 83 (1978) 1050.

(12) S. M. Faber and J. S. Gallagher, in *Annual Reviews of Astronomy and Astrophysics*, ed. G. Burbridge, D. Layzer, and J. G. Philips, 1979.

FIGURE CAPTIONS

Figure 1: A diagram in the O(10) model which leads to neutrino masses. A lepton number violating effective interaction is generated by exchange of the heavy fermion χ.

Figure 2: A hypothetical one loop diagram which might generate a lepton number violating interaction. Superheavy particles are circulating in the loop.

Figure 3: A two loop diagram which, in the simplest O(10) model, gives mass to the χ. Circulating in the loop are ordinary quarks and leptons and superheavy fermions.

Figure 4: Galaxies are believed to consist of a visible region (stars) surrounded by a much larger and more massive halo of dark matter.

*Research supported in part by the National Science Foundation under Grant No. PHY77-22864.

(1)

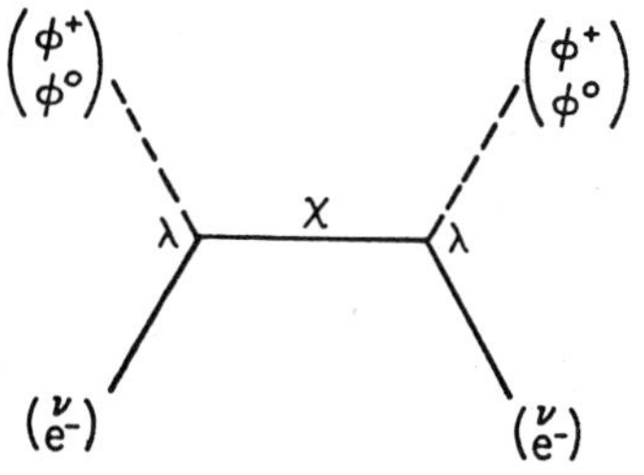

(2)

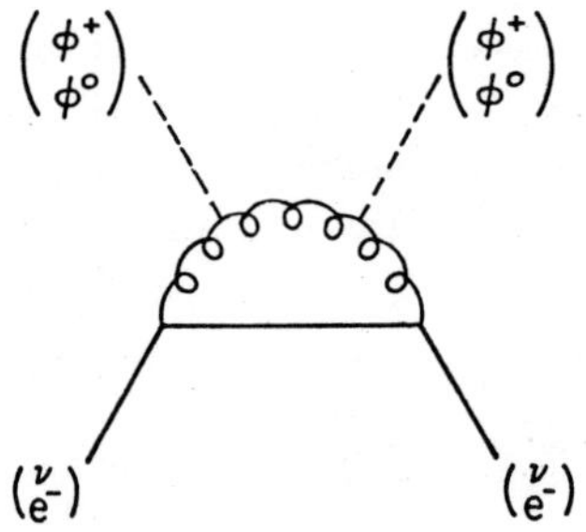

$\begin{pmatrix}\phi^+\\ \phi^0\end{pmatrix}$
$\begin{pmatrix}\phi^+\\ \phi^0\end{pmatrix}$
$\begin{pmatrix}\nu\\ e^-\end{pmatrix}$
$\begin{pmatrix}\nu\\ e^-\end{pmatrix}$

(3)

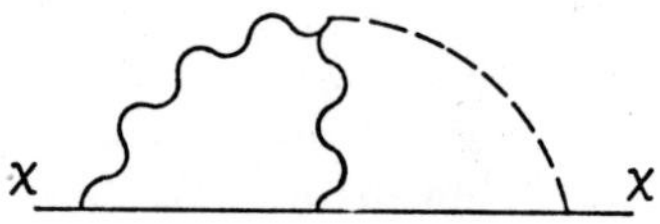

χ
χ

(4)

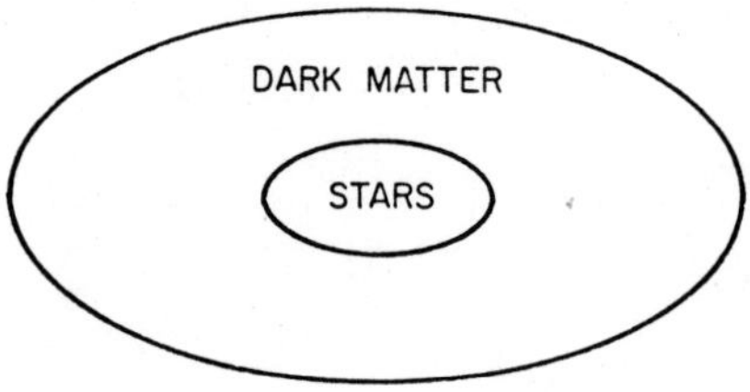

DARK MATTER
STARS

SUPERGUT*

John Ellis

CERN
Geneva, Switzerland

1. INTRODUCTION

In this talk is summarized some recent work[1,2] in which we have been trying to obtain a Grand Unified Theory (GUT) of particle interactions from an extended supergravity theory.[3] It is well known[4] that the phenomenologically appealing GUTs ($\geq SU(5)$) are too large to be contained in conventional supergravities ($\leq SO(8)$). A possible way of avoiding this problem is suggested by the work of Cremmer and Julia[5] which revealed concealed local unitary symmetries ($\leq SU(8)$) in supergravity theories. One may conjecture[5] that the composite gauge connections of these symmetries may become dynamical through some as yet unknown non-perturbative mechanism, and also the states appearing in the same supermultiplet.[1] These would be candidates for the "fundamental" particles appearing in GUTs such as quarks, leptons, and Higgs bosons, which would then all be seen as composite on a scale of order the Planck length $\approx (10^{19}\ \text{GeV})^{-1}$. The composite supermultiplet is highly chiral and the states cannot all acquire masses through simple matching of different helicities with the same internal symmetry properties,[1] as is done by the conventional Higgs and super-Higgs mechanisms. However, we are encouraged by Veltman[6] to believe in a theorem which guarantees that the only possible low energy spectrum must form a renormalizable theory. Such a theory must be based on a subset of the supermultiplet states with gauge bosons and other particles of spins $\leq 1/2$, which is free of anomalies with respect to some subgroup of the maximal unitary group $SU(8)$. We find an attractive renormalizable subtheory in which $SU(8)$ breaks down to $SU(5)$ at or near the Planck mass, and which is vector-like with respect to $SU(3)_{color} \times U(1)_{e.m.}$ so that all particles except the neutrino can acquire masses. The theory has at most three generations of $\underline{\bar{5}} + \underline{10}$ fermions, and may have Higgs fields in the $\underline{5}$ and $\underline{24}$ representations of $SU(5)$ which are massless at the tree level.

2. BACKGROUND

We all know that GUTs have some phenomenological successes such as the calculation[7-9] of $\sin^2 \theta_W \approx 0.2$ and the estimate $m_b \sim 5$ to $5\frac{1}{2}$ GeV,[8-10]

---

*Joint work with Mary K. Gaillard, LAPP, Annecy-le-Vieux, France; Luciano Maiani, CERN, Geneva, Switzerland and University of Rome; Bruno Zumino, CERN, Geneva, Switzerland.

valid if there are just three generations of light fermions. Simple GUTs predict exciting observable consequences such as nucleon decay and possibly neutrino oscillations. However, GUTs are incomplete and ugly in that they do not include gravity and have many free parameters--at least 23 in the simplest, minimal SU(5) GUT of Georgi and Glashow.[11] By way of contrast, extended supergravities[3] are complete and beautiful, including gravity as well as particle interactions and having no free parameters apart from one over-all scale. However, supergravities make no contact with phenomenology,[12] as none of the observed "fundamental" particles has any apparent superpartners, and it is not at all clear that the internal symmetry group is gaugeable in any of the larger supergravities $(8 \geq N > 4)$. In fact, even the largest global internal symmetry group SO(8) is too small[4] to contain such "observed" states as the $W^{\pm}$ bosons, μ and b quark. The largest supergravity contains the particles shown in Table 1, where X,Y,Z and W are antisymmetrized SO(8) indices. The phenomenological defects enumerated above have stymied progress in applying supergravity to the real world.

Spin	2	3/2	1	1/2	0
Multiplicity	1	8	28	56	70
Field	e^a_μ	$\psi_{\mu X}$	$V_{\mu[XY]}$	$f_{[XYZ]}$	$A_{[XYZW]}$

Table 1

3. HIDDEN SYMMETRIES

Some time ago, Cremmer and Julia[5] noticed in the course of constructing the SO(8) supergravity by dimensional reduction that it possessed surprising hidden symmetries--a global non-compact E_7 group and a local SU(8) group. To get a flavor of how these work, let us focus on the scalar fields listed in Table 1. In the analysis of Cremmer and Julia,[5] these are assigned to a 133 dimensional basic representation of E_7, 63 of whose degrees of freedom can be removed by SU(8) gauge transformations to leave us with 70 physical degrees of freedom, those indicated in Table 1. More explicitly, the scalar fields are represented by a 56×56 matrix with E_7 indices, M,N,... and SU(8) indices A,B,... :

$$S = \begin{pmatrix} U_{[AB]}{}^{[MN]} & V_{[AB][MN]} \\ \bar{V}^{[AB][MN]} & \bar{U}^{[AB]}{}_{[MN]} \end{pmatrix}$$

where U and V are 28×28 submatrices subject to certain constraints so that S only has 133 free parameters. One can construct a composite SU(8) vector out of two S matrices by contracting the E_7 indices M,N:

$$(\partial_\mu S)S^{-1} = \begin{pmatrix} 2Q_\mu{}^{[C}{}_{[A}\,\delta^{D]}{}_{B]} & P_{\mu[ABCD]} \\ \bar{P}_\mu^{[ABCD]} & 2\bar{Q}_{\mu[A}{}^{[C}\,\delta^{D]}{}_{B]} \end{pmatrix} \tag{2}$$

The $P_{\mu[ABCD]}$ quantity in Eq. (2) is the derivative of a scalar, while $Q^C_{\mu A}$ generates the SU(8) transformations on the scalar fields which can be used to reduce the number of free parameters to the 70 of Table 1. Of course, the composite $Q^C_{\mu A}$ field does not have a kinetic term in the original SO(8) Lagrangian, and hence would not appear to propagate as a physical state.

4. POSSIBLE ANALOGY WITH CP^{N-1} MODELS

Cremmer and Julia[5] noticed an apparent similarity between the supergravity models and CP^{N-1} models in two dimensions.[13] These models start with N complex scalar fields A_i subject to the constraint

$$\sum_{i=1}^{N} |A_i|^2 = 1 \qquad \text{(cf. the matrix S)} \tag{3}$$

and a Lagrangian of the form

$$\mathscr{L} = -\sum_{i=1}^{N} (\partial_\mu - iV_\mu)A_i^*(\partial_\mu + iV_\mu)A_i \tag{4}$$

which has a U(1) gauge invariance (cf. SU(8) above)

$$A_i(x) \to e^{i\Lambda(x)} A_i(x)\ ; \qquad V_\mu \to V_\mu - \partial_\mu \Lambda(x) \tag{5}$$

The equations of motion derived from (4) show that

$$V_\mu = \frac{i}{2} \sum_{i=1}^{N} A_i^*(x)\overleftrightarrow{\partial_\mu} A_i(x) \qquad \text{(cf. Eq. (2))} \tag{6}$$

which does not appear to propagate according to the Lagrangian Eq. (4). However, it has been shown[13] that quantum effects in the two dimensional CP^{N-1} model in fact cause the U(1) gauge field to propagate as a real, dynamical physical particle.

The natural conjecture, made already by Cremmer and Julia,[5] is that perhaps also in four dimensional supergravity the adjoint vector fields $Q^C_{\mu A}$ become dynamical physical states. It is natural also to conjecture[1] that the composite fields in the supermultiplet containing $Q^C_{\mu A}$ also become dynamical. This is indeed the case for the fermion superpartner of the U(1) vector field in a supersymmetric version of the CP^{N-1} model.[1] With this philosophy all "fundamental" particles with low masses--except presumably the graviton--would be made out of supergravitational preons. It should be remembered that the infra-red properties of two dimensional models are very different from those in four dimensions. In particular, supergravities seem to be infra-red finite in perturbation theory.[3] But this should not worry us, because it is presumably non-perturbative infra-red effects which lead to confinement in QCD, and we may also hope that non-perturbative phenomena will also work the trick of generating the desired dynamical bound states.

5. THE COMPOSITE SUPERMULTIPLET

The next job is to determine the supermultiplet containing the adjoint vector field $Q^C_{\mu A}$. This is not yet known with certainty, but some known special cases indicate[1] a probable general structure. In the case of $N = 1$ supergravity a massive supermultiplet with spins $(2, 2 \times 3/2, 1)$ decomposes to $(2, 3/2) + (3/2, 1) + (1, 1/2) + (1/2, 2 \times 0)$ when we take the $m = 0$ limit. Similarly, when $N = 2$ a massive supermultiplet $(2, 4 \times 3/2, 6 \times 1, 4 \times 1/2, 0)$ contains a supermultiplet $(2 \times 3/2, 4 \times 1, 2 \times 1/2)$ in which occur three vector fields in an adjoint representation of SU(2), plus an extra one reminiscent of a U(1) factor. Known examples from superconformal gravity support the natural conjecture[1] that the general supermultiplet of interest starts with the supercurrent and is

$$\left(\frac{3}{2}\right)^A , \quad (1)^A_B , \quad \left(\frac{1}{2}\right)^A_{[BC]} , \quad (0)^A_{[BCD]} , \quad \dots , \quad \left(\frac{3-N}{2}\right)^A \tag{7}$$

where one SU(N) lower index is added for each reduction of 1/2 unit of spin, and they are all totally antisymmetrized. This antisymmetrization means that the procedure must end with spin $3/2 - N/2$.

As an example, consider the largest $N = 8$ supergravity whose relevant composite supermultiplet is presumably that in Table 2. The representations in parentheses are those obtained by tracing the upper index in Eq. (7) with one of the lower ones. In contrast to $N \leq 6$, the traced vector field in Table 2 does not correspond to an apparent unitary symmetry of SO(8) supergravity.[5] We conjecture that it and its parenthetic partners are not dynamical at low energies, though their analogues in smaller supergravities might well be.

Spin	3/2	1	1/2	0	-1/2	-1	-3/2	-2	-5/2
Representation	8	63 (+1)	216 (+8)	420 (+28)	504 (+56)	378 (+70)	168 (+56)	36 (+28)	$\overline{8}$

Table 2

6. IDENTIFICATION OF ORDINARY PARTICLES

If one seeks to identify composite fields in the supermultiplets of Eq. (7) with known "fundamental" particles, one is rapidly convinced[1,2] of the necessity to consider $N = 8$ supergravity. The smallest GUT is based on the group SU(5), and any superunified theory should presumably include some non-trivial generation group, SU(2). We therefore need a dynamical supergravity group $G_E \geq SU(5) \times SU(2)$, and the available G_E are U(5) from SO(5), U(6) from SO(6), and SU(8) from SO(8), which is presumably identical with the SO(7) theory. We are therefore forced into SU(8) as the minimal available G_{E_8}. But this has rank 7, and perhaps we can fit a larger GUT into it. The only rank 5 group suitable for a GUT is SO(10), and this is not a subgroup of SU(8). One can also easily satisfy oneself[1,2] that there is no suitable rank 6 GUT group contained in SU(8)/SU(2). Hence, the pattern $G_E = SU(8) \to SU(5)_{GUT}$ is the only way to go, and we look at a decomposition of the supermultiplet of Table 2 into $SU(5) \times SU(3)_{generation}$:

$$\text{spin } \frac{1}{2} \begin{cases} \underline{216} = (10,\bar{3}) + (\bar{5},\bar{3}) + (\overline{45},1) + (1,3) + (5,1) + (1,\bar{6}) + (24,3) + (5,8) \\ \underline{\overline{504}} = (45,3) + (40,\bar{3}) + (24,1) + (15,1) + (10,1) + (10,8) + (\overline{10},6) \\ \qquad + (\overline{10},\bar{3}) + (5,3) + (\bar{5},\bar{3}) \end{cases} \tag{8}$$

These certainly include the minimal three generations of $\bar{5} + 10$ fermions and much else besides.

$$\text{spin 0: } \underline{420} = (24,\bar{3}) + (5,3) + (5,\bar{6}) + (45,3) + (\overline{40},1) + (\overline{10},\bar{3}) + (\bar{5},1) + (10,1) + (10,8) + (1,\bar{3}) \tag{9}$$

These certainly include $\underline{24}$, $\underline{5}$ and $\underline{45}$ representations of Higgs fields which are those generally used[11,9] in breaking SU(5) symmetry down to $SU(3) \times SU(2) \times U(1)$.

However, the identification of ordinary particles encounters severe problems associated with chirality. The spin 1/2, 3/2 and 5/2 representations in Table 2 contain many states which cannot[1] be given masses by conventional mechanisms because all the corresponding helicity states with identical transformation laws under the exact symmetry group $SU(3)_{color} \times U(1)_{e.m.}$ do not exist. Furthermore, the SU(8) symmetry has anomalies. Shown in Table 3 are different fermion contributions to the net SU(8) anomaly:

Helicity	3/2	1/2	-1/2	-3/2
Group Factor	-1	3	-75	-55
Spin Factor	3	1	-1	-3
Product	-3	3	75	165

Table 3

Omitted from Table 3 is the contribution to the anomaly from spin 5/2. It is not at all clear that a consistent interacting field theory exists for spin 5/2, and hence it is not clear that calculating its anomaly makes any sense. However, it seems unlikely that any rationale could be given for assigning to spin 5/2 an anomaly of -240, as required to cancel the contributions in Table 3.

We have tried various approaches to avoiding these problems. One possibility was to try using a different supermultiplet, or combination of supermultiplets. But the same chirality problems always recur, and after all, the multiplet Eq. (7) is the only one that is strongly motivated. We also[1] tried abandoning the conventional embedding of $SU(3)_{color} \times U(1)_{e.m.}$ through SU(5) in SU(8), but this did not enable us to give masses to unwanted states. One might give a vector-like SU(5) theory. But no consistent ghost-free supergravity with $N > 8$ exists. We now[2] evade these problems with a new philosophy which is to use only a subset of the full supermultiplet of Table 2, motivated by:

7. VELTMAN'S "THEOREM"

Veltman[6] has stressed to us the probable existence of a theorem which we understand as follows. In a theory with composite states whose masses are much less than their inverse sizes ($O(m_P)$ in our case), whose mutual interactions are describable by perturbation theory at low energies $E \ll (\text{size})^{-1}$, these interactions *must* be renormalizable. If not, non-renormalizable infinities could only be regulated by the bound state size, and this cut-off would appear in calculations of masses and vertex functions, giving masses of $O(\text{size})^{-1}$ and/or breakdowns of perturbation theory which should require the offending states causing non-renormalizability to have masses $O(\text{size})^{-1}$. An example of this "theorem" is afforded by technicolor,[14] with a characteristic bound state size $O(1\ \text{TeV})^{-1}$, while there should be pseudo-Goldstone bosons (PGBs) with masses $\ll 1$ TeV. The interactions of these PGBs would presumably be correctly described by a renormalizable $SU(3) \times SU(2)_L \times U(1)$ gauge theory at energies $E \ll 1$ TeV. States which lead to non-renormalizability are believed to have masses $O(1\ \text{TeV})$ or more--e.g., the massive technirho with spin 1 and all states with spins > 1.

So our assumption[2] is that the full symmetry of the SU(8) supermultiplet of Table 2 is broken at a scale $O(m_P) \sim 10^{19}$ GeV, and that the only states appearing at low energies are the gauge vectors and states with spin $\leq 1/2$. There is a further restriction that the low energy fermions should be a subset of those in Eq. (8), which are anomaly free, and hence renormalizable, with respect to the low energy invariance subgroup of SU(8). We have[2] two arguments that SU(8) should break directly to SU(5) at or near the Planck mass.

8. SYMMETRY BREAKING

The full SU(8) symmetry could, in principle, be broken by some component of the $\underline{420}$ of Higgs fields in Table 2 acquiring a vacuum expectation value. If we make an SU(7) decomposition of the $\underline{420}$ we find $\underline{224} + \underline{140} + \underline{35} + \underline{21}$. This set does not include a singlet of SU(7), which must therefore be broken. Similarly, the SU(6) breakdown $\underline{420} \to \underline{105} + 2\,\underline{84} + \underline{35} + 2\,\underline{20} + 4\,\underline{15} + 2\,\underline{6}$ reveals no SU(6) singlet. The first time we encounter a singlet is in the SU(5) decomposition, as can be seen from Eq. (9). This means[7] that SU(5) is the maximal possible symmetry after Higgs breaking. To see whether breaking can occur, we can try to construct[2] the supersymmetric $(\underline{420})^4$ Higgs potential. This potential may in general have zeros for non-trivial values of some components of the Higgs fields. It requires further analysis to see whether an SU(5) invariant direction is one of these, and an analysis of radiative corrections and possibly non-perturbative effects would be necessary for determining if such an SU(5) invariant direction is indeed the one selected. But at least we have seen[2] that SU(5) is likely to be the maximal possible symmetry at energies $\ll m_P \simeq 10^{19}$ GeV.

Another indication in this direction is provided by the analysis[2] of anomalies. The Veltman's "theorem" criterion of renormalizability has emboldened us to discard all fermions with spins $> 1/2$. The remaining spin 1/2 fermions should be free of all anomalies in the sub-Planck symmetry group SU(5). A maximal subset of SU(5) anomaly-free fermions which is vector-like with respect to $SU(3)_{\text{color}} \times U(1)_{\text{e.m.}}$ is

$$(\underline{45}+\overline{\underline{45}}) + 4\ \underline{24} + 9(\underline{10}+\overline{\underline{10}}) + 3(\underline{5}+\overline{\underline{5}}) + 3(\overline{\underline{5}}+\underline{10}) + 9\ \underline{1} \qquad (11)$$

Most of these states are vector-like with respect to SU(5) and so can acquire masses $O(10^{15}$ to $10^{19})$ GeV. We are left[2] with a maximum of three chiral generations of $\overline{\underline{5}}+\underline{10}$ SU(5) fermions.

Anomaly-free representations of larger subgroups SU(6), SU(7), or SU(8) also exist. However, if they are to be vector-like with respect to $SU(3)_{color} \times U(1)_{e.m.}$ they must also be vector-like for the whole SU(6) or SU(7) group. For example, the maximal set of SU(6) anomaly-free fermions is

$$(\underline{84}+\overline{\underline{84}}) + 2\ \underline{35} + 4\ \underline{20} + 2(\underline{15}+\overline{\underline{15}}) + (\underline{6}+\overline{\underline{6}}) \qquad (12)$$

which has fewer helicity states than the maximal SU(5) set (11). Conversely, if we tried to embed the SU(5) anomaly free set in an SU(6) theory, we would find that it had SU(6) anomalies. With a purely vector-like representation like Eq. (12), it is difficult to understand why any fermions have masses much less than the grand unification mass. We therefore conclude[2] that SU(5) is the maximal subgroup of SU(8) with a phenomenologically acceptable set of anomaly-free fermions.

Let us assume then that SU(8) breaks to SU(5) at or near the Planck mass, and that this is done by an SU(5) singlet coming from an adjoint $\underline{35}$ of the SU(6) subgroup. What are the masses of the remaining scalar fields? It seems likely[2] that at least some of the irreducible SU(5) representations of Higgs may be massless at the tree level, and that not all of these need be Goldstone bosons. In this case, it may[15] be possible to obtain subsequent stages[9,11] of SU(5) symmetry breaking down to $SU(3) \times SU(2) \times U(1)$ and $SU(3)_{color} \times U(1)_{e.m.}$, respectively, through radiative corrections[16] to the Higgs potential. It remains to be seen whether this program can actually be carried out.

9. CONCLUSIONS

So far, it seems possible that a suitable GUT may be obtained from dynamically broken supergravity. The unique possiblity that we see[1,2] is to take the SU(8) symmetry of SO(8) supergravity and have it break down to an SU(5) GUT. Analyses of the Higgs system and fermion anomalies both suggest[2] that this breakdown should occur at or near to $m_P \simeq 10^{19}$ GeV. We find[2] that the maximal residual GUT would contain three generations of $\overline{\underline{5}}+\underline{10}$ fermions, and may also contain Higgs fields which are massless at the tree level and hence suitable for further symmetry breaking. Thus, there is at most one Super GUT: much more work is necessary before we know whether there is in fact less than one! The problems for the future are many.

Acknowledgments. We would like to thank D.V. Nanopoulos, P. Sikivie, and M. Veltman for encouragement and instructive discussions. We would also like to thank P.H. Frampton, A. Yildiz and other members of the organizing committee for the opportunity to present these results at such a pleasant meeting.

REFERENCES

1. J. Ellis, M.K. Gaillard, L. Maiani, and B. Zumino, LAPP Preprint TH-15/CERN Preprint TH-2481 (1980).

2. J. Ellis, M.K. Gaillard, and B. Zumino, CERN Preprint TH-2842/LAPP Preprint TH-16 (1980).

3. See, for example, "Supergravity", *Proc. of the Supergravity Workshop at Stony Brook*, Sept. 1979, P. van Nieuwenhuizen and D.Z. Freedman (eds.), North-Holland, Amsterdam, 1979.

4. See, for example, B. Zumino, *Proc. Einstein Symposium*, Berlin, 1979, Lecture Notes in Physics 100, 114, H. Nelkowski, A. Herman, H. Poser, R. Schrader, and R. Seiler (eds.), Springer-Verlag, Berlin, 1979, and references therein. The first attempt to embed low energy gauge theories in SO(8) supergravity using the basic supermultiplet was reported by M. Gell-Mann, talk at the 1977 Washington Meeting of the American Physical Society.

5. E. Cremmer and B. Julia, *Phys. Lett.* 80B, 48 (1978), and *Nucl. Phys.* B159, 141 (1979). A previous attempt to relate this work to phenomenology has been made by T. Curtright and P.G.O. Freund, Ref. 3.

6. M. Veltman, private communication. Arguments similar in spirit have probably occurred to many people, such as G. 't Hooft, *Cargèse Summer Institute Lecture Notes* (1979); G. Kane, G. Parisi, S. Raby, L. Susskind, and K.G. Wilson, private communications. We understand that A. Kabelschacht has an ingenious proof of this theorem which is unfortunately too long to be given in the margin here.

7. H. Georgi, H.R. Quinn, and S. Weinberg, *Phys. Rev. Lett.* 33, 451 (1974).

8. M.S. Chanowitz, J. Ellis, and M.K. Gaillard, *Nucl. Phys.* B128, 506 (1977).

9. A.J. Buras, J. Ellis, M.K. Gaillard, and D.V. Nanopoulos, *Nucl. Phys.* B135, 66 (1978).

10. D.V. Nanopoulous and D.A. Ross, *Nucl. Phys.* B157, 273 (1979).

11. H. Georgi and S.L. Glashow, *Phys. Rev. Lett.* 32, 438 (1974).

12. See, for example, P. Fayet, "New frontiers in high energy physics", *Proc. Orbis Scientiae*, Coral Gables, 1978 (Plenum Press, New York, 1978), p. 413; G. Barbiellini et al., DESY Preprint 79/67 (1979).

13. A. D'Adda, P. Di Vecchia, and M. Lüscher, *Nucl. Phys.* B146, 63 (1978); E. Witten, *Nucl. Phys.* B149, 285 (1979).

14. L. Susskind, *Phys. Rev.* D20, 2619 (1979).

15. H. Georgi and A. Pais, *Phys. Rev.* D16, 3520 (1977).

16. J. Ellis, M.K. Gaillard, A. Peterman, and C.T. Sachrajda, *Nucl. Phys.* B164, 253 (1980), and references therein.

FERMION MASSES IN UNIFIED THEORIES*

Howard Georgi
Lyman Laboratory of Physics, Harvard University, Cambridge, MA 02138

Abstract

I conduct a tour of a bestiary of exotic models.

1. INTRODUCTION

Alas, I have no theory of the fermion mass matrix. Instead, I will talk about many models, none entirely satisfactory but each with some interesting features. All the models will follow the general rules for naturalness of mass relations involving explicit Higgs scalars. I will not discuss technicolor models.[1] There are two reasons for this omission:

(1) I don't know how to build any such model that is even close to being realistic.

(2) I regard the success of the prediction for $\sin^2\theta_W$ as (modest) evidence in favor of explicit Higgs scalars, or at least as a very strong constraint which makes the building of a satisfactory technicolor theory even harder.

I will also adhere to a set of principles which I think that any sensible unified theory should satisfy. I would like to say that my high moral principles allow me to consider only theories which are sensible in that the only particles which are light compared to 10^{15} GeV are light for some known reason. Unfortunately, I do not know any such theories with enough structure to describe the world. Instead I will do what I think is the next best thing. I will restrict to theories in which the only mysteriously light particles are the Higgs particles which spontaneously break the low energy gauge symmetries. Usually, this will be only a single SU(2) doublet, the Higgs multiplet of the standard SU(2) x U(1) model.

In each of the models described below, I will be trying to explain some apparent regularities of the fermion mass matrix in the context of a simple and beautiful structure with a small enough number of parameters to retain some predictive power. Usually, this will be too much to ask.

2. $E(6) \to SU(3)_C \times SU(2)_L \times SU(2)_N \times U(1)$

This model is essentially the one described earlier by Gursey.[2] I find it interesting because it almost yields an explanation of the multiplicity of quarks and leptons.

It is often easiest to describe E(6) in terms of the $SU(3)_{color} \times SU(3)_{left} \times SU(3)_{right}$ subgroup. The fermion fields are in 27's. The left-handed quarks are in a (3,3,1)

$$\begin{pmatrix} u \\ d \\ b \end{pmatrix}_L \tag{2.1}$$

The left-handed antiquarks are in a $(\bar{3},1,\bar{3})$

$$(u^c,\ d^c,\ b^c)_L . \tag{2.2}$$

The left-handed leptons are in a $(1,\ \bar{3},\ 3)$,

$$\begin{bmatrix} N_\tau & \tau^+ & e^+ \\ \tau^- & \nu_\tau & E^0 \\ e^- & \nu_e & T^0 \end{bmatrix}_L . \tag{2.3}$$

The weak $SU(2)_L$ is the subgroup of the $SU(3)_L$ that acts on the first two rows in (2.1) and the first two columns in (2.3). The $SU(2)_N$ (N for neutral) is the subgroup of $SU(3)_R$ which acts on the last two columns of (2.2) and the last two rows of (2.3).

This is a topless model with a fourth charge - 1/3 quark, the ℓ. The b and ℓ do not mix with the light quarks. b decay and CP violating interactions are mediated by the $SU(2)_N$ gauge bosons. All b decays are semileptonic.[3] So why are there two 27's of fermions?

An answer is almost available in the lepton sector. The same Higgs couplings which give rise to quark masses give rise (unless there are unnatural cancellations) to neutrino masses of the same order of magnitude. In particular, we expect terms like

$$m_u(\nu_e\ E^0 - \nu_\tau\ T^0). \tag{2.4}$$

This is a disaster, because it leads to a ν_e mass of order m_u unless something happens to remove the E^0 from the low energy theory.

Now the E^0 is part of an $SU(2)_N$ doublet, but it is a singlet with respect to all the other low energy gauge groups, SU(3), $SU(2)_L$, and U(1). We would expect any $SU(3) \times SU(2)_L \times SU(2)_N \times U(1)$ mass term to be large, of order 10^{15} GeV. With only a single 27, we cannot make such an invariant, because the mass term must be a symmetric combination of two $SU(2)_N$ doublets, which has no $SU(2)_N$ singlet. However, with two families, the $SU(2)_N$ singlet coupling exists and we expect the E^0 to be superheavy.

Unfortunately, this is too much of a good thing. Not only does the E^0 become superheavy, but the T^0 as well. We need the T^0 to couple to N_τ.

Without it, the τ^- and its neutrino are degenerate.[3]

I see no natural way out of these difficulties.

3. O(10): VERSION 1

O(10) models are very attractive from the point of view of understanding the fermion mass matrix because it is possible to incorporate the zeroth order relations

$$m_b = m_\tau, \quad 3m_s = m_\mu, \quad m_d = 3\, m_e$$

$$\text{and} \qquad \tan^2\theta_c = m_d/m_s. \tag{3.1}$$

The idea has been explained elsewhere[4] and can be summarized in the "Yukawa coupling matrix"

$$\begin{bmatrix} 0 & 10 & \\ 10 & 126 & \\ & & 10 \end{bmatrix} \tag{3.2}$$

meaning that the b and τ masses and the d-s and e-μ mass mixing come primarily from a 10 of Higgs mesons while the s and μ mass comes from a 126, giving rise to the factors of 3 in Eq. (3.1). Some slots in the matrix are blank in Eq. (3.2) because it is not obvious how to extend the structure to a complete theory incorporating b decay and CP violation.

In an earlier talk by Ramond[5], you heard about a possible extension of Eq. (3.2) of the following form (with three different 126 Higgs fields):

$$\begin{bmatrix} 0 & 10 + 126_1 & 0 \\ 10 + 126_1 & 126_2 & 126_3 \\ 0 & & 10 + 126_1 \end{bmatrix} \tag{3.3}$$

The striking thing about this coupling matrix is the presence of 10 and 126_1 couplings in the same slot. The idea was that their VEV's could be constrained so that, for example, the 10 contributed to d quark and lepton masses while the 126_1 gave rise to large Majorana masses for the right-handed neutrinos. A very similar idea has been suggested by Pakvasa.[6]

When I first heard this idea, I was sure it would not work. I know from bitter experience how hard it is to implement a mass relation like this which depends not only on the form of the Higgs couplings but actually on the form of the VEV's. I assumed that this model would fail the test of naturalness. However, I was only partially right. True, the original form of the model in which the O(10) symmetry is broken with a 45 of Higgs (in addition to the 10 and 126) is no good. The renormalizable coupling of two 45's, 10 and $\overline{126}_1$ cannot be forbidden and spoils the naturalness of the VEV.

But there is another possibility. Instead of a 45 of Higgs one can use a 54. This has a similar effect. The 54 VEV breaks O(10) to SU(4) x SU(2) x SU(2) and then the SU(5) singlet component of the 126 (responsible for the ν_R mass) breaks it further down to SU(3) x SU(2) x U(1).

The appropriate VEV's can then be written schematically as

$$\begin{aligned} &\langle 10\rangle^D \neq 0, \quad \langle 10\rangle^U = 0 \\ &\langle 126_{1,3}\rangle^D = 0, \ \langle 126_{1,3}\rangle^U \neq 0, \ \langle 126_{1,3}\rangle^{\nu_{R,L}} \neq 0 \\ &\langle 126_{1,3}\rangle^D = 0, \ \langle 126_2\rangle^U = \langle 126\rangle^{\nu_{R,L}} = 0. \end{aligned} \tag{3.4}$$

The superscripts label the various neutral components of the Higgs field. For example, the 126 has 4 independent neutral components $\langle 126\rangle^D$, $\langle 126\rangle^U$, $\langle 126\rangle^{\nu_{R,L}}$ associated with D quark and charged lepton masses (D), U quark and Dirac neutrino mass (U), the very large Majorana mass of the right-handed neutrino (ν_R), and the very small Majorana mass of the left-handed neutrino (ν_L). This is not exactly the structure advocated by Ramond, but it has many of the same features *and it is natural.*

In this model, we can estimate one contribution to the Majorana mass of the right-handed neutrinos. The heaviest is the $\nu_{\tau L}$, about $m_t/(G_F^{1/2}\langle 126_1\rangle^{\nu_R})$. This is small for any reasonable t quark mass. But there is another contribution (from $\langle 126_{1,3}\rangle^{\nu_L}$) which cannot be estimated with any certainty.

There is nothing much wrong with this model except that it has a lot of parameters and a lot of 126's.

4. O(10): VERSION II

Can we do something similar with only a single 126? Almost, with couplings of the following form:

$$\begin{bmatrix} 0 & 10 & 126 \\ 10 & 126 & 0 \\ 126 & 0 & 10 \end{bmatrix}. \tag{4.1}$$

This does not require a 54 for naturalness. We can break the O(10) symmetry with a 45. This is somewhat better because the 45 can break the O(10) symmetry down to SU(4) x SU(2) x U(1). As shown by Nanopoulos et al.[4], this allows the best fit to the b/τ mass ratio. This also produces a τ neutrino mass of order $10^{-10}\ m_t$, which is interesting.

The amusing features of this model have to do with CP violation and can be summarized in two comments:

A. The only reasonable phase which can appear in the mass matrix is large, a multiple of π/3.

B. Even with such a maximal CP violation in the mass matrix, the CP violation in the K meson system is tiny, indeed too small to account for the observed effect.

The reason for A is the strong CP problem. If the Yukawa couplings are real and if (for some unexplained reason) the phases of all VEVs are multiples of $\pi/3$, then the determinant of the quark mass matrix is real. It is not clear that this yields a solution to the $\bar{\Theta}$ problem, but it would certainly seem to be a necessary condition.

Comment B is not so obvious. It just happens that because of the zeroes in Eq. (4.1), there is very little CP violation in the K meson system if the KM angles are relatively small.

The lack of CP violation can be remedied (at the cost of extra parameters) by adding a 120 of Higgs mesons as follows:

$$\begin{bmatrix} 0 & 10 & 126 \\ 10 & 126 & 120 \\ 126 & -120 & 0 \end{bmatrix} . \tag{4.2}$$

The minus sign reflects the antisymmetry of the coupling of the 120 to two 16's.

The trouble with this scheme is that it predicts a dangerously light t quark, less than 20 GeV. Having been burned once[4], I am loth to predict such a light t quark mass again.

5. PERMUTATION SYMMETRY

My colleagues Shelly Glashow and Alvaro De Rújula are fond of telling me that I am wasting my time messing around with relations between charged lepton masses and charge -1/3 quark masses with all the attendant factors of three when there is an even simpler relation waiting to be explained. According to these gentlemen, the lepton masses and charge 2/3 quark masses are proportional:

$$\frac{m_u}{m_e} \simeq \frac{m_c}{m_\mu} \overset{?}{\simeq} \frac{m_t}{m_\tau} . \tag{4.3}$$

I have to admit that the first equality is not ridiculous and the second is not inconsistent. It predicts a "safe" t quark mass, ~20 GeV.

It is possible to produce this mass relation naturally in an SU(3) x SU(2) x U(1) model with a permutation symmetry. The Yukawa couplings for the leptons and charge 2/3 quarks are

$$\sum_i (g\, \overline{U^i_R}\, \phi^\dagger_i\, \psi^i_L + g'\, \overline{\ell^i_R}\, \overline{\phi^\dagger_i}\, L^i_L)$$
$$+ \text{h.c.} + \text{charge} - 1/3 \text{ quarks}. \tag{5.2}$$

The ratios of masses are then the ratios of VEV's:

$$u:c:t = e:\mu:\tau = \langle\phi_1\rangle:\langle\phi_2\rangle:\langle\phi_3\rangle. \quad (5.3)$$

This scheme looks ugly for two reasons. First, some totally different structure is required for the charge - 1/3 quarks which clearly do not satisfy

$$d:s:b = \langle\phi_1\rangle:\langle\phi_2\rangle:\langle\phi_3\rangle. \quad (5.4)$$

Second, as far as I can tell, it is impossible to unify this theory. I really have no idea how to obtain Eq. (5.2) (without also getting the bad one, Eq. (5.4)) in a unified theory.

6. SU(2N+1)

I sometimes suspect that nature is trying to tell us something by arranging the quark and lepton masses much smaller than the W and Z masses. It might be interesting, for example, to find a class of models in which all quark and lepton masses are radiative corrections. It is easy to find such models in the context of the SU(N), N > 5 models I proposed some time ago as a possible direction for attack on the flavor problem.[7]

In the notation of Ref. 7, the idea is as follows. The gauge group is SU(2N+1), N > 2. The fermions are a series of antisymmetric tensor representations of alternate handedness, for example

$$[0]_L, [1]_R, [2]_L \ldots [N]. \quad (6.1)$$

The number of each type must be consistent with the anomaly constraint and must yield the desired number of light families (presumably 3). The Higgs comprise N - 5 [1]'s and one adjoint.

The VEVs of the N - 5 [1]'s break the gauge symmetry down to SU(5). But they give mass only to superheavy fermions. I assume that the model is constructed so that the Yukawa couplings of the [1]'s to the fermions break all the chiral symmetries.

Under the SU(5) subgroup, the adjoint contains a 24, 5's, and 1's. The VEV of the 24 breaks SU(5) down to SU(3) x SU(2) x U(1), as usual. But the 5's can also develop VEV's which break SU(2) x U(1) → U(1).

The light fermions are massless in tree approximation, because the Higgs adjoint, whose VEV breaks SU(2) x U(1) and gives rise to quark and lepton masses, does not couple to the fermions. But quark and lepton masses are induced as radiative corrections due to diagrams such as those shown in Fig. 1. If all couplings are of order g_G (the unifying gauge coupling), then those quark and lepton masses which arise from one loop diagrams such as Fig. 1 are of order $\alpha_G M_W$.

The family hierarchy structure is not obvious in models of this kind, though it might be possible to have some light fermions which get their mass only in two or more loops. A more serious problem with models of this type is that it is hard to find any without inventing an enormous number of superheavy fermions. I like the abstract idea much better than any of the realizations I have found.

7. SUBMINIMAL O(10)

The c quark mass is much larger than the s quark mass. The t quark mass is unknown, but it seems to be quite a bit larger than the b quark mass. Perhaps we should look for a model in which charge 2/3 quark masses are naturally much larger than charge -1/3 quark masses. One amusing possibility is a subminimal O(10) model.

We heard Ed Witten[8] talk earlier about a minimal O(10) model with 16's of fermions and only a 45, a 16 and a 10 of Higgs particles. My model is subminimal in that I do without the 10. The VEV's of the 45 and 16 can break O(10) down to SU(3) x U(1). The 16 contains an SU(5) 5 which breaks SU(2) x U(1). But the Higgs 16 cannot couple directly to the fermion 16's.

To get the fermion masses started, I introduce three (one for each 16 of fermions) O(10) singlet fermions. These have Majorana masses and also couple the Higgs and fermion 16's together, giving mass to the right-handed neutrinos.

The charge 2/3 quarks develop mass in one loop from diagrams like those in Fig. 2. But the charge - 1/3 quarks and charged leptons get mass only in two loops. Thus we expect

$$m_U \simeq \alpha_G M_W, \; m_D \simeq \alpha_G^2 M_W. \tag{7.1}$$

Really we want some much more bizarre mass relations. Something like

$$\begin{aligned} &m_c, m_b \sim \alpha_G M_W, \; m_s \sim \alpha_G^2 M_W \\ &m_u, m_d \sim \alpha_G^3 M_W, \; m_t \sim \alpha_G M_W \text{ or } M_W. \end{aligned} \tag{7.2}$$

Getting such a complicated structure in standard games looks very, very hard.

*Research supported in part by the National Science Foundation under Grant No. PHY77-22864 and by the Alfred P. Sloan Foundation.

REFERENCES

1. See L. Susskind's talk and references therein.
2. See F. Gursey's talk and references therein.
3. H. Georgi and S. L. Glashow, Nucl. Phys. B, to be published.
4. H. Georgi and D. V. Nanopoulos, Nucl. Phys. B159, 16 (1979).
5. R. Ramond's talk.
6. S. Pakvasa, University of Hawaii Preprint (1979).
7. H. Georgi, Nucl. Phys. B156, 126 (1979). See also G. Barbieri and D. V. Nanopoulos, Phys. Lett. 91B, 369 (1980); and CERN preprint 28-70 (1980).
8. E. Witten's talk.

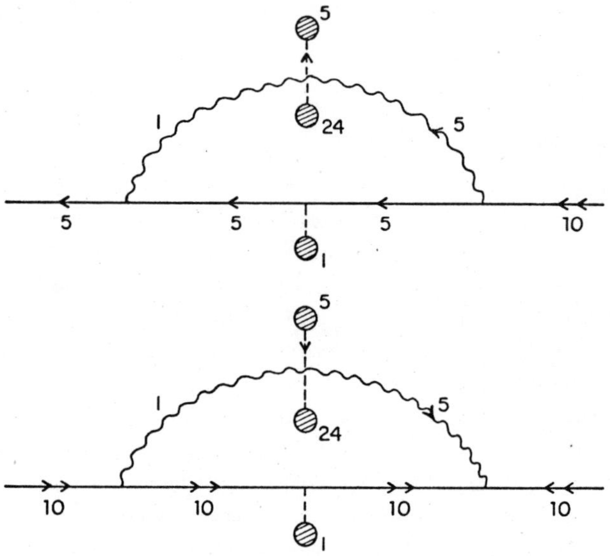
5
1
24
5
5
5
5
10
1
5
1
24
5
10
10
10
10
1

Fig 1

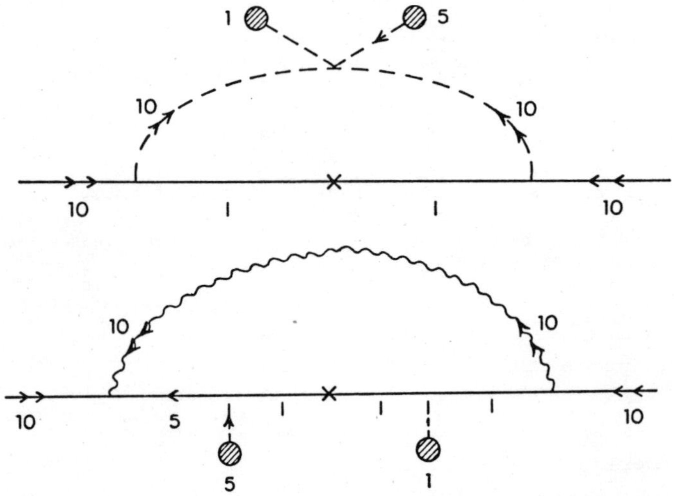
1
5
10
10
10
1
1
10
10
10
10
5
1
1
1
10
5
1

Fig. 2

THE SOUDAN MINE EXPERIMENT: A DENSE DETECTOR FOR BARYON DECAY*

Marvin L. Marshak

School of Physics
University of Minnesota
Minneapolis, MN 55455

Abstract

We describe the progress to date on construction of a baryon decay detector in the Soudan Mine in northeastern Minnesota. This first phase of this detector has a mass of 40 tons (3×10^{31} nucleons) of taconite concrete and steel ionization tubes. The first operation of this detector is scheduled for Summer, 1980.

1. INTRODUCTION

In the past year, several experiments have been designed to search for baryon decay with a lifetime of order 10^{30} to 10^{33} years. An overview of the field requires discussion of the current experiments, plus an earlier experiment in South Africa,[1] which has been used to set the existing experimental lower limit on the baryon lifetime. Table 1 lists the primary characteristics of these experiments. Both the South African and the Homestake[2] experiments use non-confining detectors, which reduces their sensitivity and makes their results branching ratio dependent. The Ohio[3] and Utah[4] experiments use large water Cherenkov detectors, while the Soudan (Minnesota)[5] and Mont Blanc[6] experiments use ionization detectors in conjunction with a dense, high-Z mass. The remainder of this report will describe the design of the Soudan detector and report on its progress to April, 1980.

Table 1. A comparison of baryon decay detectors

Characteristic	S. Africa	Homestake	Ohio	Utah	Mt. Blanc	Soudan
Mass (tons)	64*	500	10,000	1000	150	40**
Depth (mwe)	8700	4500	1500	1400	5000	2000
Decay Material	rock	water	water	water	iron	Fe/O
Detection Method	scint.	------- Cherenkov light --			Ionization tubes	
Time Frame	done	now	------------- Future -----------			

*64 ton-years was the measured sensitivity.

**This is the mass of the first phase.

2. DESIGN AND PROPERTIES OF THE SOUDAN DETECTOR

The Soudan detector is designed to maximize the amount of information about baryon decays per unit cost. This cost ratio is important, because ultimately the size of any detector, and thus the lifetime sensitivity, will be fixed by the funding resources available. This economy of design is accomplished by using conventional detection techniques, by constructing the detector from conveniently handled materials indigenous to the mine area, and by placing minimal requirements on the underground space needed by the experiment. They key to this latter aspect is the use of a dense, high atomic number material for the fiducial mass.

A schematic of the detector is shown in Fig. 1. The sensitive elements are proportional tubes, constructed from 2.5 cm i.d. steel pipe with 1.5 mm walls. The tubes use a 50 μ gold-plated tungsten wire strung between two injection-molded plastic endcaps. With a 90 percent argon--10 percent carbon dioxide mixture (permissible for use underground), these tubes have a typical plateau of 200 volts at 2 kV. The tubes are embedded in taconite concrete, a mixture of Portland cement, water, and concentrated iron ore. The hardened concrete is about 50 percent iron and 45 percent oxygen by weight. Its specific density is typically 3.3. For ease of assembly, the detector is currently constructed in modules of 8 tubes, 290 cm long by 32 cm wide by 4 cm high, with a total mass of about 80 kg. In the mine, the modules will be stacked as shown in the figure, in order to obtain two 90 degree stereo views of each event.

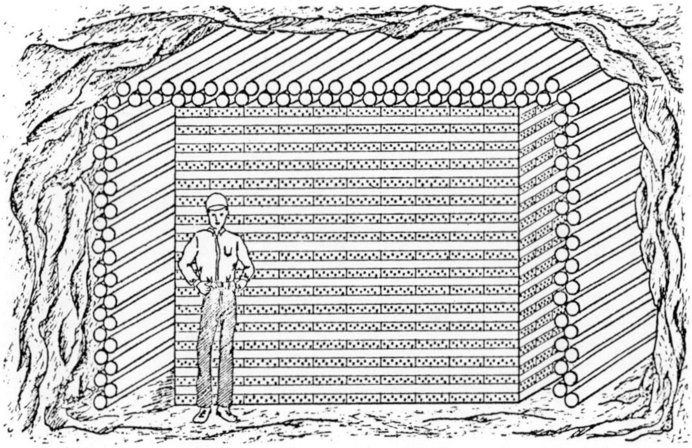

Figure 1. A conceptual sketch of the Soudan detector in the mine. The large proportional tubes surrounding the detector are one possibility for an active shield.

The design of a baryon decay detector can be examined in terms of two different sets of criteria. One is to maximize the information obtained about events, independent of any specific decay mode. The criteria in this set include energy resolution, spatial and angular resolution, and particle and charge identification. The second set is to minimize the logistical difficulties associated with the placement of the detector in a deep underground location. Mastering these problems is a key to the construction of a real, working detector in a reasonable amount of time.

a. Physics Considerations

From the physics point of view, the Soudan detector is not conceptually different from the instrumentation which is planned for the Mont Blanc tunnel. Both devices are track-sensitive, total absorption calorimeters, which utilize a dense, high-Z mass. There are several attributes of a dense calorimeter which are particularly useful in distinguishing various expected modes of baryon decay.

1. Position and Angular Resolution: The position resolution of the Soudan (Minnesota) detector is ± 1.3 cm. The typical width of a 400 MeV electromagnetic shower at the shower maximum is 3-4 times this spatial resolution. The angular resolution on showers is thus set by the shower fluctuations, not by the limited resolution of the detector. The angular resolution on muon tracks of several hundred MeV kinetic energy is smaller than 25 mr.

2. Energy Resolution: The energy resolution of a sampling ionization detector is different for particles which generate an electromagnetic shower and those which do not. In the former case, the resolution is determined by the graininess of the sampling and the critical energy of the material. The critical energy of taconite concrete is about 30 MeV.[7] For showers, the energy resolution of the Soudan detector is expected to be 25 percent for a 500 MeV electron.[5] Muon energies are best established by range. In the Soudan detector, the energy resolution for a 500 MeV muon is ± 10 percent. Determination of the energy of charged pions is more difficult because of the possibility of strong interaction, but even in this case, the high-Z material yields a higher probability of stopping before a strong interaction occurs.

3. Particle and Charge Identification: The ease of distinction between showering and non-showering particles depends on the ratio of the particle energy to the critical energy. The taconite concrete will cause a multi-level cascade at electron and photon energies important for baryon decay. The good spatial resolution of the Soudan detector will also permit the separation of photons from electrons in many cases. This good resolution can also make possible the identification of pions by their strong elastic scatters and muons by their lack of elastic scatters.

The Soudan detector has some possibilities for the identification of charge. Positive pions and muons can be identified about 60 percent of the time by their decay chain $\pi \rightarrow \mu \rightarrow e$. In a high-Z material, most of the negative muons are captured, which is a rare occurrence in a light material. The ferrimagnetic property of taconite concrete also makes possible the construction of a detector with magnetic analysis.

4. Directionality: The one apparent disadvantage of ionization calorimeters is that they do not directly measure the direction that a particle is traveling. The good spatial resolution of these detectors and the possibility for measuring specific ionization does provide indirect means for distinguishing the beginning from the end of particle tracks. The direction of showering particles is given by the increase in the width and the decrease in the specific ionization at the end of the shower. The end of non-showering tracks can be distinguished by increases in the multiple scattering and the specific ionization. The good pattern recognition capabilities of this type of device have been demonstrated in several accelerator neutrino experiments.

Figure 2 shows a typical Monte Carlo-generated event as it would be observed in the Soudan detector. The actual tracks are shown in the figure at the left, while the hit and ionization patterns are shown to the right. Typically, 15 to 20 tubes will be hit in a baryon decay to a mode such as $e\pi$ or μK.

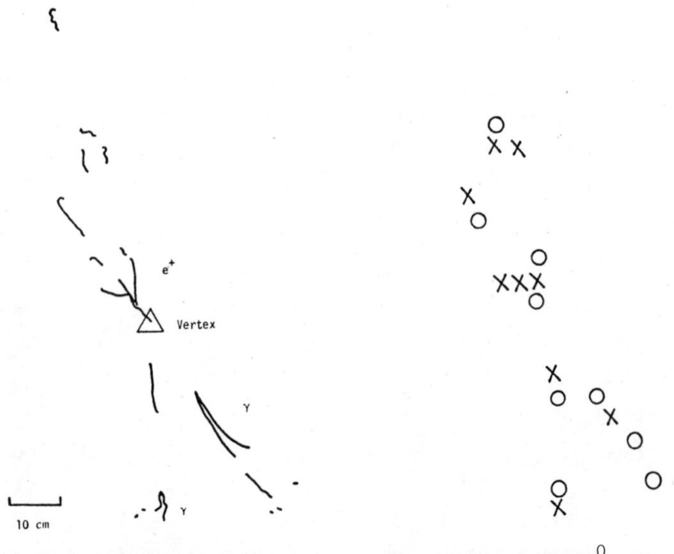

Figure 2. A Monte Carlo simulation of a proton decay to $e\pi^0$ event in the Soudan detector. The figure at the left shows one view of the charged tracks in the event. The right figure shows the pattern of tubes hit in the Soudan detector. The circles are tubes hit in this view, while the crosses represent tubes hit in the other view.

5. Absorption: A detector constructed from any material other than hydrogen suffers from the problem that the products of baryon decay may be absorbed before they exit from the nucleus. This problem has been studied in some detail[8] and two observations are relevant. First, there is little difference between iron and any other common element in the percentage absorption. Second, events may still be recognizable in a track-sensitive ionization detector, even if a pion is absorbed. For example, an excess of electrons of energy about 400 MeV above the neutrino-induced spectrum would still be indicative of baryon decay. It is important to remember that none of the detectors proposed has the resolution of a bubble chamber. The discovery of baryon decay is unlikely to rest on a single event.

b. Logistical Considerations

To minimize backgrounds, a baryon decay detector must be located deep underground (approximately 700 m for the Soudan detector). At these depths, cavities are subject to immense gravitational pressures (about 200 atmospheres at Soudan). As a result, it is difficult to construct and support cavities with large rock spans. On the other hand, there are large numbers of existing, stable tunnels at various depths. A typical tunnel cross-section is 3 by 3 m, although most tunnels can be increased to about 6×6 m at a reasonable cost. The length of tunnels varies, but several kilometers is not uncommon. The ability to fit into a standard or modified tunnel cross-section would be a significant advantage to any detector.

Another problem for a baryon decay detector is the large number of uncertainties involved and the resulting possibilities for cost escalation. These questions can mostly be answered by the construction of a prototype of sufficient size to adequately test the detection scheme but small enough to minimize possible risks. The experience gained with the prototype could be used to avoid problems on a large scale.

A third practical criterion is that, because of the size of the detector, the design should avoid the use of extreme state-of-the-art techniques. Ideally, it should not require few nanosecond timing measurements over a large number of channels or special construction techniques. The size of the baryon detector is already state-of-the-art; other technical complications could prove disastrous.

The Soudan design meets all of these criteria. The radiation length including gas-filled tubes is 8.5 cm, which means that a 5 m span contains about 60 radiation lengths. In these terms, a tunnel-sized taconite detector has the same fiducial percentage as a low-Z detector with a span of 20 m. By these same arguments, the 40 ton Soudan first phase has the same percentage ability to confine events as a detector composed of 3 kilotons of a light material. The dense detector can therefore be prototyped on a much smaller scale.

The Soudan detector does not require state-of-the-art electronics. Indeed, the Soudan design is entirely conventional, similar to the design of neutrino detectors which have been proven at many accelerators. The major novel feature of the Sourdan design is the use of taconite concrete. The main advantage here is cost, although there is the ancillary benefit of keeping the proportional tubes rigid and straight with a small amount of machining. We note that formed iron (for example, black iron pipe) costs about $600 per ton, while the materials cost of the taconite concrete is less than $50 per ton.

3. SCHEDULE OF THE SOUDAN DETECTOR

About one-third of the Soudan detector has currently been constructed with module manufacturing proceeding at a rate of 3 tons per week. Prototype electronics has been constructed and is now being tested on a 32 tube module. Sufficient integrated circuits are on hand to assemble 4000 channels of electronics during May. Work at the mine site has been facilitated because the State of Minnesota, Division of State Parks, has employed a seven man, full-time crew at the mine since ore operations ceased in 1963. This crew maintains the mine shaft, the electrical system, the pumps, the rail system and the mine hoist. The use of the mine for public, underground tours during the summer requires a high standard for maintenance and safety in the entire mine operation. The state work crew has prepared the detector site by barring down loose rock, removing debris and washing the walls and roof. Electricity and water have been connected to the detector site and a bulkhead has been constructed to retain heated, dehumidified air in the detector area. The detector itself will be supported on a concrete pad, which is surrounded by a treated wooden decking area for the electronics instrumentation and the on-line computer. Installation of the taconite concrete modules is scheduled for May. The summer of 1980 will be devoted to debugging and initial operation of the detector.

4. SUMMARY

Given the range of uncertainty concerning the lifetime and decay modes for baryon decay, there is a necessity for several different types of experiments. The Soudan experiment is sensitive to all even remotely likely modes. It uses a modular detector, one which will start operation in the near future. The detector can easily be expanded while the installed portion continues to operate.

We are grateful for the assistance of the Minnesota Department of Natural Resources and the Eveleth Taconite Company. This work has been supported by the U.S. Department of Energy and the Institute of Technology and the Graduate School, University of Minnesota.

J. Bartelt, H. Courant, K. Heller, E. Peterson, K. Ruddick, and M. Shupe are also investigators on this project.

REFERENCES

1. J. Learned, F. Reines, and A. Soni, *Phys. Rev. Lett.* 43, 907 (1979).

2. B.M. Schwarzschild, *Physics Today* 33, 19 (1980).

3. Irvine-Michigan-Brookhaven Collaboration, Proposal for a nucleon decay detector, Preprint.

4. Harvard-Purdue-Wisconsin Collaboration, A multi-kiloton detector to conduct a sensitive search for nucleon decay, Preprint.

5. H. Courant, et al., "A Dense Detector for Baryon Decay", University of Minnesota Preprint.

6. Frascati-Milano-Torino Collaboration, Nucleon Stability Experiment, Preprint.

7. D.H. Perkins, *Introduction to High Energy Physics*, Addison-Wesley, Reading, 1972.

8. D. Sparrow, *Phys. Rev. Lett.* 44, 625 (1980).

THE HOMESTAKE MINE PROTON DECAY EXPERIMENT*

R. Steinberg**

Department of Physics
University of Pennsylvania
Philadelphia, PA 19104

Abstract

We describe briefly the philosophy, design, and operation of the Homestake mine proton decay experiment. Preliminary data on the proton lifetime based on 127 effective days of running time are presented.

1. INTRODUCTION

Over the past several years the importance of experimental tests of nucleon stability has become universally recognized. Since an excellent review[1] of existing limits on the nucleon lifetime has recently appeared, we will explicitly mention only the most sensitive existing search for nucleon decay,[2] in which a lower limit of 10^{30} year for the total nucleon lifetime was obtained.

There exist many possible approaches to an improved search for nucleon decay. All of them require detectors of very large mass located in an environment exceedingly well shielded from cosmic rays. A further difficulty is introduced by the very large number of possible decay modes. Nucleon decay is a rare and poorly understood process with many background processes which might imitate it. There exist several recent calculations of branching ratios for nucleon decay with appreciably different rates. An experimental nucleon decay search focussed on a specific decay mode is therefore likely to suffer unknown and possibly large sensitivity losses caused by suppression of the decay mode being studied.

For this reason we have decided that the first requirement for a new proton decay experiment is that it be about equally sensitive to most of the possible decay modes. The approach we have chosen for our Homestake proton decay experiment utilizes the characteristic 2.2 μsec. delayed coincidence provided by muon decay as the primary signature of nucleon decay. Of the 12 hadrons of mass less than 938 MeV to which nucleons might decay, nine yield a substantial fraction of detectable $\mu \to e$ decays. In addition, direct $N \to \mu^+ X$ and $N \to \mu^- X$ are detectable by this method with high efficiency.

*Research supported in part by the Department of Energy under contract DE-AC02-76-ERO-3071 and by the National Science Foundation under Grant No. AST 79-08670.

**Written in conjunction with M. Deakyne, K. Lande, C.K. Lee, and S. Sutton, Department of Physics, University of Pennsylvania, Philadelphia, PA.

The second requirement for a successful proton decay experiment is that the backgrounds, essentially all of which are produced by cosmic rays, be reduced to exceedingly low levels. To meet this requirement, we have constructed our experiment in a $20\,m \times 20\,m \times 10\,m$ underground chamber located in the Homestake Gold Mine, Lead, South Dakota, at a depth of 4850 feet of rock (4400 meters water equivalent). The cosmic ray muon flux at this depth is approximately $1000/m^2\,yr$, which is about 10^7 times lower than that at the surface of the earth and 10^3 times lower than that in the competing shallow mine proton decay experiments.

2. DESCRIPTION

A. Apparatus

The Homestake detector[3] as presently operating (see Fig. 1) consists of 36 optically separated rectangular water Cerenkov modules each of which measures approximately $2 \times 2 \times 1.2\,m$ together with 34 liquid scintillation counters each measuring $1.6 \times 1.6 \times .05\,m$. The water Cerenkov modules are arranged in two groups of 6×3 modules. These two groups cover opposite walls of the chamber. The scintillation counters provide an anti-coincidence shield along the roof of the chamber. Further additions to this detector utilizing the water near the chamber floor and the remaining wall areas are under construction. The 36 operating modules together comprise an active volume of 150 tons of water or 9×10^{31} nucleons.

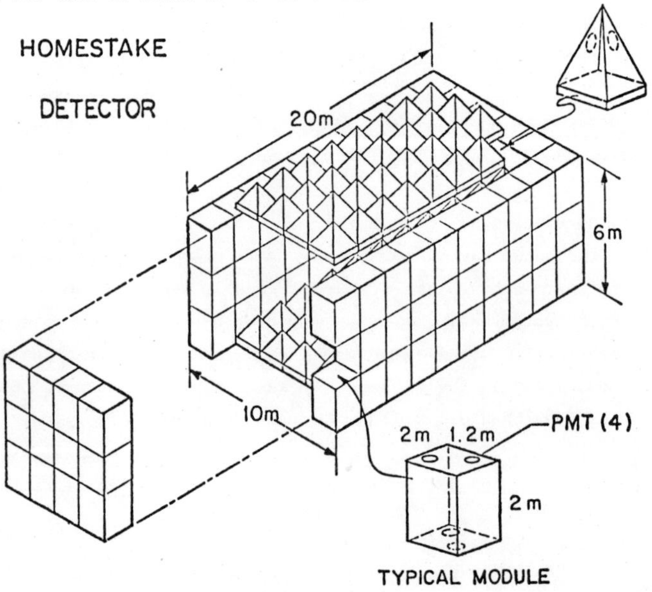

Figure 1.

Each Cerenkov module consists of highly reflective walls containing $4\ m^3$ of filtered water to which wavelength shifter has been added. Four 5 inch hemispherical photomultiplier tubes collect the Cerenkov light from each module. The use of wavelength shifter, highly reflective walls and large photomultiplier tubes allows efficient detection of electrons with energies above 10 - 15 MeV, a value which is ideal both for rejection of the large background of low energy electrons produced by natural radioactivity and for efficient detection of the electrons from muon decay (mean energy 38 MeV).

B. Decay Mode Sensitivity

In order to proceed further with any decay-mode-specific nucleon decay detector, we require theoretical predictions of the decay branching ratios. Within the SU(5) grand unified gauge model there exist at present four different calculations,[4-7] a remarkable feature of which is that no two are in quantitative agreement with regard to individual decay modes. Nevertheless, since our $\mu \to e$ detection method integrates over a large number of decay modes, the overall sensitivity of our experiment to nucleon decay is essentially independent of which calculation is correct.

In order to characterize the appearance of nucleon decay events in our detector and to determine the detection efficiency within a given set of predicted branching ratios, we must consider in detail the behavior of the nucleon decay products from the original decay point inside the oxygen nucleus (where most of the decays in a water detector will occur) to the point where they come to rest or escape from the detector.

The intranuclear propagation of nucleon decay secondaries has recently been examined by Sparrow.[8] This calculation indicates that pions are particularly subject to absorption and scattering because their energy is close to that of the (3,3) resonance. About 50% of the pions are absorbed inside the oxygen nucleus; of those that escape 1/3 undergo a scattering that results in a significant change of energy or direction. Since the ρ lifetime is less than the nuclear transit time, the ρ can be treated as two pions. For both ρ^+ and ρ^0 there is a 50% chance that a π^+ will emerge from the nucleus. Neither the η^0 nor ω^0 has an appreciable probability of decaying in the nucleus. Since the (3,3) resonance considerations do not apply to these particles and since their kinetic energy is very small, a reasonable estimate is that 90% escape the nucleus unscathed.

We have performed careful Monte Carlo simulations of the behavior of charged and neutral pions, positrons, gammas, and muons in our detector. We find that, depending on their energy, from 50 to 90% of the π^+ will come to rest within the detector and then undergo the $\pi \to \mu \to e$ decay sequence. The maximum excursion of the pions in water will be less than 1 meter from the point of nucleon decay. Indeed, the pion stopping point, and thus the location of the $\pi \rightarrow \mu$ and $\mu \rightarrow e$ decay, is typically 30 cm from the nucleon decay. From the electromagnetic cascade simulations, we find, since the radiation length in water is 36 cm, that 80 to 85% of the electromagnetic energy is deposited in a cylinder 20 cm in diameter extending typically 1 m from the nucleon decay point. The direct muon secondaries will merely travel to the end of their range without significant deviation. Since, within the SU(5) picture, the main source of direct muons in nucleon decay is from the $K^0\mu^+$ mode, muons will have a kinetic energy of 230 MeV giving them a range of about 1 meter.

In Table I we present results of the Monte Carlo calculations of the partial detection efficiences* for six of the SU(5) allowed decay modes. These calculations assume realistic detector geometry and energy detection thresholds and include the detailed particle propagation phenomenology described above. In the table e_N represents the efficiency with which a given decay mode will be detected as an event in which N modules fire. We see that the summed partial detection efficiencies (e_T) are quite high, ranging from 30 to 85%. Furthermore, approximately 90% of the events detected appear as 1- or 2-module triggers. This situation allows us to impose a 1- or 2-module event geometry as an additional signature for nucleon decay. The Monte Carlo calculations also provide information on the pulse height distributions of the Cerenkov radiation for various decay modes. We find that, for our detector, these pulse heights lie with high probability in the region of 100 to 400 MeV equivalent energy. For the 2-module events and some of the 1-module events (those with a prompt pulse in one module and the muon decay in another), we can also determine the direction of motion of the decaying particle by the relative orientation of the prompt- and delayed-pulse modules.

Since nucleon decay products should be isotropically distributed, whereas the cosmic ray background is almost exclusively downward, we can discriminate strongly against the latter by eliminating downward going trajectories. Fewer than 25% of the nucleon decays will be rejected by this cut.

In Table II we calculate the total detection efficiencies for the six SU(5) allowed proton decay modes to which we are sensitive. The values range from 15 to 52%. In Table III we show the proton decay branching ratios predicted in Refs. 4-7 and calculate the corresponding net signals per proton both for bound and for free protons by summing the branching ratios weighted

---

*By partial detection efficiency for a given decay mode, we mean the fraction of events which would be detected assuming that that decay mode yielded a 100% branching ratio to π^+/μ and 100% intranuclear π^+/μ survival probability. Total detection efficiency for a given decay mode is the product of the partial detection efficiency, the π^+/μ branching ratio for that mode, and the intranuclear π^+/μ survival probability.

Decay Mode	e_1 1 Module	e_2 2 Modules	$e_{\geq 3}$ $\geq$ 3 Modules	e_T
$p \to e^+\rho^0$	.300	.133	.025	.46
$p \to e^+\eta^0$	.405	.333	.115	.85
$p \to e^+\omega^0$	.340	.250	.065	.66
$p \to \bar{\nu}\pi^+$	.249	.053	.002	.30
$p \to \mu^+K^0$	.349	.167	.050	.57
$p \to \bar{\nu}\rho^+$	.261	.142	.049	.45

Table I. Partial detection efficiencies for various proton decay modes calculated by the Monte Carlo method.

by the total detection efficiencies of Table II. The net signal for protons in water (denoted by $B(\mu)$) is reasonably insensitive to which set of branching ratios is chosen. We will use 0.27, the average of the four values of $B(\mu)$.

C. Backgrounds

The most significant background for a nucleon decay experiment is produced by the cosmic ray muon flux. We are indeed fortunate to have access to a chamber at 4850 ft depth where the muon flux $(4 \times 10^{-9}/\text{cm}^2 \text{ sec})$ is reduced by 10^7 compared to that at the surface of the earth. At our depth the muons have a mean energy of 300 GeV and thus a mean range of 1500 m of water. In one cm^3 of water we therefore expect 0.67×10^{-5} muon stops per incident muon. Since there are 6×10^{23} nucleons per cm^3 of water, we have

$$\frac{4 \times 10^{-9} \times .67 \times 10^{-5}}{6 \times 10^{23}} = 0.45 \times 10^{-37} \quad \text{decays/nucleon sec.}$$

or one muon decay per 2.2×10^{37} nucleon sec. If these cosmic ray muon decays were all attributed to nucleon decay, they would give an apparent lifetime of 2.2×10^{37} sec. or 7×10^{29} years. In addition to these direct muon decays, there are also muon decays that result from pions produced locally by muon photonuclear interactions. Muon decays associated with this process reduce the apparent muon-induced nucleon decay lifetime from 7×10^{29} years to 2×10^{29} years.

Mode	B.R. to π^+ or μ	π^+/μ Internal Survival Probability (from Ref. 8)	Partial Detection Efficiency (from Table I) $e_1 + e_2$	Total Detection Efficiency
$p \to e^+\rho^0$	1.0	.5	.43	.21
$p \to e^+\eta^0$	.29	.9	.74	.19
$p \to e^+\omega^0$	.91	.9	.59	.48
$p \to \bar{\nu}\pi^+$	1.0	.5	.30	.15
$p \to \mu^+K^0$	1.0	1.0	.52	.52
$p \to \bar{\nu}\rho^+$	1.0	.5	.40	.20

Table II. Calculation of the total detection efficiencies for six SU(5) allowed proton decay modes.

Mode	Ref. 4	Ref. 5	Ref. 6	Ref. 7
$e^+\rho^0$	.08	.18	.21	.02
$e^+\eta^0$	.06	.03	.05	.07
$e^+\omega^0$	.11	.47	.19	.18
$\bar{\nu}\pi^+$	.28	.03	.11	.15
μ^+K^0	.17	.14	.01	.19
$\bar{\nu}\rho^+$	.14	.07	.08	.01
Net muon decays per bound proton	.23	.36	.18	.23
Net muon decays per free proton	.33	.45	.27	.27
Net muon decays per proton for $H_2O \equiv B(\mu)$	.25	.38	.20	.24

Table III. Proton decay branching ratios and net signals in the SU(5) model. The net decay entries are the sums of branching ratios weighted by total detection efficiencies.

Since the expected muon decay rate from cosmic rays is known, the observed rate for this process provides a measure of the efficiency of the decay electron detection as well as a measure of the overall volume-on time factor for the detector. Having the calibration signal one order of magnitude greater than the expected nucleon decay signal is an ideal arrangement. The calibration signals are frequent enough to insure that the apparatus is functioning, but not so intense as to overwhelm the signal. Indeed, one major concern in a very low counting rate experiment is to maintain a continuing measure of the status of the apparatus and to insure that the calibration process is sufficiently similar to the desired signal that the measure is meaningful.

The remaining background process which can simulate nucleon decay by producing muon decays is the interaction of cosmic ray neutrinos with nucleons of the detector. At lifetimes below 10^{31} years, this process is not significant.

3. OPERATIONAL STATUS AND PRELIMINARY RESULTS

We began operation of the Homestake Detector at the beginning of 1979. We present here the results of 127 days operation. During this period we observed 89 $\mu \to e$ decays in 150 m^3 of water. The observed events are characterized by three parameters: number of modules involved, magnitude of energy deposition, and presence or absence of veto counts. We display these events in Fig. 2, where the energy deposition is plotted along the x-axis and module number along the y-axis. As explained previously, module numbers greater than two are inconsistent with nucleon decay. About half of such events are also accompanied by veto pulses, a fraction consistent with that expected for background events produced by cosmic ray muons.

The region of interest for nucleon decay is shown in more detail in Fig. 3. For the two-module events and for some of the one-module events, we show arrows indicating the direction of flight of the particle producing the muon decay as determined by the prompt-delayed module configuration. Of the eight events between 100 and 400 MeV, all four for which directional information is available are moving downward. Of the remaining four, two were accompanied by veto pulses and therefore are also cosmic ray muons. The remaining two represent the sum of signal plus unvetoed background. Since the veto efficiency is estimated to be 50%, we expect a background of two events. The net signal is then $2 - 2 = 0 \pm 2$. We can therefore set an upper limit of two nucleon decay events. Using the clearly identified cosmic-ray-muon-associated $\mu \to e$ decays as a normalization, we have

$$\tau_{\text{nucleon}} \geq \frac{\#\ \text{cosmic ray muon decays}}{\#\ \text{nucleon decay muon decays}} \times 2 \times 10^{29}\ \text{yr} \times B(\mu)$$

$$\geq \frac{89}{2} \times 2 \times 10^{29}\ \text{yr} \times B(\mu)$$

$$\geq 9 \times 10^{30}\ \text{yr} \times B(\mu) \quad .$$

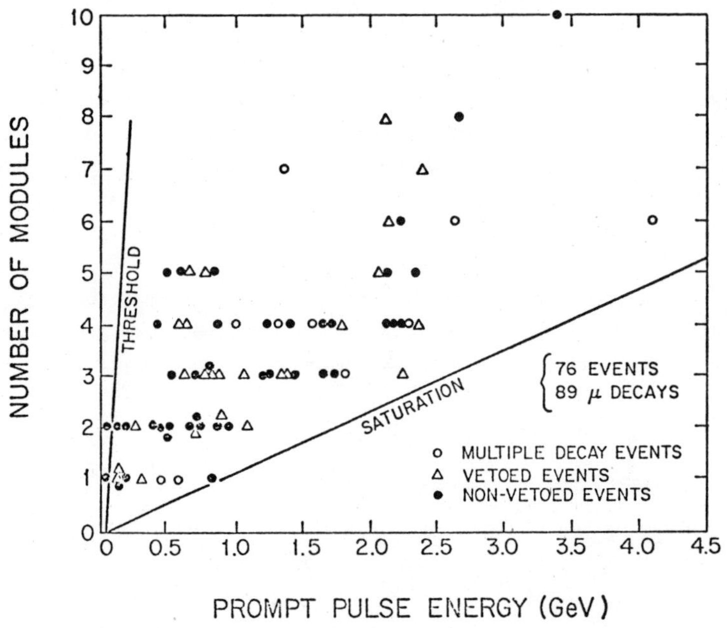

Figure 2.

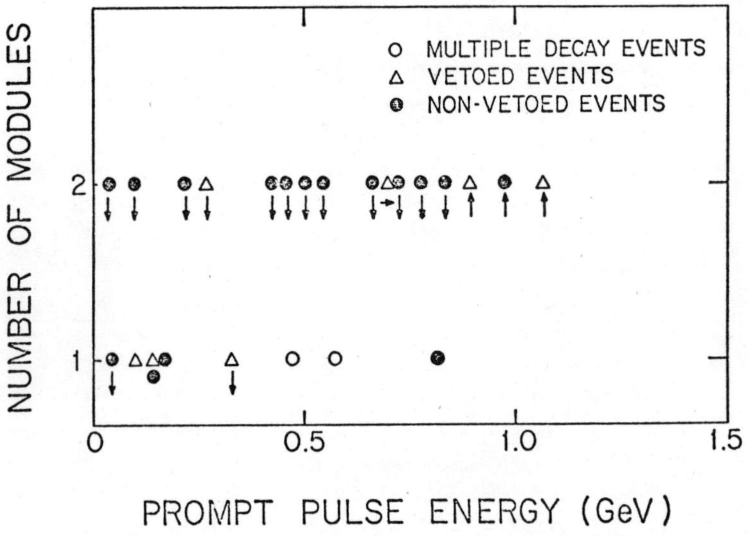

Figure 3.

We therefore find, using the results of Table III for $B(\mu)$

$$\tau_{nucleon} \geq 2 \times 10^{30} \text{ yr.}$$

Various improvements to the Homestake Detector are now being installed or assembled and will become operational during the next few months. Improvement of the veto efficiency to better than 90% will be attained after completion of the instrumentation for the bottom and remaining side areas. New pulse height measuring circuitry will improve the energy resolution and extend the dynamic range of our observations. We will also reduce the electronic dead time following a muon stop from 800 to 300 ns, thereby increasing the sensitivity of the detector by 25%. Increased running time and improved computer processing of the data will also help substantially. We expect the present Homestake Detector to achieve an ultimate sensitivity corresponding to nucleon total lifetimes beyond 10^{31} years.

It is a pleasure to thank the Homestake Mining Company for its generous cooperation.

REFERENCES

1. M. Goldhaber, P. Langacker, and R. Slansky, Preprint LA-UR-80-356, March 1980.

2. J. Learned, F. Reines, and A. Soni, *Phys. Rev. Lett.* 43, 907 (1979); F. Reines and M.F. Crouch, *Phys. Rev. Lett.* 32, 493 (1974).

3. M. Deakyne, W. Frati, K. Lande, C.K. Lee, R.I. Steinberg, and E. Fenyves, *Neutrinos '78*, Purdue University, April 1978, p. 887; M. Deakyne, W. Frati, K. Lande, C.K. Lee, and R.I. Steinberg, *Proc. Conf. on Nucleon Stability*, Madison, December 1978.

4. M. Machacek, *Nucl. Phys.* B159, 37 (1979).

5. J.F. Donoghue, Preprint CTP #824, Nov. 1979.

6. A.M. Din, G. Girardi, and P. Sorba, *Phys. Lett.* 91B, 77 (1980).

7. M.B. Gavela, A. Le Yaouanc, L. Oliver, O. Pene, and J.C. Raynal, Preprint LPTHE 80/6, March 1980.

8. D.A. Sparrow, *Phys. Rev. Lett.* 44, 625 (1980).

A MODEST APPEAL TO SU(7)

Asim Yildiz*

Research Laboratories of Mechanics
University of New Hampshire
Durham, NH 03824
and
Lyman Laboratory of Physics
Harvard University
Cambridge, MA 02138

A particular version of SU(7) symmetry without top (t) quark provides an explanation for low energy phenomenology and remains consistent with cosmological baryon generation, proton decay, neutrino mass generation and similar exotic processes by including intermediate mass scales.

Since many aspects of a particular symmetry reflect varied physical scenarios, an extended or enlarged symmetry such as SU(7) will have multiple themes. The particular version of SU(7) symmetry as unification of strong and electroweak forces which is the subject of my talk has been developed recently with my collaborators M. Claudson and P. Cox [1]. The specific questions which motivated us are directed to include the reported decays of the B-meson ($B \to J/\Psi\ K\pi$) while excluding top quarks from the theory. Such assumptions are encouraged by the combination of Goliath and negative e^+e^- annihilation (PETRA) experimental results. With these two initial ingredients a naive search among simpler symmetries was carried out by us. Our aim has been to find a simple gauge group which contains an $SU(3)_c \times [SU(2) \times U(1)]_W$ subgroup which is consistent with low energy phenomenology and our new conditions.

Embedding this subgroup in a larger and simple group reminds us that the observed fermions appear in complex representations. Since unbroken chiral symmetries provide a natural explanation for the mass scales of the observed particles, we restrict our attention to groups with complex representations [2]: SU(N), SO(4N+2) and E_6.

In the SU(N) series, first SU(5) symmetry is reinvestigated. The standard model has 10_L and $\bar{5}_L$ (ten dimensional and five dimensional left-handed) representations. Without a t-quark, the bottom quark (b) and tau lepton (τ) can be put into the same multiplet (b,τ) which can be

*Research supported in part by the National Science Foundation under Grant No. PHY77-22864, and by the Department of Energy under Grant No. ER-78-5-02-4999.

represented by 5_L and $\bar{5}_L$. This is the only possibility that uses only fundamental representations. The defect of such an arrangement is that no chiral symmetry in such a theory will keep the b, τ, ν_τ masses light. The next symmetry SU(6) [3] has the anomaly free set: $2 \times \bar{6} = 2 \times \bar{5} + 2 \times 1$ plus $15 = 10 + 5$ (indicating the SU(5) decomposition). This choice provides the right particle content; however, defects of the model are i) d,s,b,b' all mix; thus the GIM mechanism is spoiled for neutral currents; ii) neutrino masses will be on the order of lepton masses.

Before we go to higher dimensions in SU(N) family we pause to examine the most prominent two alternative models. In SO(10) the only complex representations are spinors, with equal numbers of $Q = 2/3$ and $Q = -1/3$ quarks which give no room for a quark scheme excluding the top quark. However E_6 seems to be an ideal theory for the purpose since the fundamental representation 27 contains in SU(5) dictionary $10 + \bar{5} + \bar{5} + 1 + 1$, an adequate picture for the purpose. However SU(N) family promises a much simpler symmetry picture than that of E_6 which we preferred to investigate.

SU(7) is a rank-6 symmetry, the same rank as E_6, and it possesses two basic anomaly free sets [4] (expressed in terms of SU(5) x SU(2) decomposition)

$$3 \times \bar{7} = 3(\bar{5},1) + 3(1,2)$$
$$21 = (10,1) + (5,2) + (1,1)$$

and

$$2 \times \bar{7} = 2 \times (\bar{5},1) + 2(1,2)$$
$$35 = (\overline{10},1) + (10,2) + (5,1).$$

Among many, the most satisfactory combination of the two sets is the anomaly free choice: $5 \times \bar{7} + 21 + 35$, giving

$$5 \times \bar{7} = 5 \times (\bar{5},1) + 5 \times (1,2)$$
$$21 = (10,1) + (5,2) + (1,1)$$
$$35 = (\overline{10},1) + (10,2) + (5,1).$$

Pairing of conjugate multiplets identifies superheavy fermions: $(\overline{10},1)$ $(10,1)$; $(\bar{5},1)$ $(5,1)$; $4(1,2)$. The rest form:

i) Two ordinary (e,μ) families: $(10,2) + 2(\bar{5},1)$ where the first and second terms contain $[(u,d;e),(c,s;\mu)]$ and $(d;e,\nu_e)$, $(s;\mu,\nu_\mu)$,

ii) Two new families, $(b;\tau,\nu_\tau)$ and $(b';\tau',\nu'_\tau)$ in $(5,2) + 2(\bar{5},1)$,

iii) Other light fermions in, $(1,2) + (1,1)$.

Symmetry breaking has multiple stages, with the initial breaking at 10^{15} GeV, and intermediate mass scales at 10^{11} GeV, 10^5 GeV, and finally at 10^2 GeV. The first symmetry breaking reduces SU(7) to $SU(3)_c \times SU(4) \times U(1)$. The constraints of this breaking are provided by proton decay. The mass scale 10^{15} GeV promises the decay lifetime of the proton to be $\gtrsim 10^{30}$ years. The alternative possibility of breaking to SU(5) x SU(2) is disregarded since it leads to a much faster decay of the proton. In this transition the 24 baryon number non-conserving gauge bosons and, we assume, all colored Higgs (which also provide baryon number nonconservation) get masses.

The second stage realizes a breaking of $SU(3)_c \times SU(4) \times U(1)$ to $SU(3)_c \times [SU(2) \times U(1)]_W \times SU(2)_N$. Note that the U(1) in the last expression is not the first U(1) of $SU(7)/SU(3)_c \times SU(4)$, nor the U(1) of $SU(4)/SU(2) \times SU(2)$, but a linear combination. (In fact, it is actually the U(1) of $SU(5)/SU(3) \times SU(2)$. See reference 1 for a further discussion.) There seem to be no overwhelming constraints on this second symmetry breaking, since suggestions focus on certain physical phenomena which neither experimentally nor theoretically can be expressed quantitatively so far. These are B-L (Baryon number minus lepton number) nonconserving, exotic processes such as neutrino oscillations which require light neutrinos $(\nu_e, \nu_\mu, \nu_\tau)$, mass generations in this second transition, and certain proton decay modes (especially the $p \to \pi^+ \nu$ mode) influenced by the symmetry breaking at $\sim 10^{11}$ GeV. We also observe a new $SU(2)_N$ subgroup emerging which describes a symmetry with neutral vector bosons W_N, W_N, and Z_N and notably new fermions.

$SU(3)_c \times [SU(2) \times U(1)]_W \times SU(2)_N$ will experience a symmetry breaking at about 10^5 GeV to $SU(3)_c \times [SU(2) \times U(1)]_W$ which will give masses to the new neutral vector bosons (W_N, W_N, Z_N) and new fermions which are heavier than ordinary fermions. This intermediate symmetry breaking is related to considerations of neutral currents; CP-violation; b,τ and (unobserved) b',τ' masses.

Finally, the symmetry breaking at 10^2 GeV reduces $SU(3)_c \times [SU(2) \times U(1)]_W$ to $SU(3)_c \times U(1)_{E.M.}$. This breaking is constrained by:

i) Light fermion masses

ii) Gauge boson masses

exactly as in the standard SU(2) x U(1) theory.

HIGGS IN SYMMETRY BREAKING AND SUPERHEAVY FERMIONS

The first symmetry breaking $SU(7) \to SU(3)_c \times SU(4) \times U(1)$ at 10^{14} GeV is achieved by scalars in the 48 dimensional adjoint representation of SU(7): ϕ^α_β. More specifically, the vacuum expectation value of ϕ^α_β is indicated by the diagonal (only non-zero) elements of the representative matrix:

$$\langle\phi^\alpha_\beta\rangle \sim (4,4,4,-3,-3,-3,-3)$$

Allowed Yukawa couplings for our SU(7) fermions can be shown as:

$$\bar{7} \times \bar{7} = \overline{21}_A + \overline{28}_S$$

$$\bar{7} \times 21 = 7 + \overline{140}$$

$$\bar{7} \times 35 = 21 + \overline{224}$$

$$21 \times 21 = \overline{35}_S + \overline{196}_S$$

$$21 \times 35 = \overline{21} + 224 + 490$$

$$35 \times 35 = 140_S + 490_S.$$

The scalar multiplets of dimensions 7, 21, 140, and 224 are used in the theory. Others are omitted due either to large dimensions or to not playing a useful role. We list these scalars with their tensorial indices and masses that they generate:

$21 \rightarrow \phi_{\alpha\beta}$	superheavy masses
$7 \rightarrow \phi_{\alpha}$	b,τ;b',τ' masses
$140 \rightarrow \phi^{\alpha\beta}_{\gamma}$	u,c masses and b,τ;b',τ' masses
$224 \rightarrow \phi^{\alpha\beta\gamma}_{\delta}$	d,s masses

The second symmetry breaking at $\sim 10^{12}$ GeV defines

$$SU(3)_c \times SU(4) \times U(1) \rightarrow SU(3)_c \times [SU(2) \times U(1)]_W \times SU(2)_N,$$

which is driven by the Higgs multiplet of 21 dimensions with the vacuum expectation value (VEV):

$$\langle\phi_{\alpha\beta}\rangle = V_2(\delta^6_\alpha\delta^7_\beta - \delta^6_\beta\delta^7_\alpha);\ V_2 \sim 10^{12} \text{ GeV}.$$

This VEV will provide superheavy masses for some fermions and the responsible Yukawa couplings are given by:

$$\mathcal{L}_{YH} = \frac{1}{24} f_2\, \Psi_{\alpha\beta\gamma}\, \gamma^0\, \Psi_{\delta\varepsilon}\phi_{\rho\eta}\, \varepsilon^{\alpha\beta\gamma\delta\varepsilon\rho\eta} + \frac{1}{2} h_{2i}\, \Psi_{\alpha\beta\gamma}\, \gamma^0\, \Psi^{\alpha}_i\, \bar{\phi}^{\beta\gamma}$$

$$+ \frac{1}{2} k_{ij}\, \Psi^{\alpha}_i \gamma^0\, \Psi^{\beta}_j \phi_{\alpha\beta} + \text{h.c.},$$

where i and j indices run from 1 to 5 for the five 7-plets. This yields mass values

$$M_{(10,1)(\overline{10},1)} = |f_2 V_2|$$

$$M_{(5,1)(\bar{5},1)} = \left[\sum_{i=1}^{5} |h_{2i}V_2|^2\right]^{1/2}$$

and a mass matrix

$$M_{(1,2)(1,2)} = k_{ij}V_2$$

Note that since k_{ij} is antisymmetric in five dimensions, one of its eigenvalues is zero; thus one (1,2) remains light.

$$(10,2) + 2 \times (\bar{5},1) \rightarrow (e,\mu) \text{ families}$$

$$(5,2) + 2 \times (\bar{5},1) \rightarrow \{(b;\tau,\eta_\tau),(b';\tau',\eta_{\tau'})\} \text{ families}$$

$$(1,2),(1,1) \rightarrow \text{new neutrals } N_L, N_L', \bar{N}_L$$

$(n_\tau)_L$, $(n_\tau')_L$, N_L, N_L', mix, and $(\bar{n})_L$, $(\bar{n}_\tau')_L$, $\bar{N}_L$ mix, to give ν_τ and three (eventually massive) neutral leptons. We assume also that the following scalars remain massless at this stage (identified by their $SU(2)_W \times SU(2)_N$ content)

$$\left.\begin{matrix} 2 \times (1,2) \\ (2,3) \end{matrix}\right\} \text{ from 7 and 140}$$

$$(2,2) \quad \text{from 224.}$$

Two (1,2)'s acquire VEV's at 10^5 GeV ($=V_3$) and leave one weak doublet light to get a VEV at 10^2 GeV($=V_4$), which reproduces the relation $M_W/M_Z = \cos\theta$.

INTERMEDIATE AND LIGHT FERMION MASSES

Intermediate and light fermions masses are generated from the couplings represented by

$$\mathcal{L}_{YL} = h_{1i}\, \Psi_{\alpha\beta}\gamma^0\, \Psi_i^{\alpha}\bar{\phi}^{\beta} + \frac{1}{2} h_{5i}\, \Psi_{\alpha\beta}\gamma^0\, \Psi_i^{\gamma}\phi_\gamma^{\alpha\beta}$$

$$+ \frac{1}{24} f_5\, \Psi_{\alpha\beta\gamma}\gamma^0\, \Psi_{\omega\delta\varepsilon}\bar{\phi}^{\omega}_{\rho\eta}\, \varepsilon^{\alpha\beta\gamma\delta\varepsilon\rho\eta}$$

$$+ \frac{1}{6} h_{4i}\, \Psi_{\alpha\beta\gamma}\gamma^0\, \Psi_i^{\delta}\phi_\delta^{\alpha\beta\gamma} + \frac{1}{36} f_4\, \Psi_{\alpha\beta\gamma}\gamma^0\, \Psi_{\omega\delta}\bar{\phi}^{\omega}_{\varepsilon\rho\eta}\, \varepsilon^{\alpha\beta\gamma\delta\varepsilon\rho\eta}$$

$$+ \text{h.c.}$$

which can be shown to give masses as follows:

i) The 10^5 GeV VEV's give masses in tree approximation to all light fermions except (e,μ) families and ν_τ.

ii) The 10^2 GeV VEV's give masses in tree approximation to (e,μ) families except e,d,ν_e,ν_μ while ν_τ still stays massless. Note that e,d,ν_e,ν_μ and ν_τ get masses from radiative corrections.

As an essential part of the low energy phenomenology, it is also important to discuss mixing among particle species in the model. Mixing among quarks and among leptons consists of : Cabibbo mixing of (u,c); (s,b,b') mixing (d is not included), and (μ,τ,τ') mixing (e is not involved). Also note that a mixed state $s_W \sim$ (s,b,b') appears in the weak hadronic charged current expression:

$$J^\mu = g\, (\bar{u},\bar{c})_L\, U_c\gamma^\mu \begin{pmatrix} d \\ s_W \end{pmatrix},$$

where U_c is the Cabibbo matrix. The model thus allows B-meson decay to $J/\Psi + K + \pi$ final states, without reference to any t quark. Rates in this decay are similar to those for D-mesons.

Other aspects of weak interactions can be briefly summarized as:

i) electron family interactions are the same as in SU(2) x U(1) theory,

ii) muon family shows departures due to mixings,

iii) b,b' have no weak interactions except through mixings ,

iv) τ,τ' have vector weak interactions with $n_\tau, n_{\tau'}$.

One shortcoming of the model so far is that no CP-non-conservation appears in the kaon system. This is due to the fact that the physical s-quark couplings involve no complex mixings. This, however, can be overcome by introducing an extra 7-plet of scalars, assuming that the $SU(2)_N$ doublet components of this multiplet have masses about 10^5 GeV, while the remaining components are superheavy. The $\bar{d}$-s vertex to scalars thus has a complex coefficient, and requiring that this mass scale is responsible for a milliweak kaonic CP violation is thus an important justification for the third symmetry breaking scale.

LIGHT NEUTRINO MASSES

Most probably the most promising aspect of the model is its prediction of light neutrino masses. In the standard SU(5) theory with the two mass scales, there is no natural way to obtain neutrino masses that are large enough to be potentially observable while remaining small enough to be presently unobserved. In this standard version the SU(5)-allowed couplings of these fields necessarily respect a global U(1) symmetry in addition to gauge symmetry. This symmetry can be related to baryon number minus lepton number (B-L), and B-L is still an exact symmetry after SU(5) is broken, and it forbids neutrino masses to all orders. However, this symmetry can easily be broken by adding scalar representations with corresponding couplings.

Tree level masses for neutrinos in SU(5) can arise only if the model contains a $\underline{15}$ of scalars; no other scalar can couple neutrals. If this scalar in fact occurs, the sum of all tree level mass contributions gives an effective Yukawa coupling constant times an effective $\underline{15}$-plet VEV. If it is zero, then no neutrino mass appears at tree level; if non-zero, we obtain neutrino masses. However, our criterion of naturalness implies that no mass parameter should be introduced into the theory unless it is required by phenomenology, and especially no new mass scale unless required. Hence, a $\underline{15}$-plet VEV should be comparable to the known VEV's, and thus neutrino masses should be similar to other fermion masses; a result which conflicts with experiments by five orders of magnitude.

One-loop contributions to neutrino masses do not require new VEV's; diagrams such as Figure 1 occur if either a 10 or a 15 of scalars is introduced with (B-L)-violating couplings. (The scalars propagating in the loop must be a 5 and the new multiplet; the VEV ends should be a 5 and 24 at the quark vertex and a 5 on the fermion line. Depending on the components of the various multiplets, the fermion line could be any of the massive fermions.) The neutrino mass generated by such a diagram may be estimated as

$$m_\nu \sim \frac{h^3\lambda}{\pi^2} \frac{(V_5)^2 V_{24}}{M^2},$$

where h represents a typical Yukawa coupling constant, λ represents a typical scalar quartic coupling constant, and M^2 is the heaviest mass appearing in the loop. In this case M^2 can only be of order V_5 or of order V_{24}. If $M \sim V_5$, then $m_\nu \sim V_{24}$, which is ridiculously large; even taking $h \sim 10^{-5} \sim m_e/m_W$ (which is the smallest, not a typical, known Yukawa constant) leaves $m_\nu \sim 10^{-1}$ GeV (M^2 will be $\sim\lambda\, V_5^2$, not V_5^2). If on the other hand $M \sim V_{24} \sim 10^{15}$ GeV, then taking $h \sim 10^{-1} \sim m_t/m_w$ (the other extreme for Yukawa constants) still leaves $m_\nu \sim 10^{-15}$ GeV = 10^{-6} eV, too small to observe.

Other models of violation of B-L will in general lead to the same estimates, unless something really exotic is added to the theory to break B-L. Thus, the SU(5) theory naturally yields neutrino masses only if they are too heavy to be acceptable or too light to be measurable.

In the SU(7) symmetry version that we are talking about here, a Majorana mass matrix for the $(\bar{5},1)$ multiplets is generated by the diagram of Fig. 1, where now the loop scalars are from a 7 and a 21; two of the VEV's must come from the weak symmetry breaking at 10^2 GeV; the third from the adjoint VEV which is 10^{15} GeV. The fermions inside the loop have masses controlled by the 21 VEV, which is also the relevant scale for the scalars in the loop. This VEV is not strongly constrained by established phenomenology, but estimating with the renormalization-group equations suggests a value of order 10^{11} - 10^{12} GeV. We thus arrive at a typical mass matrix element of order

$$m_5 \sim \frac{h^3\lambda}{\pi^2} \frac{(10^2\ \text{GeV})^2(10^{15}\ \text{GeV})}{\lambda(10^{12}\ \text{GeV})^2} \quad 10^{-12}\ \text{GeV} = 10^{-3}\ \text{eV},$$

where we have used $h \sim 10^{-2}$ as a typical value. If we use 10^{11} GeV for the 21 VEV estimate, or if we increase the adjoint VEV (proton decay so far only gives a lower bound, and the renormalization group equations, which fix this scale in SU(5), here have more undetermined parameters), then a larger value is quite possible.

The (1,2) neutrals also obtain Majorana mass contributions from diagrams as in Fig. 1, where the loop scalars are now from the 21 and from the 224 introduced to provide masses for lighter fermions; two of the VEV's come from the SU(2) breaking at 10^6 GeV, while the third is from the 21. The fermions here are charged leptons, one from the ordinary (10,2) multiplet, the other from the superheavy (10,1). This leads to mass matrix contributions of order

$$m_2 \sim \frac{h^3\lambda}{\pi^2} \frac{(10^6 \text{ GeV})^2 (10^{12} \text{ GeV})}{\lambda (10^{12} \text{ GeV})^2} \sim 10^2 \text{ eV},$$

which for the neutrino mass matrix must be reduced by mixing parameters. In addition, there is a contribution which couples $(\bar{5},1)$ and (1,2) neutrals via a simple radiative correction graph. Here one neutral emits a charged scalar (from the 21), becoming a muon; the muon is coupled to the VEV which gives its mass, and then absorbs the scalar becoming the other neutral state. A similar estimate of this diagram gives $(h^2/\pi^2)\, m_\mu$ times a logarithm (the loop actually gives a divergent integral; the divergence is cancelled by similar graphs with superheavy particles, giving $\log(10^{12}\text{ GeV}/10^{15}\text{ GeV}) \sim 10$); the result is

$$m_{5-2} \sim 10^4 \text{ eV},$$

which seems likely to dominate the mass matrix.

The result of these various processes will be a complicated mass matrix for neutral fermions. Its eigenvalues will be impossible to predict closely, especially since few of the coupling constants involved are experimentally accessible. However, if our estimates are reasonable and if this model is correct, neutrino masses of KeV order are to be expected. Furthermore, since these processes do not have any reason to respect family structure, neutrinos generated (or observed) in weak interactions will be mixtures of the mass eigenstates, and interesting oscillations will take place, on a scale also probably measured in KeV.

SUMMARY

This unification model with SU(7) symmetry seems to have certain virtues:

i) The model without t-quark is compatible with B-meson decay.

ii) The theory possesses intermediate mass scales which are needed for consistency with low energy phenomenology and these scales suggest observable levels for exotic processes, and neutrino masses.

iii) Standard unification results persist: baryon number generation, proton decay, and experimentally verified values of $\sin^2\theta_W$.

The shortcomings of the theory are the excessive appearances of scalar multiplets; and especially the necessity of using them for kaonic CP violation. Furthermore, the large number of Yukawa couplings endangers any quantitative predictive power of the model, especially concerning masses and mixing angles.

REFERENCES

1. Extended Unified Field Theories: An SU(7) Model, M. Claudson, P. Cox, and A. Yildiz, HUTP-80/A013.
2. For earlier contributions see: J. C. Pati and A. Salam, Phys. Rev. D 8, 1240 (1973), H. Georgi and S. L. Glashow, Phys. Rev. Lett. 32, 438 (1974); F. Gursey, Proc. Kyoto Conf., Kyoto, Japan (1975), H. Fritzsch and P. Minkowski, Ann. Phys. 93, 193 (1975).
3. A different version of SU(6) is treated by H. Georgi and M. Machacek, Phys. Rev. Lett. 43, 1639 (1979).
4. Discussion of anomalies in SU(7) is given by P. H. Frampton, Phys. Lett. 88B, 299 (1979).

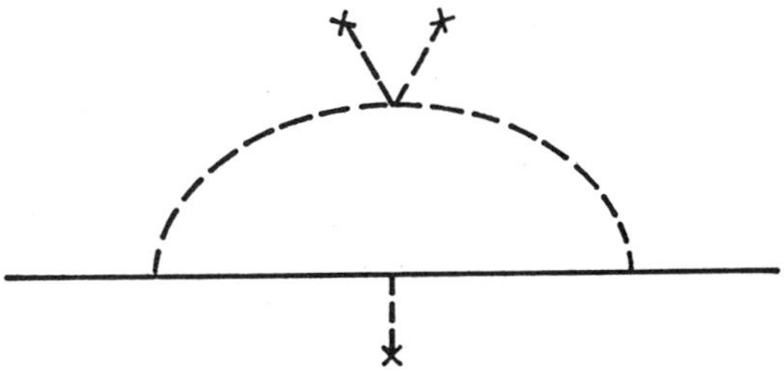

FIGURE 1

Diagram contributing to neutrino masses. Solid lines indicate fermions, dashed lines scalar mesons; lines ending in x indicate vacuum expectation value factors. The same diagram applies in several contexts; specific particle identifications are given in the text.

VERTICAL-HORIZONTAL FLAVOR GRAND UNIFICATION

Kameshwar C. Wali

Physics Department
Syracuse University
Syracuse, N.Y. 13210

This will be a very brief report on the work in progress in collaboration with Aharon Davidson. The problem of the replication of the fermionic generations and the basics of flavor dynamics have received a great deal of attention recently. As we heard from Georgi,[1] there are no satisfactory answers. Davidson and myself[2] have proposed a minimal extension of Salam-Weinberg theory, based on the group structure

$$[SU(2)\times U(1)]_{SW}\times U'(1)$$

The additional $U'(1)$ is used to classify the generations by assigning a new electro/weak hypercharge-like quantum number Y' different for the members of different generations. This way we are led to a dynamical scheme in which the mass matrices in the two charged sectors get correlated and one can obtain the quark masses, the weak mixing angles and an estimate of the CP-violation parameter in terms of Yukawa-type coupling constants and a set of vacuum expectation values for the Higgs doublets. While qualitatively very "good" results emerge from the model and there is almost certainly no difficulty in obtaining quantitatively good results, the model has too many free parameters. It is, therefore, natural to seek a grand unified theory which may limit the choice of the parameters.

We propose to investigate grand unified theories based on semi-simple group structures

$$G = \underset{\substack{\downarrow \\ \text{Vertical}}}{SU(N)_V} \otimes \underset{\substack{\downarrow \\ \text{Horizontal}}}{SU(N)_H} \tag{1}$$

The $SU(N)_V$ is the symmetry group of flavors in one generation and $SU(N)_H$ is the symmetry group in the generation space. The above group structure is a generalization of Davidson's recent proposal[3] of $SU(5)_V\times SU(5)_H$. Our main goal is to see whether from some generally accepted criteria, N can be uniquely determined and the total number of elementary fermions specified.

For this purpose, consider the left-handed components of all the fermions. Denoting them collectively by F_L, let F_L consist of

$$F_L : \sum n_{ab}(a,b) \quad , \tag{2}$$

where a and b denote the dimensionalities of $SU(N)_V$ and $SU(N)_H$, respectively. n_{ab} is the number of times the representation (a,b) occurs in the summation. We shall require:

1) That there be a single gauge coupling constant so that we have a truly unified theory. We invoke a discrete $V \leftrightarrow H$ symmetry to assure this. This means

$$n_{ab} = n_{ba} \tag{3}$$

and hence if the representation (a,b) occurs, the representation (b,a) also occurs with the same numerical coefficient.

2) That there be no "superfluous" replications of generation, which is, after all, one of the basic motivations to consider a group bigger than $SU(5)$. This implies

$$n_{a,b} = 0,1 \quad . \tag{4}$$

3) That only the usual leptons, quarks, and anti-quarks (color singlets, color triplets and anti-triplets) appear in F_L. This restricts the representations a,b to be

$$1, N, N^2, \ldots, N^{N-1} \quad ,$$

where N^k is the k-fold, totally anti-symmetrized fundamental representation.[4]

4) That only complex representations with respect to $SU(3)_c \times SU(2) \times U(1)$ occur in F_L. Hence

$$n_{ab} \cdot n_{ab} = 0 \quad . \tag{5}$$

5) Renormalizability and hence the anomaly-free condition

$$\sum_{a,b} n_{ab} A(a) D(b) = 0 \quad , \tag{6}$$

where $A(a)$ is the anomaly associated with the representation a and $D(b)$ is the dimensionality of the representation b. For generalized fundamental representations of the type N^k,

$$A(N^k) = \frac{(N-3)!}{(n-k-1)!(k-1)!}(N-2k) \quad ,$$

and

$$D(N^k) = \frac{N!}{(N-k)!\,k!} \quad . \tag{7}$$

6) Asymptotic freedom and hence the inequality

$$\sum T(a)D(b)n_{ab} < 11\,N \quad , \tag{8}$$

where $T(a)$ is the trace of the product of the normalized generators given by

$$T(N^k) = \frac{(N-2)!}{(N-k-1)!\,(k-1)!} \quad .$$

The above stated assumptions have become more or less standard assumptions in considering a grand unified theory. If we can find solutions that satisfy the above criteria, we would obtain the total number of fermions classified into generations and the representation into which they fit. We can indeed do so. We can right away place a lower and an upper bound on N. From the requirement (6),

$$N \geq 5$$

and from the constraint (8)

$$N \leq 10 \quad .$$

Hence we have to examine only a limited number of cases. Let me illustrate by giving a few examples of solutions, relegating the details to be published elsewhere.

1) $N = 5$

This is the minimal $V - H$ grand unified scheme first studied by Davidson.[3] There are two solutions for F_L,

$$F_L^{(1)} : \; [(10,1) + (1,10)] + [(\bar{5},1) + (1,\bar{5})] \quad ,$$

which corresponds to the familiar Georgi-Glashow $SU(5)$ single generation scheme.[5] Extending it to $SU(5)_V \times SU(5)_H$ introduces additional singlets, but does not provide a generation structure. Such single generation solutions will occur for any N. The only other solution is

$$F_L^{(2)} : \; [(10,\bar{5}) + (\bar{5},10)] + (5,5) \quad .$$

This we consider as a non-trivial solution. Using Georgi's[6] ideas on separating the light fermionic generations from the heavy ones, we can easily show that the above solution embodies five light generations. Using linear combinations of the two solutions, we can increase or decrease the number of such light generations by one.

2) $N = 9$

In this case, besides the single generation solution, we have the interesting solution

$$(\bar{9},\bar{9}) + [(84,1) + (1,84)] \quad ,$$

which when analyzed in terms of its SU(5) contents, has exactly three light fermionic generations. This solution corresponds to that of Frampton (also Frampton and Nandi),[7] but without the superfluous "replication", namely, nine copies of the representation $\bar{9}$.

The above two examples illustrate the general nature of the solutions. Note that they involve more than one irreducible representation. In fact, one of the drawbacks of the simple SU(5) scheme is the contrived nature of the combination of two representations $\bar{5}$ and 10 to describe the single generation of the fermions. The question that naturally arises is *whether there is an* N *for which there is a single irreducible, complex and anomaly-free representation for which all our conditions are satisfied.* First of all, we observe that within the framework of SU(n) schemes, the anomaly free condition implies

$$A(N^k) = 0 \qquad \text{or} \qquad N = 2k \quad .$$

Hence, the representation is real. Thus our requirements cannot be satisfied within any simple SU(N) schemes. In the case of $SU(N) \times SU(N)$, the irreducible representations of relevance are

$$(a;b) + (b;a)$$

and the anomaly free requirement (6) reduces to

$$A(a)D(b) + A(b)D(a) = 0 \quad . \tag{9}$$

The condition (9) *combined with the asymptotic freedom constraint* (8) *leads to the unique value for* $N = 7$. The solution for this case is given by

$$F_L : (21,\bar{7}) + (\bar{7},21) \quad .$$

By analyzing the above representation into its SU(5) content, we find that the solution contains seven generations of light fermions. Simple SU(7) has its nice features as emphasized by the previous speaker,[8] but it does not restrict the number of generations. The semi-simple unification scheme based on $SU(7)_V \otimes SU(7)_H$ does provide such a restriction and is uniquely chosen from all possible N's, if we demand that all the fermions belong to a single irreducible representation.

REFERENCES

1. H. Georgi, Talk at this workshop.

2. A. Davidson and K.C. Wali, *Phys. Rev.* D21, 787 (1980). See also *Phys. Rev.* D20, 1195 (1979).

3. A. Davidson, *Phys. Lett.* 90B, 17 (1980).

4. M. Gell-Mann, P. Ramond, and R. Mansky, *Rev. Mod. Phys.* 50, 721 (1978).

5. H. Georgi and S.L. Glashow, *Phys. Rev. Lett.* 32, 438 (1974).

6. H. Georgi, *Nucl. Phys.* B156, 120 (1979).

7. P. Frampton, *Phys. Lett.* 89B, 352 (1980); P. Frampton and S. Nandi, *Phys. Rev. Lett.* 43, 1460 (1979).

8. A. Yildiz, Talk at this workshop.

NEUTRINO FLUCTUAT NEC MERGITUR: ARE FOSSIL NEUTRINOS DETECTABLE?*

A. De Rújula**

Center for Theoretical Physics
Laboratory for Nuclear Science and Department of Physics
Massachusetts Institute of Technology
Cambridge, MA 02139

Abstract

This is a brief report on work with S.L. Glashow.[1] The question is whether light (few eV to ~ 100 eV) neutrinos, left over from the big bang, are detectable. The answer is, perhaps. If the weak currents of leptons, like those of quarks, are not diagonal in mass eigenstates, a neutrino will decay into a lighter neutrino and a monochromatic photon. The corresponding photon line may be detectable provided: 1) Neutrinos are heavy enough to participate in galaxy clustering, 2) Neutrino lifetimes are, as in some weak interaction models, short enough.

1. FOSSIL NEUTRINOS

In the standard big bang cosmology photons outnumber baryons in our present universe by a factor $\sim 10^8$. These photons have been detected in their present disguise as $\sim 2.7°$ K red-shifted background radiation. Light ($m < 1$ MeV) weakly interacting neutrinos with lifetimes long on the scale of the Hubble time must also have survived in quantities that, for each neutrino flavor, outnumber baryons by ~ 8 orders of magnitude. A neutrino "flavor" is defined as a mass eigenstate. If uniformly distributed in the universe, the neutrino number density is predicted to be of the order of 100 cm^{-3} for each neutrino flavor.[2] Sufficiently heavy neutrinos ($m \gtrsim 20$ eV) may cluster on a galactic scale.[3] In this case, their number density in our neighborhood may be 3 to 4 orders of magnitude higher. A large cathedral may contain as many as 10^{18} neutrinos and yet, it is extraordinarily difficult to detect them. Our purpose is to describe a scenario in which not all hope is lost.

*This work is supported in part through funds provided by the U.S. Department of Energy (DOE) under contract DE-AC02-76ER03069.

**On leave from CERN, Geneva, Switzerland.

2. COSMOLOGICAL LIMITS ON NEUTRINO MASSES

In 1972 Cowsik and McClelland[2] pointed out that if the missing nonvisible mass (necessary to account for a universe that is close to closed) is attributed to neutrinos, there is an *upper* limit on the sum of their masses of order 40 eV. More recently much study has been devoted to the possibility that neutrinos are responsible for a possible missing mass in objects smaller than the universe.[3] The smallest objects of interest are galactic halos. The study of 21 cm emission by hydrogen in these halos gives an astonishing result:[4] the velocity of the rotating gas is of order 200-300 km/sec and *constant* (independent of the distance to the galactic center) out to distances an order of magnitude bigger than the visible galaxy, where the signal becomes too faint to detect. Newtonian mechanics $(Gm(r)/r^2 = v^2/r)$ implies that there is nonluminous mass in the halo with a mass density distribution $\rho(r) = r^{-2}$, and a total mass bigger than the visible mass. Some attribute this mass to defunct stars[5] that one way or another managed to populate the outskirts of galaxies. More fascinating is the possibility that the halo mass (or a fraction thereof) be due to neutrinos sufficiently heavy to have participated in galaxy clustering. This possibility has been discussed by Tremaine and Gunn,[3] who conclude that the *minimum* necessary neutrino mass (in the sense of $(\Sigma_\nu m_\nu^4)^{\frac{1}{4}}$, with ν running over flavors) is of the order of 24 eV.[3] To be fair, we must point out that these authors dismiss this possibility on grounds that the neutrino mass density (averaged over the universe) should be smaller than the visible density in galaxies (similarly averaged). We side with the experts who do not understand this constraint, and for whose proof of existence you must take our word.

Our attitude is to assume, as an ansatz, that neutrinos with masses in the general ballpark of 20 eV (give or take a factor of five) may be gravitationally bound to our own Milky Way and to other friendly neighborhood galaxies, like Andromeda. The neutrinos are partially or totally responsible for the large invisible mass in the galactic halos. We will presently compute neutrino lifetimes in the channel

$$\text{heavier neutrino } (m_1) \to \text{lighter neutrino } (m_2) + \text{photon} \qquad (2.1)$$

and conclude that there exist scenarios where neutrino-decay photons from the galactic halos may be detectable. In all scenarios the lifetimes of neutrinos, often dominated by channels other than Eq. (2.1), are a comfortable number of orders of magnitude larger than the Hubble time. Thus the heavier neutrinos are still around, awaiting discovery.

A very crucial point is that the photons of Eq. (2.1), originating close enough not to be significantly red-shifted, are monochromatic to within one part in a thousand. The escape velocity of neutrinos in our galaxy is of order $v/c \sim 10^{-3}$, and this would be the fractional width of the photon line in their radiative decay. The central photon energy in reaction (2.1) is

$$E_\gamma = \frac{m_1^2 - m_2^2}{2m_1} \quad .$$

For purposes of illustration we will consider the case $m_\nu = m_1 \gg m_2$, $E_\gamma \sim m_\nu/2$. For $m = 20$ eV the photons are VUV, and for $E_\gamma < 13.6$ eV, the galactic halo is quite transparent. *What we propose is a search for monochromatic* UV *photons coming from starless, dustless directions in our galactic halo, or from the halo of nearby galaxies.* The signal would be one or several lines that are absent from the atomic and molecular tables. The likelihood of not seeing anything, as well as the stakes, are high.

3. THREE ESTIMATES OF NEUTRINO LIFETIMES

In all models to be discussed, the weak interactions are as in the conventional $SU(2) \times U(1)$ weak theory. Our first model is the most pessimistic (longest neutrino lifetime) and the most standard, i.e., a model with the largest number of prejudices. Let there be three generations of leptons, and let the weak currents be the ones implied by the conventional left-handed doublets:

$$\begin{pmatrix} \nu_e \\ e \end{pmatrix}_L \quad \begin{pmatrix} \nu_\mu \\ \mu \end{pmatrix}_L \quad \begin{pmatrix} \nu_\tau \\ \tau \end{pmatrix}_L \tag{3.1}$$

If neutrinos have masses, the "weak eigenstates" ν_e, ν_μ, ν_τ need not coincide with the mass eigenstates ν_i, $i = 1,3$. The two bases are related by a unitary matrix U

$$(\nu_e, \nu_\mu, \nu_\tau) = (\nu_1, \nu_2, \nu_3)U \quad . \tag{3.2}$$

In this model radiative decays of neutrinos occur via the diagrams:

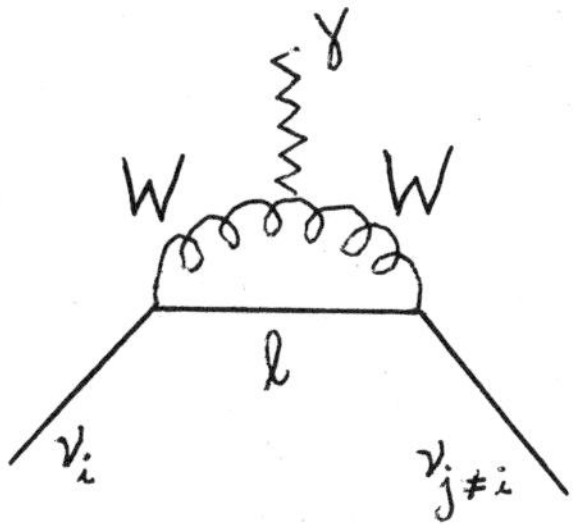

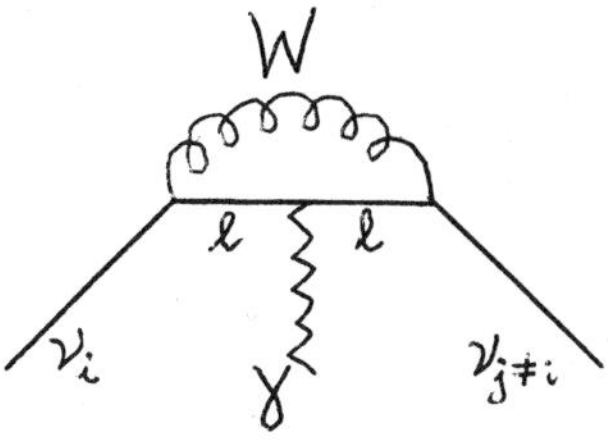

where ℓ is a charged lepton running over the three possibilities e, μ, τ. To present a simple expression for the neutrino lifetime, assume that, as is perhaps indicated by down to earth experiments,[6] ν_e and ν_τ mix with an angle β_1 ($\nu_\tau = \cos\beta_1 \nu_3 + \sin\beta_1 \nu_1$) and ν_μ does not mix significantly. The $\nu_3 \to \nu_1 \gamma$ lifetime is:[7]

$$\frac{1}{\tau} = \Gamma = \frac{G_F^2 m_\nu^5}{512\,\pi^4}\,\alpha\,(\sin^2 2\beta_1) I^2 \tag{3.3a}$$

$$I \equiv \frac{m_\tau^2}{M_W^2}\left(\ell n\,\frac{M_W^2}{m_\tau^2} + O(1)\right) - \frac{m_e^2}{M_W^2}\left(\ell n\,\frac{M_W^2}{m_e^2} + O(1)\right) \tag{3.3b}$$

Here we have assumed $m_\nu = m(\nu_3) \gg m(\nu_1)$. The factor I^2 is the "GIM" suppression: the decay is forbidden for $m_\tau = m_e$. Numerically, the lifetime implied by Eqs. (3.3) is rather long:

$$\tau_1 = 3.8 \times 10^{27} \text{ years } \frac{1}{\sin^2 2\beta_1}\left(\frac{30 \text{ eV}}{m_\nu}\right)^5 \tag{3.4}$$

Here the subindex 1 refers to our first model.

Our second model is a small departure from standard prejudices. Suppose there is an extra generation of leptons: $(\nu_\sigma, \sigma^-)_L$; and ν_σ mixes with someone among $(\nu_e, \nu_\mu, \nu_\tau)$. Equations (3.3) are still the lifetimes for the decay of the heavier neutrino, but now m_τ is substituted by m_σ in Eq. (3.3b), and we expect $m_\sigma > 18$ GeV $\sim 10 m_\tau$. Equation (3.3b) is only correct for $m_\sigma^2 \ll M_W^2$. For $m_\sigma \approx M_W$, $I = O(1)$ and there is no suppression by small mass ratios. For $m_\sigma^2 \gg M_W^2$ the lifetime is again suppressed by a power $(M_W/m_\sigma)^4$. The maximum lifetime in this model is of order:

$$\tau_2 = \left(\frac{G_F^2 m_\nu^5}{512\,\pi^4}\,\alpha \sin^2 2\beta_2\right)^{-1}$$

$$= 5 \times 10^{22} \text{ years } \frac{1}{\sin^2 2\beta_2}\left(\frac{30 \text{ eV}}{m_\nu}\right)^5 \tag{3.5}$$

A third model is based on the possibility that there exist more neutrinos than charged leptons. Just to give an example, consider the structure

$$\begin{pmatrix} \frac{\nu_1 + \nu_2}{\sqrt{2}} \\ \\ e \end{pmatrix}_L \begin{pmatrix} \nu_\mu \equiv \nu_4 \\ \\ \mu \end{pmatrix}_L \begin{pmatrix} \sin\beta_3 \left(\frac{\nu_1 - \nu_2}{\sqrt{2}}\right) + \cos\beta_3 \nu_3 \\ \\ \tau \end{pmatrix}_L$$

$$\cdot \left(\cos\beta_3 \frac{\nu_1 - \nu_2}{\sqrt{2}} - \sin\beta_3 \nu_3\right)$$

It is easy to convince oneself that in models of this kind the decays $\mu \to e\gamma$, $\tau \to e\gamma$ are conveniently GIM-suppressed, while decays as $\nu_3 \to \nu_1\gamma$ are not. The lifetime for the latter decay is

$$\tau_3 = \left(\left(\frac{25}{36}\right) \frac{G_F^2 m_\nu^5}{512\,\pi^4} \alpha \sin^2 2\beta_3\right)^{-1}$$

$$= [6 \times 10^{22} \text{ years}] \frac{1}{\sin^2 2\beta_3} \left(\frac{30 \text{ eV}}{m_\nu}\right)^5 \tag{3.6}$$

In models of type 3 a heavy neutrino also decays into three lighter ones with a shorter lifetime:

$$\tau(\nu_3 \to 3\nu\text{'s}) \sim \left(\frac{G_F^2 m_\nu^5}{192\,\pi^3} \cos^4\theta_W\right)^{-1}$$

$$= 6 \times 10^{19} \text{ years} \left(\frac{30 \text{ eV}}{m_\nu}\right)^5 \tag{3.7}$$

The branching ratio $[\nu_3 \to \nu_1\gamma]/[\nu_3 \to 3\nu\text{'s}]$ is small, but the overall lifetime is still long enough not to have depleted the universe of the heavier neutrinos.

We have not been able to construct sensible models of elementary neutrinos of mass ~ 30 eV and lifetimes in the interesting channel $(\nu' \to \nu\gamma)$ much shorter than 10^{22} years. With such lifetimes, at most one radiative decay would occur in a large cathedral every 500 centuries. Clearly, we need a bigger source, such as a galactic halo.

4. THE FLUX OF PHOTONS FROM NEUTRINO DECAYS IN GALACTIC HALOS

The gas velocities in our galactic halo are ~ 250 km/sec, independent of distance to the galactic center. This implies an energy density of material

$$\rho(r) = Ar^{-2} \tag{4.1a}$$

$$A = 1.3 \times 10^{75} \text{ eV/kpc}. \tag{4.1b}$$

Not all of this mass density can be neutrinos, for after all, the "visible" mass of the galaxy is of order 10^{77} eV. Also, we do not expect the neutrino density to rise toward the center as Eq. (4.1a) for $r < r_0 = 8.5$ kpc, our distance to the galactic center. To make an estimate of the flux of photons from neutrino decays in the halo, we assume there are no neutrinos inside a sphere of radius r_0, subtract the visible mass and ascribe the missing mass necessary to explain the gas velocity curves to a spherical halo of neutrinos at $r > r_0$. With this or similar geometries, the number of photons per second incident on a sphere of 1 cm^2 cross section at the position of the earth is of order:

$$n_\gamma [cm^{-2}\ s^{-1}] = 5 \times 10^{29} \frac{1}{\tau(\text{sec})} \left(\frac{30 \text{ eV}}{m_\nu}\right) \tag{4.2}$$

For $\tau = \tau_1$, the lifetime in the first model we discussed,

$$n_\gamma \simeq 4 \times 10^{-6} \left(\frac{m_\nu}{30 \text{ eV}}\right)^4 \sin^2 2\beta_1 \text{ cm}^{-2} \text{ s}^{-1} \quad . \tag{4.3}$$

Even for $m_\nu = 100$ eV and maximal mixing, this is a rather smallish signal. For $\tau = \tau_2 \simeq \tau_3$, corresponding to the other models of neutrino decay, the photon flux increases to

$$n_\gamma \simeq 0.3 \left(\frac{m_\nu}{30 \text{ eV}}\right)^4 \sin^2 2\beta_3 \text{ cm}^{-2} \text{ s}^{-1} \tag{4.4}$$

that is, a monochromatic photon every few seconds, for $m_\nu \sim 30$ eV. For $m_\nu = 100$ eV and maximal mixing, this increases to a few photons per second. It is not inconceivable to extract such fluxes of UV photons, monochromatic at the one per thousand level, from the background in the UV region. The angular distribution of photons is not in general isotropic, and depends on the details of the neutrino density distribution in the galaxy. Let θ be the angle between the line of observation and the center of the galaxy. The angular distribution of photons, down to angles at which $\rho(r)$ may still be assumed to behave as r^{-2}, is

$$\frac{dn_\gamma}{d\cos\theta} = \frac{\pi - \theta}{\sin\theta} \quad . \tag{4.5}$$

As one may expect, it pays to look as close as possible toward the galactic center.

The detection of neutrino-decay photons from halos of nearby galaxies has the advantage of directionality, that may compensate for the disadvantage of looking at a distant source. From the halo of Andromeda, 650 kiloparsecs distant, we would expect a photon flux

$$n_\gamma[\text{model 1}] \sim 6\times 10^{-9} \left(\frac{m_\nu}{30\ \text{eV}}\right)^4 \sin^2 2\beta_1 \ \ \text{cm}^{-2}\ \text{s}^{-1} \tag{4.6a}$$

$$n_\gamma[\text{models 2,3}] \sim 5\times 10^{-4} \sin^2 2\beta_2 \left(\frac{m_\nu}{30\ \text{eV}}\right)^4 \text{cm}^{-2}\ \text{s}^{-1} \tag{4.6b}$$

This is not a large flux, even in the most optimistic scenario $m_\nu \sim 100$ eV, $\beta_2 \sim 45°$.

Another scenario is the one where neutrinos, perhaps because they are too light, have not participated in individual galaxy clustering. They may be in halos of clusters of galaxies, or even uniformly distributed in the universe. In this latter case, photons from neutrino decays will reach us from the universe as a whole, but will be red-shifted when the decay occurred far away. The advantage of a monochromatic line is lost, and substituted by an energy distribution characteristic of the model of the universe. The total number of photons (integrated up to the maximal energy $m_\nu/2$) is of the order of magnitude

$$n_\gamma \sim \frac{\rho}{\tau}\frac{c}{H_0} \quad , \tag{4.7}$$

where H_0 is Hubble's constant. For $\rho = 100\ \text{cm}^{-3}$, $H_0 = 50\,\text{km s}^{-1}\ \text{Mpc}^{-1}$, and $\tau = \tau_2 \simeq \tau_3$

$$n_\gamma \simeq 1 \left(\frac{m_\nu}{30\ \text{eV}}\right)^5 \text{cm}^{-2}\ \text{s}^{-1} \quad . \tag{4.8}$$

If $m_\nu = 6$ eV, $n_\gamma \sim 3\ \ 10^{-4}\ \text{cm}^{-2}\ \text{s}^{-1}$, a small flux of visible nonmonochromatic light, hard to disentangle from the background. If for some reason heavier neutrinos are uniformly distributed in the universe, the flux is larger and the background smaller: perhaps the hope for seeing the red-shifted photon radiation from neutrino decay is not completely lost.

We conclude that the flux of photons from radiative decays of neutrinos bound to our galaxy or nearby galaxies is small, even in optimistic scenarios.

But the UV photon line or lines are very monochromatic and could conceivably be extracted from the background. Unfortunately, we do not know the masses of neutrinos and the precise location of the line or lines. In the standard cosmology, the lines would correspond to photon energies anywhere up to half the maximum acceptable neutrino mass. The hope to detect neutrino radiative decay increases with the fourth power of neutrino mass. We have not found it impossible to corner cosmologists into accepting $m \sim 100$ eV, or even a bit more.

Acknowledgements. We are indebted to C. Canizares, R. Jaffe, C. Papaliolios, E. Purcell, M. Rees, and G. Steigman for discussions. We are also indebted to P. Frampton and A. Yildiz for discussions, and for organizing the Workshop in whose pleasant atmosphere the considerations in this note developed.

REFERENCES

1. A. De Rújula and S.L. Glashow, to be published.

2. R. Cowsik and J. McClelland, *Phys. Rev. Lett.* 29, 669 (1972).

3. S. Tremaine and J.E. Gunn, *Phys. Rev. Lett.* 42, 407 (1979); D.N. Schramm and G. Steigman, to be published; G. Steigman, in *Proc. Conf. on Astrophysics and Particle Physics: Common Problems*, Academia dei Lincei, Rome, Feb. 1980.

4. For a review, see S.M. Faber and J.S. Gallagher, *Ann. Rev. Astron. Astrophys.* 17, 135 (1979).

5. M. Rees, Loeb Lectures, Harvard University, April 1980.

6. A. De Rújula et al., CERN Preprint TH2788, Nov. 1979.

7. T. Goldman and J.G. Stephenson, *Phys. Rev.* D16, 2256 (1977); S. Pakvasa and K. Tennakone, *Phys. Rev. Lett.* 28, 1415 (1972).

EXPECTATIONS FOR BARYON AND LEPTON NONCONSERVATION*

Steven Weinberg

Lyman Laboratory of Physics, Harvard University
and
Harvard-Smithsonian Center for Astrophysics
Cambridge, Massachusetts 02138

Are baryon and lepton conservation actually violated in nature? At this moment, we do not know. Nevertheless, it seems a good idea to try to anticipate the details of baryon or lepton nonconserving processes, so that we can at least know what to look for, and what may be learned if it is found. In this talk, I will outline the expected properties of baryon and lepton nonconserving processes, taking as a guide just the strong and electroweak SU(3) × SU(2) × U(1) gauge symmetries and some plausible dimensional analysis. I will also describe a recent calculation of superheavy particle masses, and will have a few comments on the constraints imposed by cosmology on the possible modes of nucleon decay. But before getting into details, I would like to try to set the stage, by describing how our views of baryon and lepton conservation have been shaped by changes in our views about symmetries in general.

We used to believe in all sorts of fundamental conservation laws. In addition to those that follow from exact symmetry principles like Lorentz invariance, CPT, and electromagnetic gauge invariance, there were a great many approximate conservation laws, for isospin, strangeness, chiral SU(3) × SU(3), C, P, T, and so on. But with the advent of modern gauge theories, our attitude to these conservation laws has changed. In quantum chromodynamics, one can prove that the strong interactions *must* conserve charge conjugation, strangeness, charm, bottom, etc., and (assuming small u, d, s masses) isospin and chiral SU(3) × SU(3) as well. The color gauge symmetry plus renormalizability simply do not allow the Lagrangian of the strong interactions to be complicated enough to violate these conservation laws. (The same applies to P and T conservation, apart from problems concerning instantons which have not yet been resolved.) And in the gauge theory of weak and electromagnetic interactions, the V and A semileptonic currents must be linear combinations of the Noether currents of SU(3) × SU(3). Thus all the old approximate conservation laws have been demoted, from fundamental principles that must be assumed in advance, to dynamical consequences that can be derived from gauge symmetries and renormalizability.

One is tempted to conclude that the only truly fundamental conservation laws left in physics are the gauge symmetries: Lorentz invariance and SU(3) × SU(2) × U(1). But then what about baryon and lepton conservation? They

*Research is supported in part by the National Science Foundation under Grant No. PHY77-22864.

appear to be exact and unbroken, but they are surely not exact unbroken gauge symmetries, because we do not see the effects of a long-range vector field coupled to baryon or lepton number.[1] The peculiar position of baryon and lepton conservation in today's physics suggests that baryon and lepton conservation may go the way of the other non-gauge symmetries, and turn out to be only approximate consequences of the gauge symmetries and renormalizability.

It is not hard to see how this may happen. The SU(3) × SU(2) × U(1) gauge symmetries do not allow any renormalizable baryon or lepton number violating interactions among "ordinary" particles: quarks, leptons, $W^{\pm}$, Z^0, and scalar doublets. (With all ordinary boson fields assigned baryon and lepton number zero, the only renormalizable baryon violating interactions would involve a pair of quarks, or a quark and lepton or antilepton, either by themselves or coupled to a boson. But these fermion pairs can only be color triplets or sextets, and the ordinary bosons are all color singlets or octets. The only remaining renormalizable lepton violating interactions would have to involve a pair of leptons, but in the absence of SU(2) × U(1)-neutral neutrinos, these also would be forbidden, by SU(2) × U(1).) This does not mean that baryon and lepton number are necessarily strictly conserved, but only that their violation would require the presence of new particles with exotic SU(3) × SU(2) × U(1) transformation properties. Any such particles would presumably have to be very heavy, or else they would already have been seen, so the resulting baryon or lepton nonconservation would have to be correspondingly weak. The SU(3) × SU(2) × U(1) gauge symmetries, by explaining the approximate conservation of baryon and lepton number, have opened up the possibility that these conservation laws are not exact.

Fortunately, it is possible to go pretty far in analyzing the varieties of possible baryon and lepton nonconservation, without knowing anything about the superheavy exotic particles. When we "integrate out" the superheavy particles, we obtain an effective Lagrangian involving only the ordinary particles. The effective Lagrangian contains an infinite number of arbitrarily complicated interactions, subject only to the constraint of SU(3) × SU(2) × U(1) invariance, and if we do not make specific assumptions about the superheavy particles we have to regard the coefficients of these interactions as an infinite set of unknown parameters. Nevertheless, with a little reasonable dimensional analysis, we can draw useful conclusions. An effective interaction operator that has dimensionality $[\text{mass}]^d$ (with $\hbar = c = 1$) must have an associated coupling constant g with dimensionality $[\text{mass}]^{4-d}$. If the superheavy exotic particles all have masses of the same order, say M, then g will be of order M^{4-d}. (Other factors in g will be considered later.) Hence for $d > 4$ the effect of such a coupling will be suppressed at an energy $E \ll M$ by a factor $(E/M)^{d-4}$. The dominant interactions at low energies will be those with the smallest dimensionality. Each additional field operator or spacetime differentiation increases the dimensionality of the interaction, so the dominant interactions are those which are as simple as possible.

We have already seen that there are no interactions with $d \le 4$ (i.e., renormalizable interactions) among ordinary particles that violate baryon or lepton conservation. Thus baryon and lepton nonconservation will always be suppressed by powers of E/M, which is presumably why baryon and lepton nonconserving processes have not yet been observed.

The baryon-violating operators of minimum dimensionality are four-fermion operators QQQL or $QQQ\bar{L}$ with $d = 6$. (Here Q and L stand for any quark and lepton.) These were cataloged last year by Wilczek and Zee and myself; the complete list of SU(3) × SU(2) × U(1) invariant operators with $d = 6$ is as follows:[2]

$$(d_R, u_R)(q_{iL}, \ell_{jL})\varepsilon_{ij}$$

$$(q_{iL}, q_{jL})(u_R, e_R)\varepsilon_{ij}$$

$$(q_{iL}, q_{jL})(q_{kL}, \ell_{\ell L})\varepsilon_{ij}\ \varepsilon_{k\ell}$$

$$(q_{iL}, q_{jL})(q_{kL}, \ell_{\ell L})(\varepsilon\vec{\tau})_{ij}(\varepsilon\vec{\tau})_{k\ell}$$

$$(d_R, u_R)(u_R, e_R)$$

$$(u_R, u_R)(d_R, e_R)\ ,$$

with

$$q_{iL} = \begin{pmatrix} u_L \\ d_L \end{pmatrix} \qquad \ell_{iL} = \begin{pmatrix} \nu_L \\ e_L \end{pmatrix}$$

In the abbreviated notation used here, u, d, ν, and e stand for generic quarks and leptons; generation and color indices are omitted; i, j, k, ℓ are SU(2) indices; ε_{ij} is the antisymmetric SU(2) tensor with $\varepsilon_{12} = 1$; and (a,b) is the Lorentz scalar product $(\overline{a^c}b) = (\overline{b^c}a)$ of two spin-½ fields a and b. (It should be understood that the quarks in each interaction may be of the same or different generations; thus each d may independently be a d or s or b.)

Inspection of these operators immediately reveals a number of general properties of nucleon decay processes:[2,3]

(A) $\Delta B = \Delta L$

e.g. $p \to e^+\pi^0$ and $n \to e^+\pi^-$ but $n \not\to e^-\pi^+$

(B) $\Delta S/\Delta B \le 0$

e.g. $p \to K^+\bar{\nu}$ but $n \not\to K^-e^+$

(C) $\Delta I = 1/2$

e.g. $\Gamma(p \to e_R^+\pi^0) = \frac{1}{2}\Gamma(n \to e_R^+\pi^-) = \frac{1}{2}\Gamma(p \to \bar{\nu}\pi^+) = \Gamma(n \to \bar{\nu}\pi^0)$

$\Gamma(p \to e^+\pi^0) = \frac{1}{2}\Gamma(n \to e^+\pi^-)$

$$\Gamma(p \to e_R^+ X) = \Gamma(n \to \bar{\nu} X)$$

$$\Gamma(n \to e_R^+ X) = \Gamma(p \to \bar{\nu} X)$$

(D) Universal Antilepton Polarization

If we also assume that the nucleon decay interaction arises from the exchange of a superheavy *vector* boson, then only the first two of the $d=6$ operators can contribute. In this case the antilepton polarization (measurable for μ^+, if not for e^+) is universal, depending only on whether $\Delta S = 0$ or $\Delta S = 1$. Indeed, in this case specific models can be distinguished *only* by $\bar{\nu}/e^+$, e^+/μ^+, and $\Delta S = +1 \big/ \Delta S = 0$ ratios and by parity violating effects like lepton longitudinal polarizations. Within the general context described here, all other ratios like $\Gamma(p \to e^+\pi^0)/\Gamma(p \to e^+\rho^0)$ will be the same in all models.

What about the lifetime for proton decay? If proton (or bound neutron) decay is due to exchange of a superheavy particle of mass M, then we expect the coupling constant to be of order (recall that $d=6$)

$$g \approx e^2/M^2$$

and the lifetime then on dimensional grounds should be of order[4]

$$\tau_p \simeq 1/g^2 m_p^5 \simeq M^4/e^4 m_p^5 \ .$$

This gives $\tau_p = 10^{30}$ years for $M = 7 \times 10^{14}$ GeV. More detailed calculations[5] lower the value of M that gives this proton lifetime, to $(2 \text{ to } 6) \times 10^{14}$ GeV in place of 7×10^{14} GeV. Taking this for definiteness as 4×10^{14} GeV, we have

$$\tau_p \simeq 10^{30} \text{ years} \times \left(\frac{M}{4 \times 10^{14}\,\text{GeV}}\right)^4 \ .$$

Hence τ_p will probably be in the experimentally interesting range[6] of 10^{30} to 10^{33} years if M is between 4×10^{14} GeV and 2×10^{15} GeV.

This is of course an enormous mass, but there are at least two grounds for taking seriously the possibility that such particles actually exist. First, we know that new degrees of freedom must enter at energies of order 10^{19} GeV or less, because at these energies gravitation becomes a strong interaction, and we can no longer ignore its ultraviolet divergences. Second, the strong coupling constant g_s is much larger than the electroweak couplings g, g' at ordinary energies and decreases only like $1/\sqrt{\ln E}$ as the energy E increases, so we must go up to enormous energies before we reach a scale at which strong and electroweak couplings can be comparable.[7]

The last point was developed quantitatively in Ref. 7 for a broad class of "grand unified" models.[8] Suppose that there is a simple group G which breaks down at an energy of order M to a direct product $SU(3) \times SU(2) \times U(1)$ [perhaps with extra factors G' which commute with $SU(3) \times SU(2) \times U(1)$.] Suppose also that all (or almost all) fermions either fall into generations $[q_L\, u_R\, d_R\, \ell_L\, e_R]$

like the observed quarks and leptons, or else are SU(3) × SU(2) × U(1)-neutrals. Reference 7 showed that in all such models, M is given by

$$\ln \frac{M}{m} = \frac{\pi}{11}\left[\frac{1}{\alpha(m)} - \frac{8}{3\alpha_s(m)}\right] + O(1) \simeq 35$$

and the Z^0-γ mixing parameter is

$$\sin^2\theta = \frac{1}{6} + \frac{5}{9}\frac{\alpha(m)}{\alpha_s(m)} + O(\alpha) \simeq 0.2$$

where m is some ordinary mass, conveniently chosen to be of order 100 GeV, and $\alpha(m)$ and $\alpha_s(m)$ are the electromagnetic and strong fine structure constants measured at m. In particular, these results hold also for models [e.g. of the $SU(4)^4$ type] which do not contain SU(5), SO(10), etc., even as gauge subgroups.

One would like also to calculate the O(1) terms in $\ln M/m$ and the $O(\alpha)$ terms in $\sin^2\theta$. Methods for carrying out such calculations have been developed by Goldman and Ross and Marciano.[9] In their approach renormalized couplings are defined *à la* Gell-Mann and Low as functions of mass scale in terms of off-shell Greens functions, and the Appelquist-Carazzone theorem[10] is invoked to justify the neglect of heavy exotic particles at energies $E \ll M$. As far as I know, there is nothing wrong with this approach, but it is complicated to carry out.

Recently I have developed a somewhat different approach to this sort of calculation:[11] the superheavy particles are integrated out at the beginning of the calculation, and we work with the resulting effective Lagrangian.[12] There is no need here to use the Appelquist-Carazzone theorem, so we can make use of the calculational simplicity of the "minimum subtraction" definition of coupling constants.[13] This in turn introduces a further simplification: the non-renormalizable couplings in the effective theory have no effect on the renormalization group equations for the renormalizable couplings,[14] so we can work with the renormalizable part of the effective Lagrangian and ignore the rest.

In order to use this method to calculate the O(1) terms in $\ln M/m$ and the $O(\alpha)$ terms in $\sin^2\theta$, one follows a three-step plan:

(1) First integrate out the superheavy particles using *one-loop* Feynman diagrams. (There is a technical problem here, of maintaining the gauge invariance of the effective Lagrangian despite the need to fix a gauge in integrating out the superheavy gauge bosons, but this can be dealt with by a method akin to the old "background field method."[15]) For arbitrary group G with gauge coupling $g(\mu)$ that breaks up into subgroups G_i with couplings $g_i(\mu)$, the general result is that (for μ comparable with M)

$$g_i(\mu) = g(\mu) + \frac{g(\mu)^3}{96\pi^2}\left[-21\,\mathrm{Tr}\left(t_{iV}^2 \ln\left(\frac{e^{-1/21}M_V}{\mu}\right)\right) + 8\,\mathrm{Tr}\left(t_{iF}^2 \ln\frac{\sqrt{2}M_F}{\mu}\right) + \mathrm{Tr}\left(t_{iS}^2 \ln\frac{M_S}{\mu}\right)\right] .$$

The three trace terms here come from loop diagrams involving superheavy vector bosons, spin-½ fermions, and scalar bosons (excluding Goldstone bosons), respectively; t_{iV}, t_{iF}, and t_{iS} are the corresponding matrix representations of any one of the generators T_i of G_i (using a canonical normalization with totally antisymmetric structure constants); and M_V, M_F, and M_S are the corresponding mass matrices. Note the factor 21 in the first term; this indicates that it is the vector boson masses that are the most important in such calculations.

(2) The *two-loop* renormalization group equations must then be integrated down to μ of order $m \simeq 100$ GeV, using the above result as an initial condition at $\mu \approx M$.

(3) The results for $g_i(m)$ must then be compared with experiment, taking radiative corrections into account to *one-loop* order.

This program has been carried through by L. Hall.[16] In addition to the assumptions described above, Hall for definiteness also assumes that the superheavy vector bosons are degenerate with common mass M_V; the scalar doublet (ϕ^+,ϕ^0) comes from any representation of G, all of whose other members are degenerate with superheavy mass M_H; the scalar fields whose vacuum expectation values break G form a degenerate representation of G with common superheavy mass M_S, aside from Goldstone bosons; and there may in addition be any number of superheavy fermions and additional scalars degenerate with M_V. His result can be expressed as

$$\ln\left(\frac{M_V}{80\text{ GeV}}\right) = \left[36.06 - \frac{0.750}{\alpha_s(80\text{ GeV})}\right] - \left[0.59\right] + \left[0.045 + 0.015\ \ln\frac{M_V}{M_H} + 0.045\ \ln\frac{M_V}{M_S}\right] .$$

The three bracketed terms here arise respectively from (1) the old result of Ref. 7, corrected[9] by including a scalar doublet and using $\alpha(80\text{ GeV}) = 1/128.2$, (2) two-loop corrections to the renormalization group equations used between $\mu \approx m$ and $\mu \approx M_V$, and (3) the one-loop terms in the calculation of the effective Lagrangian. It should be noted that the coefficients of the $\ln M_V/M_H$ and $\ln M_V/M_S$ terms are very small, due to the enhancement of the vector boson loops over the scalar boson loops by the factor of 21 that has been previously mentioned. As a result, we can take $M_H \approx M_S \approx M_V$, without much risk of serious numerical error. (For the same reason, the results are not very sensitive to one's assumptions about the superheavy scalars.) Using a two-loop calculation of $\alpha_s(m)$ in terms of a four-quark QCD scale factor Λ, Hall finds in this way that for a wide range of possible values of Λ,

$$M_V = 1.5 \times 10^{15}\,\Lambda\ .$$

The scale factor Λ is believed to be of the order of several tenths of a GeV, so it is quite plausible that M_V may be in the range of 4 to 20 times 10^{14} GeV which would give a proton decay rate fast enough to observe and yet slow enough not to have been already observed.[6]

Apart from nucleon decay, I know of only one other observable effect that might be produced by exotic particles with masses of order 10^{15} GeV. There is an SU(3) × SU(2) × U(1)-invariant operator[17] that violates lepton number and has the relatively low dimensionality $d = 5$:

$$\tfrac{1}{2}(\ell_{Li}\,\ell_{Lj})\phi_k\,\phi_\ell\,\varepsilon_{ik}\,\varepsilon_{j\ell}\ .$$

If this operator appears in the effective Lagrangian with coupling constant g_5, then the spontaneous breakdown of SU(2) × U(1) will produce a neutrino mass matrix

$$m_\nu = g_5 \langle\phi^0\rangle^2\ .$$

(There is no need for the lepton doublets here to be of the same generation, so g_5 and m_ν are in general non-diagonal matrices.) According to our previous estimates, we expect g_5 to be of order 1/M, but it may involve coupling constant factors as well. For instance, in any theory with superheavy Majorana SU(3) × SU(2) × U(1) singlets N_R, we may expect Yukawa couplings

$$\frac{m_f}{\langle\phi^0\rangle}\,\phi_i\,(\overline{N}_R\,\ell_{jL})\varepsilon_{ij}$$

with m_f some mass matrix whose elements are comparable with ordinary quark and lepton masses. The N_R-exchange tree graph then generates an $\ell_L\ell_L\phi\phi$ effective interaction, with coupling $g_5 = (m_f/\langle\phi^0\rangle)^2/M_N$, so the neutrino mass matrix is

$$m_\nu = m_f^2/M_N\ .$$

For example, with $M_N = 10^{15}$ GeV and $m_f \approx 1$ MeV to 10 GeV, one finds neutrino masses of order 10^{-12} eV to 10^{-4} eV. Masses near the high end of this range could produce significant solar neutrino oscillations.[31]

The neutrino mass found in SO(10) models[18] arises in just this way. It is usual in these models to calculate the neutrino mass by diagonalizing a ν_L-N_R^C mass matrix of form

$$\begin{pmatrix} 0 & m_f \\ m_f & M_N \end{pmatrix}$$

but it is equivalent to think of the mass as arising from a $\phi\phi\nu_L\nu_L$ interaction generated by N_R exchange, and the results are the same.

It should be emphasized that these neutrino mass estimates depend crucially on the assumption that there really are elementary scalar fields. In theories based on dynamical symmetry breaking, the ϕ's would have to be replaced with bilinear $\overline{F}F$ operators formed from quarks or leptons that carry the quantum numbers of "extra-strong" or "technicolor" or "hypercolor" interactions.[19] The lowest-dimensional lepton-nonconserving interaction would then be of the form $\ell_L\ell_L\overline{F}F\overline{F}F$. This has $d = 9$, so its coupling constant is at most of order $1/M^5$, and it gives a neutrino mass at most of order $(300\text{ GeV})^6/M^5$. For $M = 10^{15}$ GeV, the neutrino would have a negligible mass, at most of order 10^{-51} eV!

Whether or not there really are exotic particles with masses of order 10^{15} GeV, we clearly cannot now rule out the possibility that there are also other "medium superheavy" exotic particles with masses in the enormous range from 100 GeV to 10^{15} GeV, in which case observable levels of baryon and lepton nonconservation could arise from effective interactions with $d > 6$. In fact, even assuming that there is a simple grand unified group which suffers a spontaneous breakdown at $M \approx 10^{15}$ GeV, there are at least two possible natural mechanisms for also producing such medium-superheavy masses:

(1) In theories in which a spontaneous symmetry breaking arises from the vacuum expectation values of elementary scalars, it is not uncommon to find that the zeroth order solutions respect certain "accidental" mass relations, which are then broken by higher-order corrections. If these accidental mass relations require some of the fermion masses to vanish, then higher-order corrections would give them masses of order M times some low power of α. ("Accidental" mass relations can in general arise because the scalars belong to only a limited set of irreducible representations of the symmetry group, so that the zeroth order masses satisfy linear relations whatever values the scalar vacuum expectation values may take.[20] It is also possible for such mass relations to arise because any quartic polynomial function of the scalar fields in the Lagrangian would satisfy accidental symmetries, which are not entirely broken in zeroth order, and which constrain the scalar field vacuum expectation values.[21] Witten has given an interesting example[22] of the first type; the scalar fields in an SO(10) model are limited to just the $\underline{10}$, $\underline{16}$, and $\underline{45}$ representations and in consequence the right-handed $SU(3) \times SU(2) \times U(1)$-singlet neutrino gets no mass in zeroth order from the spontaneous breakdown of SO(10), but gets a mass of order $(\alpha/\pi)^2 M$ from two-loop corrections. In this case, the light neutrino mass that is generated by exchange of the heavy neutrino is larger than the estimate given above, by a factor of order $(\pi/\alpha)^2$.)

(2) As already mentioned, it is possible that the grand gauge group breaks at a scale M not into $SU(3) \times SU(2) \times U(1)$, but into $G' \times SU(3) \times SU(2) \times U(1)$, where G' is any group that commutes with $SU(3) \times SU(2) \times U(1)$. If G' contains simple factors larger than SU(3), then the corresponding gauge coupling constants can reach large values at energies very much larger than the 500 MeV or so where the QCD coupling becomes strong. For instance, if SU(L) breaks spontaneously at $M = 10^{15}$ GeV into $SU(N) \times SU(3) \times SU(2) \times U(1)$, and we neglect fermion and scalar terms in the renormalization group equations, then g_N becomes of order unity at energies of order $(\Lambda_{QCD}/M)^{3/N} M$, or 300 GeV for $N = 4$; 7×10^5 GeV for $N = 5$; 2×10^7 GeV for $N = 6$; etc. Spontaneous symmetry breaking can generate masses comparable to the scales at which any factors of G' become strong. Indeed, a medium superheavy scale of order 300 GeV *must* arise in this sort of way[19] if the $SU(2) \times U(1)$ symmetry is to be broken dynamically, rather than by the vacuum expectation values of scalar fields.

Let us suppose then that medium superheavy particles do exist. What sorts of baryon and lepton nonconservation could they produce? This has been explored lately in a variety of specific models.[23] In what follows, I will describe a recent model-independent operator analysis;[24] a similar analysis has also been carried out by Weldon and Zee.[25]

Medium superheavy particles with masses $M \ll 10^{15}$ GeV can produce effective baryon or lepton nonconserving interactions which have $d \geq 7$, and hence are suppressed by three or more factors of E/M, and yet are strong enough to be observable. To catalog all the $SU(3) \times SU(2) \times U(1)$-invariant operators of some high dimensionality is tedious, but the work can be considerably lightened by introducing a convenient multiplicative quantum number, which I call F-parity. It has the values:

quarks and leptons:	+1
antiquarks and antileptons:	-1
bosons ($W^{\pm}$, Z^0, γ, ϕ):	-1
spacetime derivatives:	-1

The multiplicative conservation of F-parity is a simple consequence of Lorentz invariance and weak SU(2). [Proof: it may be easily checked that these F-values are equal to $(-)^{2A+2T}$, where A is the first of the two integers or half-integers that defines the transformation of the field under the homogeneous Lorentz group,[26] and T is the weak SU(2) isospin. Lorentz invariance and weak SU(2) require that $(-)^{2A}$ and $(-)^{2T}$ are multiplicatively conserved, respectively.] We have already seen an example of F-parity conservation at work: the $\Delta B = \Delta L$ operators QQQL have F even, and are allowed, while $\Delta B = -\Delta L$ operators $QQQ\bar{L}$ have F odd, and are forbidden.

Now let's look at the possible "abnormal" baryon or lepton violating interactions that might be produced by medium superheavy particles of mass $M \ll 10^{15}$ GeV.

<u>$\Delta B = -\Delta L$ ($n \to e^- \pi^+$, etc.)</u>

To conserve F-parity, we have to add an odd number of boson fields or spacetime derivatives to the F-odd operators $QQQ\bar{L}$. The minimum dimensionality operators then either contain one scalar field, or one derivative. A complete list of these $d = 7$ operators is as follows:

$$\varepsilon_{ij}\,\varepsilon_{k\ell}\,(\overline{q^c_{Li}}\,q_{Lj})\,(\overline{\ell_{Lk}}\,d_R)\,\phi^\dagger_\ell$$

$$(\overline{q^c_{Li}}\,q_{Lj})\,(\overline{\ell_{Lj}}\,d_R)\,\phi^\dagger_i$$

$$(\overline{d^c_R}\,d_R)\,(\overline{e_R}\,q_{Li})\,\phi^\dagger_i$$

$$\varepsilon_{ij}\,(\overline{d^c_R}\,d_R)\,(\overline{\ell_{Li}}\,u_R)\,\phi^\dagger_j$$

$$\varepsilon_{ij}\,(\overline{d^c_R}\,u_R)\,(\overline{\ell_{Li}}\,d_R)\,\phi^\dagger_j$$

$$(\overline{d^c_R}\,d_R)\,(\overline{\ell_{Li}}\,d_R)\,\phi_i$$

$$(\overline{d^c_R}\,D_\mu\,d_R)\,(\overline{\ell_{Li}}\,\gamma^\mu\,q_{Li})$$

$$(\overline{\ell_{Li}\, D_\mu\, d_R})(\overline{d^c_R}\,\gamma^\mu\, q_{Li})$$

$$(\overline{d^c_R}\, D_\mu\, d_R)(\overline{e_R}\,\gamma^\mu\, d_R)$$

(We continue to omit color and generation indices, but we are now using a conventional Dirac notation, with c indicating the Lorentz invariant complex conjugate. Also, D_μ is the SU(3) × SU(2) × U(1)- gauge-covariant derivative.)

We (tentatively) expect the terms involving scalar fields to dominate here, because $\langle\phi^0\rangle = 247$ GeV, and derivatives introduce factors of only 1 GeV or so. The operators involving ϕ happen to vanish if all charge $-\frac{1}{3}$ quarks in the operator are of the same generation, so we may expect $\Delta S = +1$ processes like $p \to e^- K^+ \pi^+$ and $n \to e^- K^+$ to dominate over $\Delta S = 0$ processes like $n \to e^- \pi^+$ by factors of order 247^2. On the other hand, if there are no elementary scalars or if operators involving them are specially suppressed, then the only $d = 7$ operators available are those involving derivatives D_μ, and $\Delta S = 0$ and $\Delta S = 1$ modes should be comparable.

<u>$\Delta B = -\frac{1}{3}\,\Delta L$ $(n \to \nu\nu\nu,\ n \to \nu\nu e^-\pi^+$, etc.)</u>

The operators $QQ\overline{LLL}$ have odd F-parity, so we must include at least one additional boson field or spacetime derivative. A complete list of these $d = 10$ operators is as follows:

$$\varepsilon_{ij}\,\varepsilon_{k\ell}(\overline{d_R}\,\ell_{Li})(\overline{d_R}\,\ell_{Lj})(\overline{d_R}\,\ell_{Lk})\phi_\ell$$

$$\varepsilon_{ij}\,\varepsilon_{k\ell}(\overline{d_R\, d^c_R})(\overline{d_R}\,\ell_{Li})(\overline{\ell^c_{Lj}}\,\ell_{Lk})\phi_\ell$$

$$\varepsilon_{ij}\,\varepsilon_{k\ell}(\overline{d_R\, d^c_R})(\overline{d_R}\,\ell_{Li})(\overline{\ell^c_{Lj}\,\ell_{Lk}})\phi_\ell$$

It is noteworthy that these operators lead only to $\nu\nu e^-_L$ decays like $n \to \nu\nu e^-_L\pi^+$ and $p \to \nu\nu e^-_L\pi^+\pi^+$, but not to $\nu\nu e^-_R$ or $\nu\nu\nu$ modes like $n \to \nu\nu e^-_R\pi^+$ or $n \to \nu\nu\nu$. To produce these suppressed modes, we must go to $d = 12$ operators, such as:

$$\varepsilon_{i\ell}\,\varepsilon_{jm}\,\varepsilon_{kn}(\overline{u_R}\,\ell_{Li})(\overline{d_R}\,\ell_{Lj})(\overline{d_R}\,\ell_{Lk})\phi_\ell\,\phi_m\,\phi_n$$

$$\varepsilon_{ij}(\overline{d_R}\,\ell_{Li})(\overline{d_R}\,\ell_{Lj})(\overline{d_R}\,\gamma_\mu\, e_R)(\overline{d_R}\,\gamma^\mu\, u_R)$$

The first of these operators can produce $\nu\nu\nu$ modes. However, if there were no elementary scalars, then the $\nu\nu\nu$ modes could be produced only by operators with $d \geq 18$, and even the $\nu\nu e$ modes would require operators (like the last one above) with $d \geq 12$.

<u>$\Delta B = +\frac{1}{3}\,\Delta L$ $(p \to e^+\bar\nu\bar\nu$, etc.)</u>

There are two SU(3) × SU(2) × U(1)-invariant $\Delta B = \frac{1}{3}\,\Delta L$ operators; schematically, they are

$$\ell_L\,\ell_L\,\ell_L\, q_L\, u_R\, u_R \quad \text{and} \quad \ell_L\,\ell_L\, e_R\, u_R\, u_R\, u_R\ .$$

However, these vanish by Fermi statistics if all u_R quarks are of the same generation, so they cannot produce charm-conserving reactions like proton decay. F-parity conservation requires that we add an even number of boson fields or spacetime derivatives. The minimum dimensional operators that can produce $\Delta B = +\frac{1}{3}\,\Delta L$ nucleon decays are therefore $d = 11$ operators, such as

$$\ell_L \ell_L \ell_L q_L u_R d_R \phi^2$$

plus other operators involving spacetime derivatives.

$\Delta B = 2$ ($nn \to 2\pi$, etc.)

The lowest dimensional $\Delta B = 2$ operators have $d = 9$; they are of the form

$$d_R d_R d_R u_R q_L q_L \quad \text{and} \quad d_R d_R q_L q_L q_L q_L \, .$$

In order to estimate the rate of nucleon decay that would be produced by a medium superheavy boson of mass M, I will for definiteness take the effective coupling for an operator of dimension d involving n fields to be of order

$$g_{nd} \approx e^{n-2} M^{4-d}$$

(The factor e^{n-2} is what would be expected from tree diagrams whose vertices have couplings typical of electroweak interactions. This factor is not very important numerically in what follows, so I will not attempt to give a more precise estimate.) Using standard phase space estimates,[27] we can determine the values of M that would give nucleon decay partial lifetimes between 10^{30} and 10^{33} years. The results are given in the following table.

Process	With Elementary Scalars		No Elementary Scalars	
	d	M (GeV)	d	M (GeV)
$N \to \ell^+ X$ or $\bar{\nu} X$	6	$(4 \text{ to } 20) \times 10^{14}$	6	$(4 \text{ to } 20) \times 10^{14}$
$N \to \ell^- X$ or νX	9	$(2 \text{ to } 10) \times 10^{10}$	9	$(4 \text{ to } 20) \times 10^{9}$
$N \to \nu\nu \ell^-_L X$	10	$(3 \text{ to } 7) \times 10^{4}$	12	$(7 \text{ to } 13) \times 10^{3}$
$N \to \nu\nu \ell^-_R X$	12	$(9 \text{ to } 18) \times 10^{3}$	12	$(2 \text{ to } 4) \times 10^{3}$
$N \to \nu\nu\nu X$	12	$(9 \text{ to } 18) \times 10^{3}$	18	$(2 \text{ to } 3) \times 10^{3}$
$N \to \ell^+_R \bar{\nu}\bar{\nu} X$ or $\bar{\nu}\bar{\nu}\bar{\nu} X$	11	$(2 \text{ to } 4) \times 10^{4}$	11	$(4 \text{ to } 8) \times 10^{3}$
$NN \to X$	9	$(4 \text{ to } 10) \times 10^{6}$	9	$(4 \text{ to } 10) \times 10^{6}$

Rough estimates of the exotic particle masses required to give various nucleon decay modes a partial lifetime from 10^{30} to 10^{33} years. (Here ℓ is e or μ, and X is any set of mesons.)

It is striking that most of the "abnormal" nucleon decay modes would require exotic particles of relatively low mass.

I mentioned earlier that exotic particles with $M \approx 10^{15}$ GeV can only produce observable neutrino masses if there are elementary scalar fields. The situation is very different if there exist medium superheavy particles, with masses $M \ll 10^{15}$ GeV and exotic SU(3) × SU(2) × U(1) properties. A $d = 9$ $\Delta L = 2$ effective $\ell_L \ell_L \bar{F}F\bar{F}F$ interaction will have coupling (roughly) of order e^4/M^5, and will give a neutrino mass matrix roughly of order

$$m_\nu \approx e^4 (300\ \text{GeV})^2/M^5 \ .$$

This is in the interesting range of 10^{-2}eV to 10^2eV for M of order 9 to 60 TeV.

If there are medium superheavy exotic particles with masses small enough to produce "abnormal" nucleon decays or neutrino masses at observable levels, then how can we avoid these particles producing unacceptably high rates for the "normal" $\Delta B = \Delta L$ nucleon decays? Perhaps surprisingly, this is not a difficult problem. First, tree graphs can produce the $\Delta B = \Delta L$ decays only if there exist SU(3)-triplet vector or scalar bosons of certain types,[28] so tree graphs are no problem if we simply exclude bosons of these types. Even so, medium superheavy exotic particles that are introduced to produce "abnormal" nucleon decays in tree approximation can produce "normal" $\Delta B = \Delta L$ nucleon decays through loop diagrams, but even though these contributions are less suppressed by $(E/M)^2$ factors, in the cases that have been examined these loop graphs involve so many factors of coupling constants and $1/8\pi^2$ that they give negligible nucleon decay rates.

There is a more serious problem, having to do with the survival of baryons in the early universe. In theories of cosmological baryon production,[29] it is essential that the universe must *not* be in thermal equilibrium during the period during which baryon-nonconserving reactions occur, because in that case any baryon number that could be produced would be immediately destroyed. And in the same way, any baryon number that is produced at very early times (or put into the universe at the beginning) would be wiped out if baryon non-conserving interaction rates become comparable with the expansion rate of the universe at any subsequent time.

The expansion rate of the universe is of order $(kT)^2/m_{PL}$ where m_{PL} is the Planck mass 1.2×10^{19} GeV. At temperatures above the mass M of the exotic particles responsible for baryon nonconservation, the baryon nonconserving reaction rates are of order $f^2 kT$, where f is a product of whatever dimensionless coupling constants appear in the matrix element. Hence the expansion rate of the universe is faster than the rate of baryon nonconserving reactions at all temperatures $kT \geq M$ if and only if $M > f^2 m_{PL}$. Even if f is quite small, this condition will probably not be satisfied for any of the medium superheavy exotic particles that might produce observable rates of abnormal proton decay modes; the above table shows that such particles must have masses below (and in most cases far below) $10^{-8} m_{PL}$. Hence any such medium superheavy particles would have annihilated or decayed at a time when baryon nonconserving interactions were much faster than the cosmic expansion rate, so that equilibrium would have been maintained through the era of baryon nonconservation.

Under these circumstances, the only way that a cosmic baryon excess could survive to the present is for the baryon nonconserving interactions of the medium superheavy particles to conserve some linear combination $B + aL$ of baryon and lepton number. In this case, the equilibrium phase-space density of particles of type i is

$$n_i(p) = \left[\exp\left(\frac{E_i(p) - \mu(B_i + aL_i)}{kT}\right) \pm 1\right]^{-1}$$

where μ is the chemical potential for $B + aL$. For small chemical potentials, the excess of particles of type i over their antiparticles is then proportional to $B_i + aL_i$, so the final baryon-lepton ratio is

$$B/L = \sum_i B_i(B_i + aL_i) \Big/ \sum_i L_i(B_i + aL_i) \; .$$

With the usual pattern of quarks and leptons, this gives

$$B/L = \frac{4}{9a} = \begin{cases} 4/9 & B + L \text{ conserved} \\ \pm 4/3 & B \pm \tfrac{1}{3} L \text{ conserved} \; . \end{cases}$$

For example, if initially B = L, and B + aL is conserved, then the effect of these reactions is to change the baryon number only by a factor

$$\frac{1 + a}{1 + \frac{9}{4} a^2} = \begin{cases} 8/13 & B + L \text{ conserved} \\ 16/15 & B + \tfrac{1}{3} L \text{ conserved} \\ 8/15 & B - \tfrac{1}{3} L \text{ conserved.} \end{cases}$$

I was going to conclude here that it is only possible to have one class of "abnormal" baryon nonconserving nucleon decays at observable levels - that is, either $N \to LX$, *or* $N \to LLLX$, *or* $N \to \bar{L}\bar{L}\bar{L}X$, etc., but only one of these. However, Pati[30] has pointed out to me that if baryon nonconservation is due to the spontaneous breakdown of symmetries, then it may become ineffective not only when kT is much less than the symmetry-breaking scale M, but also for $kT \gg M$. In this case, the intermediate superheavy particles of any given mass scale must still conserve some linear combination of B and L, but it is possible to have different linear combinations conserved at well separated mass scales. As the cosmic temperature fell below each successive mass scale, the baryon-lepton ratio would readjust itself, but a baryon excess would survive.

In any case, it is hard to see how any cosmic baryon excess could survive if there are $\Delta B = 2$ processes like $NN \to 2\pi$ at observable levels. The table above shows that this would require the existence of exotic particles with $M \approx 10^6$ GeV whose interactions violate baryon conservation without preserving any linear combination of B and L except L itself. At these "low" temperatures, baryon-violating reactions would have been much faster than the expansion of the universe, and would have wiped out any pre-existing baryon excess. The observation of $\Delta B = 2$ processes would pose a serious problem for modern cosmological theory.

References

1. T. D. Lee and C. N. Yang, Phys. Rev. 98, 101 (1955). It is of course possible that baryon and/or lepton conservation could be exact gauge symmetries, but be spontaneously broken. However, in order for baryon or lepton conservation then to survive as exact global symmetries, it would still be necessary for the Lagrangian also to have an exact global symmetry, which could combine with the gauge symmetry, in the manner described by G.'t Hooft, Nucl. Phys. B35, 167 (1971).
2. S. Weinberg, Phys. Rev. Lett. 43, 1566 (1979); F. Wilczek and A. Zee, Phys. Rev. Lett. 43, 1571 (1979). Renormalization effects are considered by L. F. Abbott and M. B. Wise, preprint SLAC-PUB-2487.
3. M. Machacek, Nucl. Phys. B159, 37 (1979).
4. The proton lifetime was estimated in this way (except for the dependence on e) by H. Georgi, H. Quinn, and S. Weinberg, Phys. Rev. Lett. 33, 451 (1974).
5. A. Buras, J. Ellis, M. K. Gaillard, and D. V. Nanopoulos, Nucl. Phys. B135, 66 (1978); C. Jarlskog and F. J. Yndurain, Nucl. Phys. B149, 29 (1979); M. Machacek, Ref. 3; A. Din, G. Girardi, and P. Sorba, Phys. Lett. 91B, 77 (1980); J. Ellis, M. K. Gaillard, D. V. Nanopoulos, and S. Rudaz, preprint LAPP-TH-14; M. B. Gavela, A. Le Yaouanc, L. Oliver, O. Pène, and J. C. Raynal, preprint LPTHE80/6.
6. The first experiment designed to set a limit on the proton lifetime was that of F. Reines, C. L. Cowan, Jr., and M. Goldhaber, Phys. Rev. 96, 1157 (1954). For an early theoretical discussion, see G. Feinberg and M. Goldhaber, Proc. Nat. Acad. Sci. 45, 1301 (1958). The lower bound of 10^{30} years is set by the experiment of F. Reines and M. Crouch, Phys. Rev. Lett. 32, 493 (1974); J. Learned, F. Reines, and A. Soni, Phys. Rev. Lett. 43, 907 (1979).
7. H. Georgi, *et al.*, Ref. 4. Also see A. Buras, *et al.*, Ref. 5.
8. Leading grand unified models include those of J. C. Pati and A. Salam, Phys. Rev. D8, 1240 (1973); Phys. Rev. Lett. 31, 661 (1973); Phys. Rev. D10, 275 (1974); S. Glashow and H. Georgi, Phys. Rev. Lett. 32, 438 (1974); H. Georgi, in *Particles and Fields - 1974*, ed. by C. E. Carlson, A.I.P. Conference Proceedings No. 23 (American Institute of Physics, New York, 1975); H. Fritzsch and P. Minkowski, Ann. Phys. (N.Y.) 93, 193 (1975); H. Georgi and D. V. Nanopoulos, Phys. Lett. 82B, 392 (1979); F. Gürsey, P. Ramond, and P. Sikivie, Phys. Rev. Lett. 36, 175 (1976); F. Gürsey and P. Sikivie, Phys. Rev. Lett. 36, 775 (1976); P. Ramond, Nucl. Phys. B110, 214 (1976); etc. As proposed by Pati and Salam, these models involve baryon and lepton nonconservation because they have quarks and leptons in the same multiplets.

9. D. Ross, Nucl. Phys. B140, 1 (1978); W. J. Marciano, Phys. Rev. D 20, 274 (1979); T. Goldman and D. Ross, preprint CALT 68-704 (1979).

10. T. Appelquist and J. Carazzone, Phys. Rev. D 11, 2856 (1975).

11. S. Weinberg, Phys. Lett. 91B, 51 (1980). Also see N. P. Chang, A. Das, and J. Perez-Mercader, CCNY preprint; P. Binétruy and T. Schücker, CERN preprint; M. Yoshimura, preprints KEK-TH 11 and KEK-79-29.

12. There are problems in giving a general algorithm for the construction of the effective Lagrangian beyond the one-loop order; see B. Ovrut and H. J. Schnitzer, Brandeis preprint; Y. Kazama and Y. P. Yao, preprint UM HE 79-40; and T. Hagiwara and N. Nakazawa, preprint HUTP-80/A012. Fortunately we do not need to carry this part of the calculation beyond the one-loop order.

13. G. 't Hooft, Nucl. Phys. B61, 454 (1973), B62, 444 (1973). This prescription is used here in the modified version of W. A. Bardeen, A. J. Buras, D. W. Duke, and T. Muta, Phys. Rev. D 18, 3998 (1978). For an alternative prescription with similar properties, see S. Weinberg, Phys. Rev. 8, 3497 (1973).

14. S. Weinberg, in *General Relativity; an Einstein Centenary*, ed. by S. Hawking and W. Israel (Cambridge University Press, 1979) p. 817.

15. B. S. de Witt, Phys. Rev. 162, 1195, 1239 (1967); Phys. Rep. 19C, 295 (1975); G. 't Hooft and M. Veltman, Ann. Inst. H. Poincaré 20, 69 (1974), S. J. Honerkamp, Nucl. Phys. B48, 269 (1972); R. Kallosh, Nucl. Phys. B78, 293 (1974); M. T. Grisaru, P. van Nieuwenhuizen, and C. C. Wu, Phys. Rev. D 12, 3203 (1975).

16. L. Hall, preprint HUTP-80/A024.

17. S. Weinberg, Ref. 2.

18. M. Gell-Mann, P. Ramond, and R. Slansky, (unpublished); H. Georgi and D. V. Nanopoulos, Phys. Lett. 82B, 392 (1979), Nucl. Phys. B155, 52 (1979); T. Yanagida, Proceedings of the Workshop on The Unified Theory and The Baryon Number in The Universe (National Laboratory for High Energy Physics - KEK, 1979).

19. S. Weinberg, Phys. Rev. D 13, 974 (1976), D 19, 1277 (1979); L. Susskind, Phys. Rev. D 20, 2619 (1979). Also see E. Eichten and K. Lane, Phys. Lett. 90B, 125 (1980); S. Dimopoulos and L. Susskind, Nucl. Phys. B155, 237 (1979), Columbia-Stanford preprint (1979); E. Farhi and L. Susskind, SLAC preprint (1979); E. Eichten, K. Lane, and J. Preskill, preprint HUTP-80/A016; P. H. Frampton, Phys. Rev. Lett. 43, 1912 (1979); S. Dimopoulos, preprint ITP-649 (1979); M. Peskin, Saclay preprint; J. Preskill, preprint HUTP-80/A033; M. A. Bég, H. D. Politzer, and P. Ramond, Rockefeller preprint; P. Sikivie, L. Susskind, M. Voloshin and V. Zakharov, Stanford preprint; W. J. Marciano, Rockefeller preprint; Y. Chikashige, G. Gelmini, R. D. Peccei, and M. Roncadelli, Munich preprint; etc.

20. S. Weinberg, Phys. Rev. Lett. 29, 388 (1972).
21. S. Weinberg, Phys. Rev. Lett. 29, 1698 (1972).
22. E. Witten, Phys. Lett. 91B, 81 (1980).
23. F. Wilczek and A. Zee, preprint UPR-0135T; R. N. Mohapatra and R. E. Marshak, preprints VPI-HEP-80/1, 2; D. V. Nanopoulos, D. Sutherland, and A. Yildiz, Lett. Nuovo Cimento 28, 205 (1980); S. Glashow (unpublished); H. Georgi (unpublished); etc.
24. S. Weinberg, preprint HUTP-80/A023.
25. H. Weldon and A. Zee, to be published.
26. For a review, see e.g. S. Weinberg, Phys. Rev. 181, 1893 (1969). The use of this formalism in analyzing baryon nonconserving operators was suggested to me by E. Witten.
27. J. D. Bjorken and S. Brodsky, Phys. Rev. D 1, 1416 (1970).
28. S. Weinberg, Phys. Rev. Lett. 42, 850 (1979), and Ref. 2.
29. M. Yoshimura, Phys. Rev. Lett. 41, 281 (1978), 42, 746(E) (1979), and to be published; S. Dimopoulos and L. Susskind, Phys. Rev. D 18, 4500 (1978), Phys. Lett. 81B, 416 (1979); A. Yu. Ignatiev, N. V. Krasnikov, V. A. Kuzmin, and A. N. Tavkhelidze, Phys. Lett. 76B, 436 (1978); B. Toussaint, S. B. Treiman, F. Wilczek, and A. Zee, Phys. Rev. D 19, 1036 (1979); J. Ellis, M. K. Gaillard, and D. V. Nanopoulos, Phys. Lett. 80B, 360 (1979), 82B, 464(e) (1979); S. Weinberg, Ref. 28; N. J. Papastamatiou and L. Parker, Phys. Rev. D 19, 2283 (1979); D. V. Nanopoulos and S. Weinberg, Phys. Rev. D 20, 2484 (1979); etc. The existence of a cosmic baryon excess prompted early discussions of possible baryon nonconservation; see S. Weinberg in *Lectures on Particles and Fields*, ed. by S. Deser and K. Ford (Prentice-Hall, Englewood Cliffs, N.J., 1964), p. 482; A. D. Sakharov, Zh. Eksp. Teor. Fiz. Pis'ma 5, 32 (1967) [JETP Lett. 5, 24 (1967)].
30. J. Pati, private communication.
31. V. Gribov and B. Pontecorvo, Phys. Lett. 28B, 493 (1969); J. N. Bahcall and S. C. Frautschi, Phys. Lett. 29B, 623 (1969). For a general review of neutrino masses, see W. J. Marciano, preprint DOE/EY/2232B-207.

SYNOPSIS OF PROGRAM FOR FIRST WORKSHOP ON GRAND UNIFICATION

Thursday (April 10, 1980)

morning	afternoon
Session Chairman: Harvey Shepard	Session Chairman: Howard Schnitzer
Haaland	Slansky
Glashow	Mohapatra
Langacker	Goldman
Ruegg	Ramond
Frampton	Pati
Gursey	

Friday (April 11, 1980)

morning	afternoon
Session Chairman: Roy Schwitters	Session Chairman: Roy Glauber
Reines	Steigman
Sulak	Turner
Winn	Witten
Wilson	Ellis
Cline	Georgi

Saturday (April 12, 1980)

Session Chairman: Thomas Appelquist

Marshak

Steinberg

Yildiz

De Rújula

Susskind

Weinberg

FIRST WORKSHOP ON GRAND UNIFICATION

ORGANIZING COMMITTEE

Paul Frampton (Chairperson)

Howard Georgi

Sheldon Glashow

Kyungsik Kang

Marie Machacek

DIRECTOR OF THE WORKSHOP

Asim Yildiz

List of Participants

	Name	Affiliation
1.	Abbott, Larry	Brandeis University
2.	Albright, Carl	Northern Illinois University
3.	Andrei, Nathan	New York University
4.	Appelquist, Thomas	Yale University
5.	Barr, Stephen M.	University of Pennsylvania
6.	Bars, Itzhak	Yale University
7.	Baulieu, Laurent	Harvard University/Ecole Normale Superieure, Paris, France
8.	Brower, Richard	University of California, Santa Cruz
9.	Cahill, Kevin	Harvard University
10.	Carter, Ashton	Rockefeller University
11.	Chakrabarti, Jaypokas	City College of New York
12.	Chang, Ngee P.	City College of New York
13.	Cheng, Ta-Pei	University of Minnesota/University of Missouri
14.	Claudson, Mark	Harvard University
15.	Cline, David	University of Wisconsin
16.	Coquereaux, Robert	Harvard University/Centre de Physique Theorique, Marseille, France
17.	Cortez, Bruce	Harvard University
18.	Costa, G.	University of Virginia/Instituto Di Fisica, Padova, Italy
19.	Cox, Paul	Harvard University/University of New Hampshire
20.	Dai, Yuan-Ben	Institute of Theoretical Physics, Beijing, China
21.	Davidson, Aharon	Syracuse University/Weizmann Institute, Israel
22.	De Rújula, Alvaro	CERN, Switzerland
23.	Dimopoulos, S.	Stanford University
24.	Dokos, Nicholas	Harvard University
25.	Doria, Mauro	Yale University
26.	Dourmashkin, Peter	Massachusetts Institute of Technology
27.	Elias, Victor	University of Toronto, Toronto, Canada
28.	Ellis, John	CERN, Switzerland
29.	Farhi, Ed	Stanford University
30.	Fischler, Mark	Fermi National Laboratory
31.	Fogelman, Guy	Indiana University
32.	Foster, G. William	Harvard University

Name	Affiliation
33. Frampton, Paul H.	Harvard University
34. Gao, Han-ying	Institute of Theoretical Physics, Academic Science, Beijing, China
35. Georgi, Howard	Harvard University
36. Gipson, John	Yale University
37. Glashow, Sheldon L.	Harvard University
38. Glauber, Roy	Harvard University
39. Goldhaber, Maurice	Brookhaven National Laboratory
40. Goldman, Terrence	California Institute of Technology
41. Gonzales, Daniel	Massachusetts Institute of Technology
42. Gottlieb, Steven	Argonne National Laboratory
43. Grossman, Bernard	Harvard University
44. Guerin, Francoise	Brown University/Physique Theorique, Nice, France
45. Gunaydin, Murat	University of Bonn, West Germany
46. Guralnik, G.	Brown University
47. Gursey, Feza	Yale University
48. Haber, Howard	Lawrence Berkeley Laboratory
49. Hagelin, John	Harvard University
50. Hagiwara, Teruhiko	Brandeis University
51. Hall, Lawrence	Harvard University
52. Heller, Kenneth	University of Minnesota
53. Hendry, A. W.	Indiana University
54. Holdom, Robert	Harvard University
55. Hu, Bei-Lok	Harvard University
56. Huffman, William	Harvard University
57. Iijima, Byron	Massachusetts Institute of Technology
58. Jones, Daniel	Argonne National Laboratory
59. Kabir, P. K.	University of Virginia
60. Kane, Gordon	University of Michigan
61. Kang, Kyungsik	Brown University
62. Kaplan, Norman	Harvard University
63. Kenway, Richard	Brown University
64. Kephart, Thomas	Northeastern University
65. Kostelecky, Alan	Yale University
66. Krauss, Lawrence	Massachusetts Institute of Technology
67. Krausz, Frank	Harvard University
68. Langacker, Paul	Institute for Advanced Study, Princeton
69. Leao, Joao	Clark University/Centro de Fisica de Materia Condensada, Lisbon, Portugal.
70. Li, Bing-An	Stanford Linear Accelerator Center/Inst. of High Energy Physics, Beijing, China.

	Name	Affiliation
71.	Lo Secco, John	University of Michigan
72.	Lykken, Joseph	Massachusetts Institute of Technology
73.	Machacek, Marie	Northeastern University/Harvard University
74.	Mahanthappa, K. T.	University of Colorado
75.	Majumdar, Parthasarathi	Brandeis University
76.	Mandelbaum, G.	Munich University, West Germany
77.	Mannheim, Philip	University of Connecticut
78.	Marshak, Marvin	University of Minnesota
79.	Meshkov, Sydney	National Bureau of Standards
80.	Mohapatra, Rabindra	City College of New York
81.	Nakazawa, Nobuya	Harvard University/Kogankuin University, Shinjuka, Tokyo, Japan
82.	Nandi, Satyanarayan	Ohio State University
83.	Nelson, Charles	SUNY at Binghamton
84.	Ng, John	University of British Columbia, Canada
85.	Ng, Yee Jack	University of North Carolina, Chapel Hill
86.	Nilsson, Jan	Institute of Theoretical Physics, Goteborg, Sweden
87.	Oakes, Robert	Northwestern University
88.	Olive, Keith	Enrico Fermi Institute, University of Chicago
89.	Paranjape, Manu	Massachusetts Institute of Technology
90.	Parke, Stephen	Harvard University
91.	Pati, J. C.	International Centre of Theoretical Physics, Trieste, Italy/University of Maryland
92.	Peterson, Earl	University of Minnesota
93.	Ramond, Pierre	California Institute of Technology
94.	Reines, Fred	University of California, Irvine
95.	Rosner, Jonathan	University of Minnesota
96.	Rubbia, Carlo	Harvard University
97.	Ruegg, Henri	Stanford Linear Accelerator Center/ University of Geneva, Switzerland
98.	Salomonson, Per	Harvard University
99.	Scanio, Joseph J. G.	University of Cincinnati
100.	Schnitzer, Howard	Brandeis University
101.	Schwitters, Roy	Harvard University
102.	Segre, Gino	University of Pennsylvania
103.	Senjanovic, Goran	University of Maryland
104.	Sezgin, Ergin	SUNY at Stony Brook

Name		Affiliation
105.	Shepard, Harvey	University of New Hampshire
106.	Shore, Graham	Harvard University
107.	Slansky, Richard	LASL, Los Alamos, New Mexico
108.	Smith, Daniel	Northeastern University
109.	Soldate, Mark	Massachusetts Institute of Technology
110.	Steigman, Gary	Bartol Research Foundation/University of Delaware
111.	Steinberg, Richard	University of Pennsylvania
112.	Stone, James	University of Michigan
113.	Sulak, Lawrence	Harvard University/University of Michigan
114.	Suranyi, Peter	University of Cincinnati
115.	Susskind, Leonard	Stanford University
116.	Tsao, Hung-Sheng	Rockefeller University
117.	Turner, Michael	University of Chicago
118.	Unger, David	University of Michigan
119.	Vaughn, Michael	Northeastern University
120.	Wada, Walter W.	Ohio State University
121.	Wali, Kameshwar C.	Syracuse University
122.	Weiler, Thomas	Northeastern University
123.	Weinberg, Steven	Harvard University
124.	Weingarten, Don	Indiana University
125.	Weldon, Arthur	University of Pennsylvania
126.	Wilson, Richard	Harvard University
127.	Winn, David	Harvard University
128.	Witten, Edward	Harvard University
129.	Wu, Dan-di	Harvard University/Institute of High Energy Physics, Beijing, China
130.	Yildiz, Asim	Harvard University/University of New Hampshire
131.	Yildiz, Musa	University of New Hampshire